Research Methods in the Dance Sciences

UNIVERSITY PRESS OF FLORIDA

Florida A&M University, Tallahassee
Florida Atlantic University, Boca Raton
Florida Gulf Coast University, Ft. Myers
Florida International University, Miami
Florida State University, Tallahassee
New College of Florida, Sarasota
University of Central Florida, Orlando
University of Florida, Gainesville
University of North Florida, Jacksonville
University of South Florida, Tampa
University of West Florida, Pensacola

RESEARCH METHODS IN THE DANCE SCIENCES

EDITED BY

Tom Welsh, Jatin P. Ambegaonkar, and Lynda Mainwaring

University Press of Florida

Gainesville · Tallahassee · Tampa · Boca Raton

Pensacola · Orlando · Miami · Jacksonville · Ft. Myers · Sarasota

Cover: Left, Philip Ancheta (dancer) is a freelance actor, singer, and dancer based in Orlando, Florida. *Right,* Kehinde Ishangi (dancer) is a Somatic Educator and Assistant Professor in the School of Dance at Florida State University. Rick McCullough (photographer) is Professor Emeritus for the School of Dance at Florida State University.

Published in the United States of America

28 27 26 25 24 23 6 5 4 3 2 1

Library of Congress Cataloging-in-Publication Data
Names: Welsh, Tom (Dance professor), editor. | Ambegaonkar, Jatin P., editor. | Mainwaring, Lynda, editor.
Title: Research methods in the dance sciences / edited by Tom Welsh, Jatin P. Ambegaonkar, and Lynda Mainwaring.
Description: Gainesville : University Press of Florida, [2023]. | Includes bibliographical references and index.
Identifiers: LCCN 2022026773 (print) | LCCN 2022026774 (ebook) | ISBN 9780813069548 (hardback) | ISBN 9780813070285 (pdf) | ISBN 9780813072630 (ebook)
Subjects: LCSH: Dance—Research—Methodology. | Dance—Study and teaching. | Dance—History—Study and teaching. | BISAC: PERFORMING ARTS / Dance / Reference | SCIENCE / Research & Methodology
Classification: LCC GV1589 .R46 2023 (print) | LCC GV1589 (ebook) | DDC 792.8071—dc23/eng/20220630
LC record available at https://lccn.loc.gov/2022026773
LC ebook record available at https://lccn.loc.gov/2022026774

The University Press of Florida is the scholarly publishing agency for the State University System of Florida, comprising Florida A&M University, Florida Atlantic University, Florida Gulf Coast University, Florida International University, Florida State University, New College of Florida, University of Central Florida, University of Florida, University of North Florida, University of South Florida, and University of West Florida.

University Press of Florida
2046 NE Waldo Road
Suite 2100
Gainesville, FL 32609
http://upress.ufl.edu

Contents

Figures

Tables

Preface

The purpose of *Research Methods in the Dance Sciences* is to identify and teach concepts and practices that promote effective, rigorous empirical research in the dance sciences, including the medical and educational sciences. It is designed to provide a sound methodological foundation upon which domain-specific concepts and practices may be layered.

We are using the plural form—*Dance Sciences*—for the title of the text to emphasize that work in our interdisciplinary field grows from a collection of sciences, each of which has its own strengths and biases. We encourage readers to embrace the diversity in the field and welcome the unique contributions that each discipline provides. Some chapter authors use *Dance Science,* and others use *Dance Medicine and Science* to refer to our collective effort.

The text is written for all who want to understand, apply, and conduct empirical research with dancers, including but not limited to dancers, teachers, researchers, clinicians, and those who aspire to such professions. It is an invitation to learn how research is conducted and how research results can be used to optimize training, treatment, and other practices that support dancer health, wellness, and performance. We have three primary audiences for the text. The first is dancers interested in learning to conduct empirical research with other dancers as a means of enhancing their involvement in dance. The second is researchers and aspiring researchers who have less direct experience in dance and want to learn how to apply empirical research concepts and methods to dance. They will learn how dancers think and work by reading the dance examples in this text. Our third audience is experienced researchers who have skills in one of the specializations represented in our field. We hope they will use the text to enhance their understanding of how colleagues in other specializations work so that they can offer a greater variety of research tools to their research mentees.

We aim to help readers develop a common foundation in fundamental research skills and their application to dance science. By doing so, we hope to facilitate interdisciplinary collaboration. Our approach is intentionally inclusive. We describe aspects that are common across disciplines and resist the temptation to offer a single set of recipes that are specific to a particular domain. Readers will be able to choose among a diverse set of methodological approaches to find the best match for particular research interests. They also may be better able to find ways to integrate their work with the work of other dance medicine and science researchers. Ideally, some will be motivated to engage in collaborative research with colleagues who use different research methodologies.

Finally, an important purpose for this text is to connect readers to other resources that researchers use to develop their skills and inform their work. Interacting with the many collaborators who helped build this book has allowed us to grow as researchers and scholars. We hope your interactions with the work of our multidisciplinary team will inspire you as well.

Editors
Tom Welsh
Jatin Ambegaonkar
Lynda Mainwaring

Acknowledgments

Research Methods in the Dance Sciences has been some time in the making. The seed for the text was planted at the 1998 founding meeting of the International Association for Dance Medicine & Science (IADMS) Research Committee in Hartford, Connecticut. It took more than 12 years for that seed to germinate and another 12 years for it to grow into this multidisciplinary resource collaboratively authored by many of the most productive scholars in our field. We could not have asked for a more thoughtful, dedicated, and cooperative group of colleagues with whom to pursue this project. If you are inspired by reading any chapter in this text, let the chapter authors know they have made a difference.

We also benefited from the support of a dozen IADMS research committees and boards that provided input, guidance, and encouragement as the project unfolded. Furthermore, many of these colleagues became chapter authors. We are grateful for the targeted feedback we received from students and colleagues who worked through early drafts asking questions and sharing perspectives on what we were trying to teach. Their investment makes the text more insightful and interesting to read.

Several other professional colleagues provided essential input that kept the project moving forward when its needs exceeded our abilities. Our editorial assistant, Leah Bailey, collected and organized the documents used to build the text, performed the initial copy editing, formatted the chapters twice in different formats—one for peer review and another for the final submission. She also enhanced the presentation of figures in Chapters 1, 3, and 5. Kaitlin Morgan helped build figures for Chapter 16, and Lyric Hartley improved the quality of the figures in Chapters 1 and 5 and consulted on drafts of selected figures and cover designs. Unless otherwise noted, all other images were created by the chapter authors or volume editors.

We feel especially fortunate that the University Press of Florida (UPF) agreed to work with us on this project. As a not-for-profit publisher, they understood the challenges of producing a unique resource for our unique audience—all of you. We are particularly grateful for the expert guidance provided by our UPF editors, Stephanye Hunter, Marthe Walters, and Sally Antrobus. Melissa Stearns Hyde built the index.

Our collaborative efforts as the volume editors have been inspiring and enriching. Together we have grown intellectually from our many online meetings, dotted with lovely interactions with one another and with our families. Many treasured outcomes have emerged from the seed planted years ago.

Of course, none of us could have invested the time and energy required to bring this project to completion without the support of our families and our colleagues.

Jatin thanks his superwoman wife, Shruti, for her unwavering support over the years and looks forward to many more years. He is very grateful to his twins, Aanya and Arjun, for their patience when Baba could not spend time with them during the countless hours of writing and editing this text.

Lynda sends deep appreciation to her father Eric, for his sense of humor, sage advice, and exceptional trumpet tone that echoed in the background; to her mother, Bernice, for her encouragement and wisdom; and to her family John, Michelle, Tab, Trish, and Jim for their unyielding support. She also sends special thanks to Lyric Hartley for great technical and creative contributions.

Tom thanks his research mentors Richard Malott, Keith Miller, and Jon Baily, who "taught [him] how to think." He thanks his life partner Sue Carpenter, who listened and asked all the right questions as ideas were being explored, carried more than her share of the home-front duties, and offered welcomed patience when the demands of the project were extreme.

Finally, dear readers, we thank you for choosing to pursue greater understanding of the research process and for dedicating a portion of your life to working with dancers. We hope to meet you at a professional conference, to read your research in our journals, and to see you flourish as a future generator of new knowledge. We welcome you as a member of the international dance medicine and science community.

I

Bringing Science to Dance

1

Researching Dance

TOM WELSH

Questions to Stimulate Thinking

- How can science inform dancing?
- What does dance science research involve?
- What would it take to become a researcher?

Dance is an aesthetic activity. It is an art form in which the human body is the essential, expressive instrument. With help from their teachers, dancers build and refine their movement skills and they work to develop their body's ability to perform at levels well beyond those common in everyday living. The challenges of training the human body to achieve such high levels of performance are on par with, and sometimes exceed, those involved in elite sports. Building a science devoted to helping dancers manage the challenges of training is a worthy endeavor for a healthy and successful career in dance.

The dance sciences are a diverse collection of domains in which researchers and practitioners have a variety of backgrounds but share a common interest in promoting health, well-being, and optimum performance in dancers.

Emergence of the Dance Sciences

Over the past 60 years researchers in the sport sciences have been building a knowledge base that athletes can access for guidance in managing the demands on their body systems as they pursue progressively higher levels of human performance. The sport sciences and sports medicine are specializations that have been developed to serve athletes and those who train and treat them. Athletes, coaches, and sports medicine clinicians can use research from the sport sciences to inform their approach to training and treatment.

Over the past 25 years the dance sciences and dance medicine have experienced tremendous growth. Many dancers have come to see themselves as aesthetic athletes, and more research is being conducted with dancers as participants. Pioneers in dance medicine and science observed the development of sports medicine and science and saw it as a harbinger for building a new interdisciplinary field (Garrick, 1986; Howse & Hancock, 1988, 2000; Ryan & Stephens, 1987). In areas where there has been little research with dancers, research with athletes has been used to inform training for dancers. As more research focuses on the unique challenges that dancers face, the utility of research to dancers is likely to grow.

The maturing of the International Association for Dance Medicine & Science (IADMS), now in its third decade, and its flagship *Journal of Dance Medicine & Science,* have played a key role in promoting and disseminating the results of empirical research with dancers. Other national and regional organizations have developed around the world and are also making valuable contributions to the dance sciences. The emergence of dance science training programs at universities (see "Student's Guide" at https://iadms.org/resources/#ref1) is a product of the developing field and a stimulus for further growth. The dance sciences will continue to mature as more work is conducted with dancers on topics that improve dancers' abilities and provide information that is useful to those who train and treat dancers, including dancers themselves. As the field continues to grow, it is likely to receive greater recognition from researchers in other movement sciences. As the work being done by dance science researchers addresses dancers' needs more explicitly, dance science research is likely to receive progressively more support from dancers, teachers, choreographers, and healthcare practitioners who work with dancers.

Need for Research in the Dance Sciences

Researchers examining the dance sciences to improve dance performance, reduce injury risk, and enhance recovery have been using a variety of research methodology textbooks from other domains to inform empirical research with dancers. These textbooks (e.g., Arnold et al., 2005; Berg & Latin, 2008; Gillen, 2007; Ingersoll, 2001; Payton, 1994; Portney & Watkins, 2009; Thomas et al., 2011) provide information that can be useful to dance science researchers even though their examples and perspectives are selected and tailored for a different interest group. *Research Methods in the Dance Sciences* is designed to address the need for a domain-specific approach to teaching empirical research skills to new dance science researchers. In this text we identify, organize, and teach concepts and practices intended to promote effective research in the dance sciences. We aim to build a sound conceptual and methodological foundation upon which domain-specific and higher-level concepts and research practices may be layered. *Research Methods in the Dance Sciences* is written for dancers, teachers, aspiring researchers, clinicians, and other practitioners who want to understand, apply, and conduct empirical research with dancers.

To optimize the utility of dance science research, particularly for dancers, teachers, choreographers, and company directors, we believe more research will need to be conducted by individuals who are or have been intensively involved in dance at some point in their lives. Intensive experience with dance will help researchers identify relevant questions to address, important abilities to assess, and practical interventions to evaluate. A major purpose of this text is to teach dancers how to become effective researchers. Research students who have less experience with dance but have a special interest in improving dancer performance and health, and who are willing to learn how dancers think and work, will also find the text useful. The perspective used and the dance examples offered will help those with less experience in dance gain a deeper understanding of dancers, their priorities, and how they work. In the final section of the book (Part VI), we recommend collaboration as a strategy for merging the complementary strengths of researchers and practitioners.

Developing the skills needed to conduct insightful and informative research with dancers can allow dancers to extend their participation in a field they love in new and creative ways. Many dancers start dancing when they are young, and they train for years to master the skills required of performers. An early and prolific modern dancer, Martha Graham (1953), once said it takes about 10 years of full-time study to become a professional dancer. An expert on expertise and his colleagues (see e.g., Ericsson & Harwell, 2019) gathered self-reported measures from domains as diverse as chess, music, sports, and surgery to show that it takes thousands of hours of deliberate practice to reach elite levels of performance. Developing the repertoire needed to become a productive dance science researcher is likely to take some time, but it is a journey that can be as exciting and demanding as learning to dance.

Implications for New Dance Science Researchers

Fortunately, aspiring researchers can practice new research skills as they are developing them, just as young dancers begin performing while developing stature as performers. New researchers in many fields participate in a research mentorship or work as research interns to refine the skills they begin to acquire through course work. In established fields, graduate training programs and postdoctoral fellowships provide a rich and productive context for learning to become a capable researcher. New researchers in younger fields such as the dance sciences may need to arrange their own apprenticeships to acquire the depth of training needed to become proficient researchers. In the final chapters in this text we offer suggestions for building a research repertoire.

As a diverse and rapidly developing field, the dance sciences offer a variety of loosely defined pathways that new researchers may follow to enter the field. In addition to being able to tailor their own pathway into the field, those who enter it now may be able to play an influential role in its development relatively early in their careers. As the dance sciences mature, the pathways to competency for researchers are likely to become more defined and more formal. We might all relish the opportunity to work and learn as next-generation pioneers in a field that is still evolving. In the seventeenth century Isaac Newton suggested that scientists build their understanding of how the world works by standing on the shoulders of giants. He was pointing out that the contributions of scientists today are made possible by the foundation built by those who preceded them. We would do well to acknowledge the colleagues who nurtured our field in its early years, such as Dr. A. Ryan, who was a pioneer in the field and wrote about its early history (Ryan, 1997). These dancers, teachers, and health care practitioners learned research skills on the job and in their spare time. There were few opportunities for formal training in the dance sciences then. Now is an exciting time to join our field.

Dance Science Is Applied and Interdisciplinary

Most research in the dance sciences is applied, whether it takes place in the field or in a laboratory. This means that in addition to creating new knowledge about how dancers' bodies, (including their brains) operate, applied researchers study topics that address practical problems faced by dancers, teachers, fitness trainers, company directors, and health care practitioners. Research in the dance sciences emphasizes the creation of new knowledge

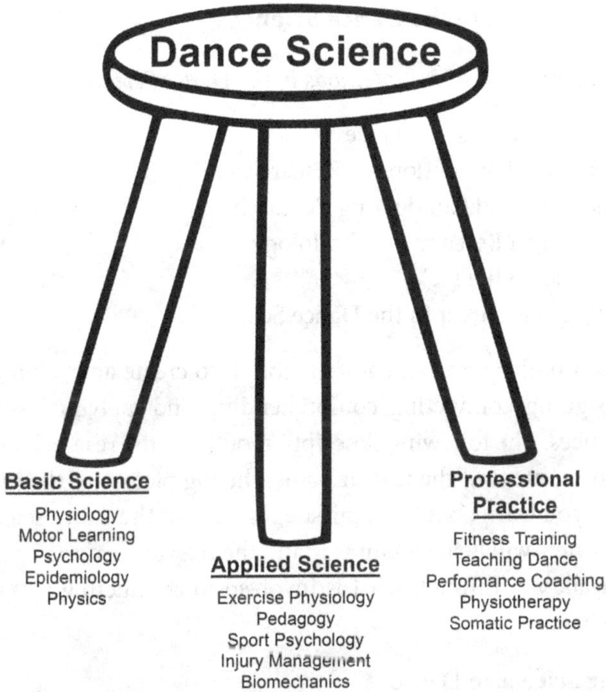

Figure 1.1. Examples of Dance Science Interdependence. The combined contributions from every leg are what make our field unique and strong.

that is clinically or socially important and pragmatically relevant to dancers and those who train and treat them.

The dance sciences, like the sport sciences, are also interdisciplinary. They integrate basic scientific disciplines such as physiology, biomechanics, and the behavioral, psychological, and medical sciences with applied domains such as dance medicine, dance fitness training, dance pedagogy, dance psychology, and somatic training for dancers (see Figure 1.1). The scope of the dance sciences is still humble enough to be included under one umbrella in most contexts. That integration encourages interaction among professionals who have diverse areas of specialization and diverse perspectives on how they might contribute to and draw from the knowledge base we use to support the work we all do. The history of the sport sciences can allow us to make informed guesses about how the dance sciences are likely to evolve.

Research Methods in the Dance Sciences

We have organized *Research Methods in the Dance Sciences* into six parts:

I. Bringing Science to Dance
II. Creating a Foundation for Research
III. Designing and Conducting Research
IV. Choosing a Research Methodology
V. Influencing the Field
VI. Pursuing a Career in the Dance Sciences

Each part builds on what comes before it to create an evolving scaffolding for designing, conducting, comprehending, and applying research in the dance sciences. The following description outlines the research process as it outlines the content of the text. It paints the big picture of the journey you will take as you learn about and pursue research in the dance sciences. Having an overview will help you understand the relevance of each part, and the chapters walk you through what is involved in conducting research in the dance sciences.

I. Bringing Science to Dance

In the remainder of this chapter we provide an overview of the research process and outline how *Research Methods in the Dance Sciences* addresses the steps involved in conducting empirical research with dancers. In *Chapter 2* we explain the discipline needed to conduct empirical research and how that discipline differs from the ways of knowing that we use in other aspects of our lives. *Chapter 3* provides a catalogue of concepts that are fundamental to empirical research in the diverse collection of domains that comprise our field. The catalogue provides an introductory mapping of fundamental concepts that are expanded upon and explained in detail in later chapters. You may want to refer back to the brief descriptions in *Chapter 3* as you work through the rest of the book.

II. Creating a Foundation for Research

In Part II we describe the foundation on which empirical research is built. Researchers identify a topic of interest, learn what has been done by previous researchers, learn about ethical considerations involved in conducting research with dancers, and choose an approach for studying their topic.

In *Chapter 4* we explain how to choose a topic to research and how to acquire the background information needed to design and pursue an empirical research study. In many research approaches a major component of this phase is reviewing the research literature. Reading what other researchers have done helps you determine what is already known about your topic and what remains to be discovered. Reading broadly at first and developing depth as you home in on a specific research question that you wish to answer allows you to make educated guesses about where the points of leverage may be. In some qualitative approaches to research, a thorough review of the literature on the topic of interest is intentionally not performed in advance so that the researcher can remain free of the biases prior reading might create. In other approaches to research, reviewing the literature allows you to develop a solid understanding of a particular topic or area (what is needed and where the gaps in knowledge are) and to sample the research methods that have been used to study topics like the one that currently interests you.

A second purpose for the literature review is to collect information that will help you choose and tailor a research methodology to address your research question. You will learn how other researchers have assessed similar dancer capacities, abilities, or preferences, and you will see what strategies other researchers have used to deliver training and administer treatment systematically so that their procedures can be replicated by other practitioners and researchers.

A third benefit of reviewing research literature is that it can help you determine how to explain why your study is important to the field. You will learn how to justify the use of considerable human resources as you design, execute, analyze, and report your research. Reviewing the work of previous researchers provides access to findings that you may build on. It also allows you to compare the results of your work with prior research findings. You will begin to see how your work may ultimately contribute to the expanding evidence base that is available to researchers and practitioners in the field.

As a heads-up for new researchers: the literature review begins early in the research process and may continue until submission of your final report (thesis, dissertation, research article, or conference presentation). It can feel as though there is always more to learn through reading reports by other researchers. *Chapters 4* and *7* include suggestions for how to keep this realization from inhibiting your progress. We suggest reading early, reading often, and reading a lot as you pursue other steps in the research process.

As the research question you wish to answer becomes clearer, you will be ready to begin deciding what types of research methodology are better suited to the study you want to conduct. Fortunately, various approaches are available to address the variety of questions researchers may want to answer. In *Chapter 5* we provide an overview of types of research that are used in the dance sciences. We also offer suggestions for choosing an approach that is appropriate for answering your research question as well as for securing resources available to you (personnel, space, equipment, funding).

It will become clear that assumptions made at earlier stages need to remain open to refinement as you learn more about your topic and how to study it. It is common, for example, for researchers to revise their research question as they continue to review the research literature and as they select a research methodology. Indeed, your research question will benefit from repeated refinement throughout the planning phases of your study. You may even see it with new clarity after your results are collected, summarized, and analyzed. Such insight is similar to discoveries made by choreographers and composers when they finally see or hear their work fully produced.

Before initiating research involving human participants, new researchers are expected to develop and demonstrate competence in research ethics. Ethics training may be provided through course work or as online independent study. After being certified in research ethics, researchers prepare applications in which they describe their research plans, highlighting interactions with their study participants. The research proposals are reviewed and ultimately approved by a human participants committee (also called an *Institutional Review Board, Research Ethics Committee,* or some combination of these terms). These committees are responsible for protecting the rights of research participants and for reducing personal, reputational, and financial risks for all stakeholders. *Chapter 6* addresses these issues.

Chapter 7 provides guidance on writing a review of the research literature that you have been reading to prepare for your study. In addition to helping you decide how to approach your study and identifying a specific, clearly stated purpose for your research, your written review of the literature offers a justification for conducting your study and establishes the context that will help readers of your report understand what you have done and why it is important.

III. Designing and Conducting Research

In Part III we explain steps in the research process that are common to most empirical approaches to research in a manner intended to encompass the diversity of approaches used in our field. Consider it an overview or a road map for conducting an empirical research study. It outlines where you are headed, helping you see how the work you are doing relates to the work of other researchers answering different questions using different research methodologies.

Chapter 8 begins with a description of how to identify a population for whom you hope the results of your study are likely to be relevant. For example, your target population might be professional ballet dancers, pre-professional modern dancers, or dance studio competition dancers between the ages of 12 and 17. Researchers commonly select a representative and accessible sample of participants from their target population (e.g., a group of dancers to whom you have access to conduct assessments and, possibly, apply training or treatment). Researchers also identify a setting (e.g., medical clinic, dance studio, fitness center, movement lab) where observations and measurements can be made. If you will be conducting intervention research, you will also need a setting in which to administer training or treatment. We describe the trade-offs that can influence these decisions and offer strategies for recruiting participants and securing your research setting.

An important challenge in any new area of research is finding or developing a way to assess outcome variables. In some research approaches, outcome variables (sometimes called dependent variables) are measured or counted. They may involve a dance skill, a physical capacity, or perceptions of an experience you want to learn more about. Strategies for defining and measuring outcome variables that are important in dance are offered in *Chapter 8*.

When looking for possible cause-and-effect relationships, researchers may also specify one or more independent variables (IVs). Some independent variables, such as gender, age, and level of experience, are not directly controlled by the researchers. Others, such as training or treatment, are applied by the researcher (and often called interventions) in order to assess their influence on study participants. The ability to replicate an intervention (repeat it with precision) is essential in experimental studies. Operational definitions, pilot testing, and a procedural integrity checklist are used to ensure the consistent application of an intervention.

In other types of research, dependent and independent variables are not pre-specified. Rather, they arise from the process of intensive observation. In *Chapter 13* we describe how qualitative researchers discover variables that are important to dancers and those who treat and train them. A well-specified research question and the type of question asked guides the choice of methodology.

Gathering, managing, and interpreting data (records of the observations) are the heart of any empirical research endeavor (Sidman, 1960). We describe general strategies for handling these phases of the research process in *Chapter 9*. How data are gathered, analyzed, and presented differs with their type and the methodological approach chosen to answer your research question. In *Chapter 9*, we provide an overview of the approaches described in detail in Part IV.

In *Chapter 10*, we explain how inferences are made from data and how conclusions are drawn. Both are affected by how you arrange the methodological components described in the previous two chapters. Planning the arrangement is called research design.

The first three parts of this text are designed to build a common foundation and a general framework for research in the dance sciences. That foundation and framework will help you see how your work relates to the work of other researchers. It will facilitate communications with colleagues whose research approach may differ from the approach you choose for your study. Having a common understanding of the research process will allow you to interact with other researchers on topics of mutual interest, potentially in a research collaboration.

IV. Choosing a Research Methodology

In Part IV, *Chapters 11–19,* we offer a representative collection of methodological approaches commonly used to answer a variety of research questions in the dance sciences (see Table 1.1). The value of one methodology over another is based on its fitness for the research question you want to answer.

When approaching a new area of study in which prior research is limited, some researchers start with a descriptive study to learn more about their topic. They then use what they learn from their descriptive study to inform how they will approach another type of research on their topic. Other researchers specialize in one type of research and look for questions that are ready to be addressed by the method they prefer to use. How you choose to work is up to you. You may even change your approach as your interests change, as long as you choose a methodology that is compatible with the research question you want to answer.

In Part IV we explain each research methodology in enough detail to provide guidance on how to conduct a study addressing research questions appropriate to each approach. We also provide references to other resources for establishing a deeper understanding of a particular research approach. As in learning to dance, learning to conduct research is a pursuit that has no specified end; we can always become a little more skillful as dancers and as researchers. Our intention in Part IV is to provide a solid introduction to a variety of approaches and to offer help with finding resources enabling you to continue developing skills in conducting research with dancers.

V. Influencing the Field

The purpose of doing research is to learn how the world works *and* to share what we learn with colleagues who may find the new understanding that our research creates useful. In *Chapter 20* we explain how to propose research

Table 1.1. Research Methodologies Described in Part IV

Descriptive Research
Qualitative Research
Survey Research
Correlational Research
Case Studies
Within-Subject Experiments
Between Group Experiments
Systematic Reviews
Epidemiology Research

to those who need to be convinced to support your work as mentors, thesis committees, sponsors, and funders. We also explain how to report research to colleagues who will use your findings to improve their own teaching, professional practice, and research.

The phases of proposing research and reporting findings may be separated by a substantial period of time and intervening research activity. However, they can be viewed as two stages of the same process—proposing describes what will be done to conduct your research, and reporting describes what was actually done, what was observed, and the implications. Keeping both stages in mind as your research unfolds will inform the decisions made along the way. It will also help you avoid overlooking key elements of the research process that will generate data that will need to be captured while the research is in progress.

Writing, presenting, and publishing a report of your research allows you to contribute to an evolving system of knowledge. *Chapter 21* describes strategies for connecting research to practice and for disseminating the findings of your research to dancers, teachers, health care practitioners, and others who may benefit from considering them. Research is a cyclical process. What we learn while conducting one study informs applied practice at the same time as setting the stage for new research that may be conducted by you or colleagues in the field.

VI. Pursuing a Career in the Dance Sciences

In the final two chapters we offer guidance on pursuing a career in the dance sciences. Just as watching a video about dance is not enough by itself to make a virtuosic dancer, reading a textbook about research is not enough by itself to make an insightful researcher. In *Chapter 22* we describe benefits and expectations for pursuing graduate studies in dance science research. In *Chapter 23*, we outline strategies that aspiring researchers can use to augment their formal training and continue the mentored practice needed to build a researcher's repertoire. In both chapters we discuss working collaboratively as a strategy for combining expertise and developing and managing a dynamic research agenda.

Welcome to the exciting undertaking of training for work in a field that has unbounded potential. We hope you enjoy learning to read, interpret, and conduct research in the dance sciences as much as we have enjoyed preparing this resource to guide your efforts.

Highlights

- Building a science devoted to helping dancers manage the challenges of training for a healthy and successful career in dance is a worthy endeavor.
- To optimize the relevance of dance science research, more research should be conducted by individuals who have been intensively involved in dance at some point in their lives.
- Research in the dance sciences emphasizes the creation of new knowledge that is clinically or socially important and pragmatically relevant to dancers.

Application Exercise

Brainstorm a list of possible research topics that interest you. Choose a topic and write a related question that you might like to answer. To think outside the box, see if you can adapt your question to another topic in your list of topics. The goal is not to settle on a single question but rather to expand the types of topics and questions you can imagine building a study to address. It is like *stretching out* before class so that your pliés, battements, and ronds de jambe feel less restricted when you do them in class. Or it is like creating many movement motifs for a choreographic project so that you have many ideas to inform the direction you will take.

References

Arnold, B. L., Gansneder, B. M., & Perin, D. H. (2005). *Research Methods in Athletic Training*. F. A. Davis.

Berg, K. E., & Latin, R. W. (2008). *Essentials of Research Methods in Health, Physical Education, Exercise Science, and Recreation*. Wolters Kluwer.

Ericsson, K. A., & Harwell, K. W. (2019). Deliberate practice and proposed limits on the effects of practice on the acquisition of expert performance: Why the original definition matters and recommendations for future research. *Frontiers in Psychology, 10*(2396), 1–19.

Garrick, J. G. (1986). Ballet injuries. *Medical Problems of Performing Artists, 1(4),* 123–127.

Gillen, C. M. (2007). *Reading Primary Literature: A Practical Guide to Evaluating Research Articles in Biology.* Pearson Education.

Graham, M. (1953). *An Athlete of God* [Radio Broadcast, *This I believe,* Producer: Edward R. Morrow]. Retrieved August 28, 2021, from https://thisibelieve.org/essay/16583/

Howse, J., & Hancock, S. (1988, 2000). *Dance Technique and Injury Prevention.* Routledge.

Ingersoll, C. G. (2001). *Research in Athletic Training.* Slack.

Payton, O. D. (1994). *Research: The Validation of Clinical Practice.* F. A. Davis.

Portney, L. G., & Watkins, M. P. (2009). *Foundations of Clinical Research: Applications to Practice* (3rd ed.). Pearson Education.

Ryan, A. J. (1997). Early history of dance medicine. *Journal of Dance Medicine and Science, 1*(1), 30–34.

Ryan, A., & Stephens, R. (1987). *Dance Medicine: A Comprehensive Guide.* Pluribus.

Sidman, M. (1960). *Tactics of Scientific Research.* Basic Books.

Thomas, J. R., Nelson, J. K., & Silverman, S. J. (2011). *Research Methods in Physical Education* (6th ed.). Human Kinetics.

2

Ways of Knowing

TOM WELSH, LYNDA MAINWARING,
AND JATIN P. AMBEGAONKAR

Questions to Stimulate Thinking

- How do dancers distinguish what is true and useful from what is wishful thinking, untrue, or even detrimental?
- How can a scientific approach to knowing be helpful to dancers and those who train and treat them?
- What special discipline is required to use an empirical approach to knowing?
- Will learning to use an empirical approach require me to give up other ways of knowing?

Science offers a rigorous and disciplined way of thinking and a systematic approach to answering questions, solving problems, and learning how our world works (Portney & Watkins, 2009; Thomas et al., 2011). In this chapter we consider the ways that thinking like a scientist is different from the ways we think while engaged in other life activities. Learning to think like a scientist requires discipline, just as learning to move like a dancer requires discipline. Knowing the basic steps of a dance genre lays the foundation for a well-choreographed performance. Similarly, knowing the basic steps for conducting a rigorous scientific investigation lays the foundation for a well-executed study that can enhance knowledge about the phenomenon under study. Of course, how the steps are performed is what creates the magic in dance and research. Learning to think like a scientist will not compromise and in fact may improve your ability to think creatively. All ways of thinking and knowing are useful in the right context. Good scientists have to think creatively, just as good choreographers do.

The ways we think are like a box of tools with each tool well suited to a particular task. We use hammers to pound nails, drills to make holes, and wrenches or spanners to tighten or loosen nuts and bolts. Having a diverse set of tools in our toolbox allows us to manage a variety of challenges. Being able to think in different ways has similar benefits. Here we look at some of the ways we think about the world, make decisions, and solve problems in everyday life, and then we consider the way scientists work.

Everyday Ways of Knowing

Although most of us rarely give it much thought, we have a variety of ways of knowing what is true and useful in our lives. We use different criteria when we are engaged in different activities. We use everyday ways of knowing in our daily lives because they work for much of what we need to know and do. Which of these ways of knowing do you use for the decisions you make and the actions you take in your daily life?

Tradition

One common way of knowing is tradition. We believe what we believe and do what we do because "that is the way it has always been done" (Portney & Watkins, 2009). Tradition has the advantage of creating common expectations among those with whom we share a personal or professional community. Tradition lets us know what to expect from each other and it provides guidance on how to behave. A disadvantage of being guided by tradition is that unless we have tested a tradition for its authenticity and functionality, we may be making assumptions and following implicit guidelines that are only partially true or not true at all. Another disadvantage is that tradition, once established, is hard to change (as any dancer who has studied a classical dance form with a well-seasoned teacher knows). While testing the utility of every tradition that guides us would be incapacitating, testing a tradition that has important and enduring implications for dancers may be worth some additional effort.

Authority

Another way of knowing is to rely on authority. We listen to "experts" and adjust what we believe and do based on how experts tell us the world works. With the dramatic expansion of access to information made possible by the internet and the sheer complexity of the world in which we live, reliance on authority makes it possible to function in society. If an authority (an individual, an organization, or the designer of an algorithm) is highly experienced and known for impartiality, what we learn from the authority can allow us to operate effectively and avoid having to spend all our time sorting out what is true and useful for ourselves.

However, most authorities have some degree of built-in bias, intended or not. If we are unaware of inherent biases, we run the risk of incorporating those biases into our own belief systems and our patterns of interacting with the world. In addition, the internet's propensity to disseminate false as well as truthful information complicates relying solely on authority to guide our actions. Even when we are inclined to believe and do what an authority figure is telling us, questioning the validity of the advice is the prudent choice, especially when it may have a lasting impact on others. Science offers a thoughtful, responsible, and functional means of testing assumptions promoted by authorities.

Intuition

A third way of knowing is intuition. Sometimes we just seem to know what is true or what needs to be done, without explanation. Dancers could not improvise or execute a new movement phrase that was demonstrated imprecisely by a choreographer without relying on intuition. Whether you believe that intuition has magical roots or that it results from our complex and intensive interactions with the world, it is important to appreciate its utility. Intuition is especially useful when we are surprised by an opportunity and an immediate reaction is required. Although intuition is useful in many contexts, like tradition and authority, it will reflect our own experience, tacit knowledge, and biases. When an immediate reaction is needed, intuition may be the tool of choice. When we have more time to consider a belief or action that we want to recommend to others, it is useful to test the veracity and relevance of our intuitions. We are not giving up on the use of intuition; we are just verifying the appropriateness of what intuition has taught us for the context in which we are applying it.

Trial and Error

A fourth way of knowing is by trial and error. If you have what seems like a good idea, try it and see how it works. If the idea is truly FINER (Feasible, Interesting, Novel, Ethical, Relevant) (Hulley et al., 2013) and effective, put it to use. However, for trial and error to be useful, we have to remain open to the possibility that what may initially seem like a good idea may turn out to work differently than expected when we actually try it. We have to be willing to test our beliefs and ready to learn from that experience.

A limitation of the trial-and-error approach is the time it would take to test every good idea we may have. Although we do not want to ignore what might turn out to be a useful idea, it may be more efficient to focus our efforts on the ideas that are most likely to be useful. Of course, deciding in advance what may turn out to be useful is no mean feat. Kuhn (1970) pointed out that some of the ideas that have led to the creation of new knowledge were initially rejected by experienced scientists who were deeply invested in traditional ways of viewing their field of study. Kuhn labeled these new ways of conceptualizing what we know as "paradigm shifts." Ironically perhaps, knowing when to try a new idea that is not consistent with current scientific tradition may itself require the judicious use of intuition.

Logical Reasoning and Emotion

Two other ways of knowing, logical reasoning and emotion, might be considered incompatible at first. Logical reasoning involves using logic to think through the way some part of the world should work based on what we already know (see discussion of inductive and deductive reasoning in *Chapter 3*). Of course, even well-informed reasoning is limited by the extent of what we know. It is possible that some of the assumptions on which our reasoning is based would turn out to be untrue or only partially true if we could take time to test them.

Emotion, in contrast, involves reacting to a situation based on how we feel about it, sometimes independent of what reasoning is telling us to do! We might even suspect our reactions are misguided but choose to act anyway because we feel that is how the world should work. Using reason and emotion to guide the choices we make both have their place in our collection of thinking tools. For example, scientists who want their work to be relevant to dancers may need to allow emotion to help them choose which problems to address and which intervention to develop and test. But scientists also need to develop the discipline to hold emotion in check while collecting and evaluating data that will help them answer their research question.

Most of us rely on these thinking tools to manage our everyday lives (see Figure 2.1). You can probably imagine other ways of knowing that extend or complement these approaches. If you, or someone you know, uses them to make decisions and determine actions, it is likely because the strategies work well enough for everyday decision-making. However, a more systematic approach to observing, evaluating, drawing conclusions, and offering advice is called for when our decisions will have an important and lasting influence on others (dancers, teachers, trainers, health care practitioners). How do we

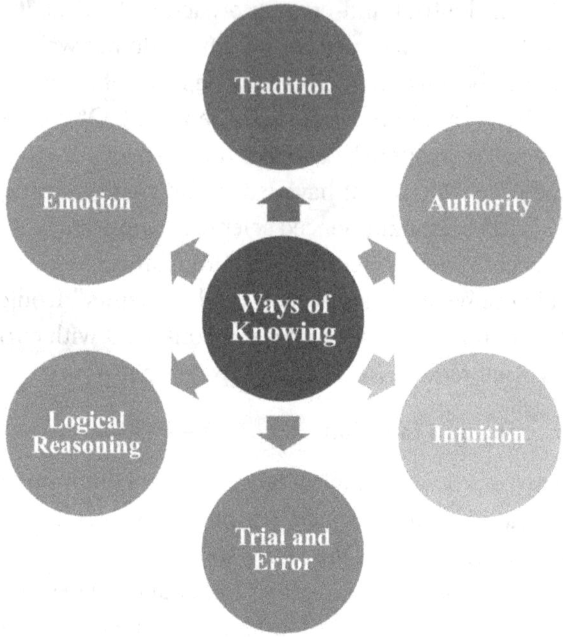

Figure 2.1. Everyday Ways of Knowing.

decide what is true and useful when the actions that follow may have a lasting effect on the dancers we teach, train, or treat?

Scientific Approach to Knowing

The scientific approach is a systematic way of learning how the world works. It allows us to confirm, refine, or revise what we think we know, and it allows us to communicate what we learn to others. Science makes use of many of the ways of knowing described earlier but it does so in a disciplined manner. It is more involved than other ways of knowing, so it may not be our first choice when we have a decision to make or an action to take in our daily lives. However, there are circumstances where the scientific approach is especially useful. We offer an overview of the scientific approach to inquiry, look at the assumptions on which it is based, and consider the discipline required to apply it.

The scientific enterprise operates when scientists employ various methods to achieve their research goals. Those methods are based on fundamental assumptions about the nature of knowledge (epistemology) and the nature of being or reality (ontology), and their associated values, beliefs, and overarching worldview. These philosophical and theoretical parameters constitute a paradigm of inquiry (Kuhn, 1970) and guide scientific thinking and research. Paradigms of inquiry have evolved over time with innovations in science and technology.

The rise of scientific thinking has its roots in the philosophers of the seventeenth, nineteenth, and early twentieth centuries. The most influential group of scientifically trained philosophers was known as the Vienna Circle. Their deliberations laid the foundation for empiricism and the systematic observation of phenomena that informs the currently dominant positivistic paradigm of inquiry. It embraces a worldview that has led to a sophisticated framework for investigating phenomena through a rigorous process known as the scientific method, which is highly prominent in the physical sciences.

As scientists across disciplines develop their approach to understanding the world, new paradigms emerge for understanding the human condition in context. Some of these paradigms focus on human experience from a holistic perspective. In this textbook we embrace all empirical approaches to research, recognizing that some research questions are better suited to one approach over another or better addressed from a combination of approaches. The research question drives the approach, and the dance sciences benefit from the diversity of methods available.

We begin with an explanation of the scientific approach to research, which is informed by the scientific method and, with some modifications, can address research across paradigms. Debate continues about whether worldviews and values from different paradigms can be blended or can coexist, and new beliefs are formulated as philosophies of science evolve (see Morgan, 2014). In Figure 2.2 we describe six steps that synthesize similar lists offered by researchers in a variety of disciplines (Berg & Latin, 2008; Kazdin, 2011; Payton, 1994; Cortney & Watkins, 2009; Sidman, 1960; Thomas et al., 2011). We have adapted our list in an effort to incorporate the diversity of approaches represented in the dance sciences.

Virtually all research begins by identifying and refining a question that a researcher wants to answer. Some research questions address a problem or deficiency, some focus on seizing an opportunity to improve an aspect in the world, and some simply feed a curiosity: "I wonder what would happen if . . ." We suggest focusing on a single, answerable question to provide a clear, specific objective for your first study, even if the steps or components required to answer the main question are elaborated as sub-questions. Studies with too diffuse a focus often end up answering no question definitively. It may be particularly important for new researchers to focus on a single question that is ready to be answered. Other interesting ideas can be saved for future research.

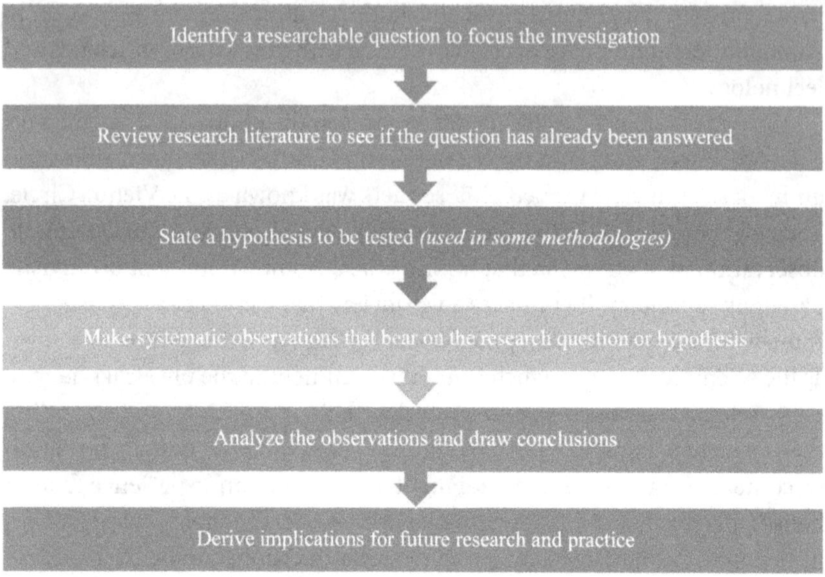

Figure 2.2. Steps in the Scientific Approach.

The second step is not often called out explicitly in descriptions of the scientific approach but it is essential and, we believe, under-appreciated by many new researchers. Before conducting a new study, researchers need to be sure their question has not already been answered by others. If an answer can be found in the literature, researchers can focus their efforts on answering the next question. Or, if their research question has not been answered clearly, they can use what they learn from reviewing the literature to design a study to answer the question more completely.

In some research approaches, the next step is to create an explicit statement that, if supported or rejected by the results, will mean the research question is answered. The statement is called a hypothesis, and it is used to provide a clear, explicit focus for a study (e.g., fatigue increases injury risk in dance). To facilitate statistical analysis, the absence of the anticipated effect is stated as a null hypothesis (e.g., fatigue has no influence on injury risk). In other approaches, researchers resist making such statements to avoid creating expectations that may bias their actions and the interpretation of results (Kazdin, 2011; Sidman, 1960; Thomas et al., 2011). Regardless of the approach, empirical researchers emphasize looking from a neutral perspective to see how the world is working, even if what they see fails to confirm what they believed to be true.

The fourth step in the scientific approach is the most interactive. It involves making systematic observations and turning those observations or measurements into the data that will be used to answer the research question. In some research approaches, the activities undertaken at this stage must be pre-planned and rehearsed so that they can be performed meticulously to ensure data integrity. In other approaches, researchers have a plan for executing the study but remain open to adapting their interactions with study participants in reaction to the evolving context. In either case, researchers document precisely what they do to make it possible for other researchers to replicate their efforts.

The fifth step requires great discipline and some deep thinking. It includes three component steps. The first is to study the data and to summarize as neutrally as possible what was observed. Researchers might summarize their observations with graphic displays, descriptive statistics, or narrative summaries and use the summaries to draw conclusions related to the research question. Then they determine whether the results agree with or differ from previous research and whether the results support any theory that may have provided impetus for the research. The final step in the research process may involve some creative thinking. Researchers identify the implications of the findings for future research and practice. At this stage they have some latitude to suggest and speculate on what they believe the results recommend. How far recommendations may extend beyond the specific context of the study is determined by the clarity and credibility of the results, the methodological approach used, and the conventions in the venue where the study findings will be presented.

Overall, these steps describe a progression that governs the scientific approach that is iterative. Researchers generally move through them in order, but it is common to revisit earlier stages as researchers work through subsequent stages, having learned more about their question and its context along the way. Researchers often clarify or refine and may even change their research question as they progress through the phases of the study.

Attitudes of Science

Just as learning to perform dance steps, phrases, and variations requires dancers to develop a disciplined approach to moving, conducting empirical research requires researchers to develop a disciplined approach to thinking. A number of attitudes have been adopted by scientists to support disciplined thinking. Based on writing by Donald Whaley (1973), we offer five attitudes of science that new researchers will need to embrace as they acquire the discipline to think like a scientist.

Suspend Judgment

Perhaps the most important attitude for researchers to develop is the ability to suspend judgment. It involves a willingness to observe impartially and assess what is being observed without attachment to a predetermined outcome. Scientists are willing to change their presuppositions if their observations suggest that a change would be prudent (Berg & Latin, 2008). Adopting this attitude means that research is not about proving a good idea or promoting a favored approach to treatment or training. Rather, research is about learning how the world actually works. It is about learning how injuries develop and resolve or how dancers learn and improve their ability to perform dance skills.

If you are mainly interested in promoting a preconceived idea of what is good for dancers, consider finding an independent researcher to answer your research question. That will allow you to avoid the risk of having your preference for a particular outcome compromise the integrity of the research. In his landmark work, *Tactics of Scientific Research,* Murray Sidman (1960) explained that "data" collected impartially and without researcher bias are never negative. Even when they do not support a preconception that we had going into a study, they teach us how the world works and that is the purpose of science.

Parsimony

Another attitude that is key to the scientific enterprise is to seek the simplest explanation available to account for the results observed. This discipline is sometimes called the law of parsimony (Whaley, 1973) or Occam's Razor (Payton, 1994). The simplest explanation of the results can facilitate integration of new findings with previous findings, and the integration of research findings may mark the highest achievement of the scientific enterprise (Sidman, 1960).

For example, you may observe that dancers learn a new dance step in fewer trials when their teacher points out the successful aspects of a performance than when the teacher calls out mistakes. Inferring that what the teacher did must have changed the dancers' mood, and that change in mood must have increased the dancers' motivation which, in turn, must have accelerated their acquisition of the new phrase would make the explanation more complicated, particularly in the absence of any direct evidence of a change in mood or motivation. The attitude of being parsimonious favors the simpler explanation. However, the level of detail deemed sufficient differs across methodologies (see Part IV).

Orderly and Dynamic

Two more attitudes that make up the foundation on which much empirical research is based are that natural phenomena are both orderly and dynamic. Although these concepts appear to be contradictory at first, taken together they suggest a synergy that can create a functional balance. If researchers are going to discover how the world works, there has to be some order to discover. If events in the world were entirely random there would be no order to reveal.

This attitude can be difficult to embrace when an event is hard to explain with concepts that are currently popular in a field. However, events that are hard to explain at one moment in history can often be explained later by researchers who create new conceptual paradigms (Einstein, 2014; Kuhn, 1970). However, assuming that phenomena are orderly does not mean that they must be static or even linear. Change is constant in our world, but presumably there is order in the way the world is changing. Some scientists seek to discover the order governing that dynamic process. Rather than focusing on the orderly and dynamic nature of phenomena, some empirical researchers emphasize the relationship between phenomena without requiring that the relationships be linear.

Hold Conclusions Lightly

The dynamic nature of scientific discovery reveals another discipline that researchers must master. Just as scientists suspend judgment while conducting their research, they also hold the conclusions drawn from research lightly (Sidman, 1960; Skinner, 1938). If natural phenomena and our understanding of them are continually evolving, we can expect at least some of what we believe today to need to change in the future.

Holding conclusions lightly lets researchers remain open to new discoveries and more functional ways of understanding the world. This attitude has implications for the use of scientific findings that some may want to use to dismiss the scientific process wholesale. We prefer simply to admit that scientific findings represent only the scientific community's current understanding of how the world may work. In *Chapter 21* we discuss ways that practitioners use research findings to inform their practice, while understanding that research findings are rarely perfectly matched to the questions we want to answer regarding practice.

We hope this description is beginning to make it clear that science is a *cyclical* and *cumulative* activity. Researchers study the reports of prior researchers to provide context for their own work and to guide their approach to answering their own research questions, which likely were suggested implicitly or explicitly by prior research. The report a researcher presents or publishes influences the work of other researchers by suggesting new strategies for studying natural phenomena and problems that are important to dancers.

A well-conceived study, carefully and thoroughly reported, will generate many new and interesting questions to be answered. It may also provide practical guidance for researchers who may wish to pursue those questions. When we engage in science, we join a community of inquisitive colleagues that transcends time and domains. We become contributors to, as well as beneficiaries of, the enterprise of science.

In the next chapter we describe a number of concepts that are fundamental to the pursuit of science. They are guided by the attitudes described in the present chapter. Together these ideas provide a foundation on which to build a common understanding of the enterprise of science. Additional details about these concepts are offered in the following chapters as appropriate.

Summary

Overall, using various strategies of inquiry such as tradition, authority, intuition, trial and error, reasoning, and emotion, independently or in conjunction with one another, are everyday ways of thinking about how the world works. They serve us well in guiding the actions we take in our daily lives. The scientific approach offers a systematic means of developing knowledge. It can be especially useful when we want to provide guidance to others. While it incorporates many of the everyday ways of knowing, the scientific approach is designed to minimize the influence of personal bias. Using a scientific approach requires disciplined thinking and the adoption of several key attitudes that guide the researcher's efforts: Suspending judgment, seeking simple explanations, looking for order even if it is dynamic, and holding conclusions lightly.

Highlights

- All ways of thinking and knowing can be useful in the right context.
- We have to be willing to test our beliefs and ready to learn from our experience.
- Seek the simplest explanation available to account for the results observed.

References

Berg, K. E., & Latin, R. W. (2008). *Essentials of Research Methods in Health, Physical Education, Exercise Science and Recreation* (3rd ed.). Wolters Kluwer.

Einstein, A. (2014). *Relativity: The Special and the General Theory.* Routledge.

Hulley, S. B., Cummings, S. R., Browner, W. S., Grady, D. G., & Newman, T. B. (eds.) (2013). *Designing Clinical Research* (4th ed.). Lippincott Williams & Wilkins.

Kazdin, A. E. (2011). *Single-Case Research Designs: Methods for Clinical and Applied Settings* (2nd ed.). Oxford University Press.

Kuhn, T. S. (1970). *The Structure of Scientific Revolutions* (2nd ed.). University of Chicago Press.

Morgan, D. L. (2014). Pragmatism as a paradigm for social research. *Qualitative Inquiry, 20*(8), 1045–1053.

Payton, O. D. (1994). *Research: The Validation of Clinical Practice.* F. A. Davis.

Portney, L. G., & Watkins, M. P. (2009). *Foundations of Clinical Research: Applications to Practice* (3rd ed.). Pearson Education.

Sidman, M. (1960). *Tactics of Scientific Research.* Basic Books.

Skinner, B. F. (1938). *The Behavior of Organisms: An Experimental Analysis.* Appleton-Century.

Thomas, J. R., Nelson, S. K., & Silverman, S. J. (2011). *Research Methods in Physical Activity.* Human Kinetics.

Whaley, D. (1973). Attitudes of science. In: R. Malott (ed.), *An Introduction to Behavior Modification.* Behaviordelia.

Additional Resources

Gladwell, M. (2015). *The Tipping Point: How Little Things Can Make a Big Difference.* Abacus.

Scalë. *FINER: FINER stands for feasible, interesting, novel, ethical, and relevant.* https://www.scalelive.com/finer.html

Strogatz, S. (2004). *Sync: The Emerging Science of Spontaneous Order.* Penguin.

3

Research Concepts

TOM WELSH, LYNDA MAINWARING,
AND JATIN P. AMBEGAONKAR

Questions to Stimulate Thinking

- What is a research paradigm?
- Can we agree to a shared purpose for dance science research?
- What key concepts will help me begin to understand research writing?

The diverse collection of domains that constitute the dance sciences is an asset because it allows complementary perspectives to be combined to enhance research and practice in our field. However, it also means we may approach our shared purpose with different assumptions about what constitutes convincing evidence and effective science.

Our intention for this chapter is to identify a common purpose for the dance sciences and to introduce new researchers to a collection of terms and concepts that are frequently used in dance science research. Having an elementary grip on these concepts will help you begin to comprehend research reports, which we encourage you to begin reading now. Researchers learn the basic concepts of science early in their careers and continue to use them, often without much conscious thought, throughout their careers. We begin this chapter with a description of paradigms of inquiry followed by a discussion of the purpose of science as it relates to dance.

Paradigms of Inquiry

The research enterprise operates when scientists employ various methods to achieve their research goals. These methods are based on fundamental assumptions about the nature of knowledge (epistemology) and the nature of being or reality (ontology), and their associated values, beliefs, and overarching worldview. Each of these sets of parameters is called a *paradigm of inquiry* and these paradigms guide scientific thinking and research.

Scientific research in all fields, including the dance sciences, often leans toward one of two commonly used research paradigms. When using a *naturalistic* approach, researchers seek to learn and describe how the world works without affecting it unnecessarily. They try to describe what they observe as clearly and as uninvasively as possible. When researchers use narrative descriptions to record observations, they may adopt the label "qualitative" for their work. When using an *analytic* approach, researchers commonly make observations or measurements under carefully controlled conditions. In some instances, they systematically manipulate variables to isolate the influence of one variable on another. This approach is often labeled "quantitative" because numerical values are used to quantify the measurements made.

We believe both perspectives can have value by allowing researchers to answer different types of research questions. Some researchers find it useful to embrace a single paradigm, while others choose different paradigms to answer different types of research questions. We encourage new researchers to learn the strategies, advantages, and challenges of a variety of approaches to empirical research before choosing to specialize in one or another. Doing so will allow you to comprehend the work of colleagues, gain insights that may influence your own work, and generate opportunities to collaborate with other researchers to achieve complementary goals.

Purpose of Science

As any dancer can confirm, it is often the most basic tasks that are the hardest to accomplish convincingly. Identifying the purpose of science may be one such task. Passer (2014) suggested that science can be defined as a process for acquiring knowledge—a way of "systematically gathering and evaluating empirical evidence to answer questions and test ideas." Marx and Hillix (1973) identified science as a *social* enterprise, the product of which is *knowledge,* acquired by using the scientific *method,* applied with a scientific *attitude.* A prolific scientist from the twentieth century described science as "a search for order, uniformity, and lawful relations among events in nature"

(Skinner, 1953), and another offered five motivations for conducting scientific experiments (Sidman, 1960):

To satisfy a researcher's curiosity (perhaps the purest motivation)
To demonstrate a new phenomenon
To evaluate a new method or technique
To explore the conditions under which a phenomenon occurs
To evaluate hypotheses and test theories

Can we glean from the writings and our own research experiences, however modest they may be so far, an underlying purpose for the type of science our field seeks to promote?

It seems clear that the purpose of science involves answering questions systematically to support the development of an accumulating knowledge base. Careful planning, focused observation, meticulous record keeping, and thorough evaluation of results are common in the sciences and essential for answering questions scientifically. In our daily lives, tradition, personal preferences, and beliefs play important roles in the decisions we make and the answers we give. Empirical observation and analysis are either as free as possible from personal biases or the researchers who conduct them openly acknowledge their unavoidable biases through a procedure called *reflexivity*, so that findings can be interpreted with that bias in mind. Both approaches help make the scientific approach to answering questions self-correcting rather than of self-promoting.

By answering questions scientifically, we can come to understand our world in ways that help us comprehend human experiences from various perspectives. It can also facilitate the detection of relationships that may permit researchers to make predictions, such as which body types among young dancers are better suited to survive and thrive in a particular form of dance training. And it can permit researchers and practitioners to influence events by, for example, strengthening dancers' core-stabilizer and lower extremity muscles, to make dancers more resistant to overuse injuries as they begin an intensive course of study in a challenging dance form. Different research methodologies emphasize different aspects of these collective intentions. We propose the following purpose for research in our interdisciplinary field:

> The purpose of dance science research is to create knowledge that can help us understand, explain, predict, and influence phenomena that relate to dancers' health, well-being, and performance.

Research Concepts

In this section we describe a number of concepts used in many of the disciplines represented in the dance sciences. They grow from the attitudes of science described in the previous chapter. You might think of them as a tourist guide to reading empirical research. Becoming aware that they exist and getting an idea of the relationships among them will help you begin to make sense of what you are reading. Similar to learning a foreign language and new culture, your understanding of these research terms and concepts will grow more informative as you use them in your interactions with dance science research. Methodological purists may take issue with some of the groupings we are making, but if you view our suggestions as a first step in comprehending these concepts, you can expect them to grow more precise and more refined as your research experience expands.

Topic, Research Question, Hypothesis

When initiating research, researchers select a topic that interests them. They look for a problem that needs to be solved or an opportunity that is ready to be seized. As they review previous research to learn what is known about a topic, problem, or opportunity, they develop a research question to focus their efforts. A *research question* is specific, limited in scope, and identifies a particular intention or purpose for the research. Research questions often appear in the *purpose* statement for a research report or presentation. Some researchers also compose a *hypothesis*—a statement that describes a relationship between variables or predicts an outcome—which can be tested to answer the research question (Portney & Watkins, 2009).

Basic and Applied Research

Basic and applied describe a continuum along which scientific research and observations of phenomena may vary. At the *basic* end of the continuum, research primarily answers a question of curiosity ("I wonder what would happen if dancers could not use their arms while performing pirouette—a spin on one leg?") or to test a theory ("How is the theory of conservation of momentum revealed while performing multiple pirouettes?" (Laws, 2002). At the *applied* end of the continuum, research usually addresses a concern that has immediate practical significance, such as: "Would giving dancers a 3-week break from training and rehearsing following an intensive performance season reduce symptoms of over-training?" (Koutedakis et al., 1999).

Most research can be placed somewhere along the basic-applied continuum, and many research questions can be modified to move research in one direction or the other. For example, "How can the theory of conservation of momentum be used by dancers to improve their performance of pirouette?" gives a basic research question immediate, practical relevance, moving it in the applied direction. Knowing that you can refine your research question to make it more applied or more basic is more important than being able to place your question at any specific point along the continuum. It also makes arguments about whether a specific study is basic or applied research more academic than pragmatic.

Laboratory and Field

A similar continuum exists between laboratory and field settings. A *lab(oratory)* environment allows researchers to measure variables and manage procedures with greater precision. That precision can lead to more definitive assessments of the variables under study. In contrast, *field* settings are environments that are similar to the environments where dancers, teachers, and other practitioners work. It may be harder to control variables precisely in the field, but what is observed in field settings may be more similar to the environments where the practices will ultimately be deployed. This may make findings from field settings more immediately useful to practitioners. Of course, researchers can adjust either type of setting to make it more like the other. As with basic and applied research, the lab-field distinction can be viewed productively as a continuum that you can navigate to match your research question.

Population, Sample, Sampling

A population is a set of individuals or entities to which researchers imagine their findings might be applicable. For example, dancers enrolled in professional training programs in a particular country might be a population of interest to dance science researchers studying the effects of pre-season conditioning. When researchers are interested in generalizing their findings to a target population, they select a representative sample of individuals to participate in their study. A variety of *sampling* procedures have been devised to accommodate a variety of research contexts.

Observations and Data

Observation is a fundamental activity of empirical science. Rather than imagining how the world might work, empirical researchers look to see how it actually is working. Researchers may directly observe a dancer performing a movement or skill, or they may test a physical or mental capacity relevant to dancing. Researchers also may ask participants to report experiences, opinions, or perspectives to get an indirect indication of a preference or trait. Or researchers might engage intensively with participants to get a deeper understanding of their experiences. Observation—whether direct, indirect, objective, or subjective—is a keystone of empirical research.

Data are records of observations made by researchers who are seeking to answer a specific research question (Payton, 1994). Qualitative data may include a narrative account of a dancer's experience with injury or a list of reoccurring themes generated by researchers who review and code the meaning in the narratives. Quantitative data generally include numerical descriptions of observations or measurements made by researchers such as frequency, percentage, duration, length, mass, force, and other measurable dimensions of the phenomenon under study.

Variables

A variable is an event, feature, or factor that is likely to change. An *outcome variable* (also called a *dependent variable* in some methodologies) is the performance observed or the capacity measured by researchers to address their research question. Identifying and detailing important outcome variables is often a key purpose in qualitative research. In quantitative research, data are gathered on one or more dependent variables and compared across participants, settings, time, or experimental conditions to look for similarities, differences, and changes. Dependent variables may involve some form of measurement, although the measurement can be as simple as counting how many times a particular event occurs. A researcher might, for example, count the number of times a dancer asks a question while learning a new movement combination. The label *dependent variable* can be abbreviated as DV when space is tight and if using an abbreviation will not confuse readers.

An *independent variable* (called a factor in some research methodologies) is a treatment, training, or other procedure (perhaps a company policy), whose influence on a dependent variable is assessed. In intervention studies the independent variable is applied by the researcher and its application is carefully controlled across conditions or groups. For example, a dance pedagogy researcher might compare the differential effects of a condition in which dancers receive only affirmations of successful execution of a dance skill with a condition in which dancers receive only critiques of errors. In other studies, differences in the independent variable may be naturally occurring or otherwise beyond the immediate control of the researchers. The independent variable is the procedure, intervention, or treatment whose influence on the dependent variable is being evaluated. The term *independent variable* can be abbreviated as IV when doing so will not confuse readers.

Potential *confounding variables* are extraneous or uncontrolled influences that have the potential to affect the dependent variable. Particularly when conducting experimental analyses, controlling extraneous variables is important for ruling out uncontrolled, perhaps undetected, influences that might confound interpretation of the results.

Methodological Integrity

Methodological integrity refers to the ways in which researchers ensure that their research findings provide an authentic representation of the phenomenon examined. One key step to ensuring methodological integrity is to have researchers *operationally define* the procedures used in their study. Operationally defining research procedures (e.g., how to qualify participants for a study, how to measure dancer performance, how to apply a treatment or training protocol) involves describing precisely what is to be done by the researcher, teacher, clinician, or dancer and how the concept under study will be measured or standardized. In some approaches the operations are left flexible intentionally but are documented in detail using an *audit trail.*

Observations and measurements must be *reliable* (consistent) and *valid* (measure what they purport to measure), and *procedures* must be implemented with *fidelity* for research findings to be *trustworthy* and *credible. Rules of evidence* or *standards of methodological integrity*—tailored to a particular research methodology—are used to guide and to evaluate the *credibility* of research finding (e.g., CASP, JARS, PEDro, and see Kratochwill et al., 2010, and other additional resources following references, this chapter). Problems of interpretation can occur if the standards appropriate for one type of research are inappropriately used to evaluate the methods used in a different type of research.

Internal validity addresses the trustworthiness in the data generated and the research procedures employed. Internal validity describes the degree to which research findings accurately represent the phenomena under study. *External validity* and *transferability* represent the degree to which the findings of a particular study can be applied to individuals or entities who were not involved in the study but are similar to the study participants. Generalization is a statistical approach to forecasting applicability based on how participants were selected for a study (i.e., were they selected at random from the population).

Replication is the ultimate test of the applicability of research findings to different participants, settings, and contexts in analytic research (Kazdin, 2011, p. 375). Are other researchers or practitioners able to obtain the same or similar results when following the procedures operationally defined in a

Table 3.1. Similar Intentions across Research Paradigms

Lay Term		Qualitative Concept		Quantitative Concept
Consistency	→	Dependability	→	Reliability
Accuracy	→	Confirmability	→	Objectivity
Applicability	→	Transferability	→	Generalizability
Believability	→	Credibility	→	Validity

research report? Some fields require replication to confirm a finding, and some research methodologies use the replication of an effect of one variable on another within a study as a means of assessing *experimental control* or *confirmability.* The disciplined application of this practice can help scientists avoid jumping to conclusions, which, you will remember, is an essential discipline in the sciences. It also allows science to be self-correcting.

In qualitative studies the focus is on revealing what may be unique details of the phenomenon under study. Particular findings in these studies may be *transferable* to other contexts but with specified limitations. The *credibility* and *authenticity* of findings are affirmed though the use of *triangulation,* which involves cross-checking research findings by using multiple sources of data, multiple methods of data collection, or more than one researcher. Table 3.1 highlights similar intentions across research paradigms.

Thinking or Reasoning

There are two general directions of thinking or reasoning. *Inductive reasoning* involves looking at a collection of observations and creating a summary of what those observations reveal or suggest. For example, you may observe that a majority of dancers experiencing an injury this season returned to work after a month or more hiatus from regular physical activity comparable to the in-season demands of rehearsing and performing dance. If you observe the same pattern across many dancers, over a series of seasons, and across a variety of dance training settings, you might use inductive reasoning to summarize the observations with a statement such as "returning to dancing after an extended hiatus in dance training appears to increase the risk of injury."

Deductive reasoning, by contrast, involves using a general rule (a theory, principle, or hypothesis) to predict how a specific event is likely to unfold. If previous research shows that large, quick changes in the volume of training increases dancers' risk of injury, a company director might predict that because Sally took the past month off for a holiday in Europe, she may be at greater risk of injury this season than she was last season, when she was

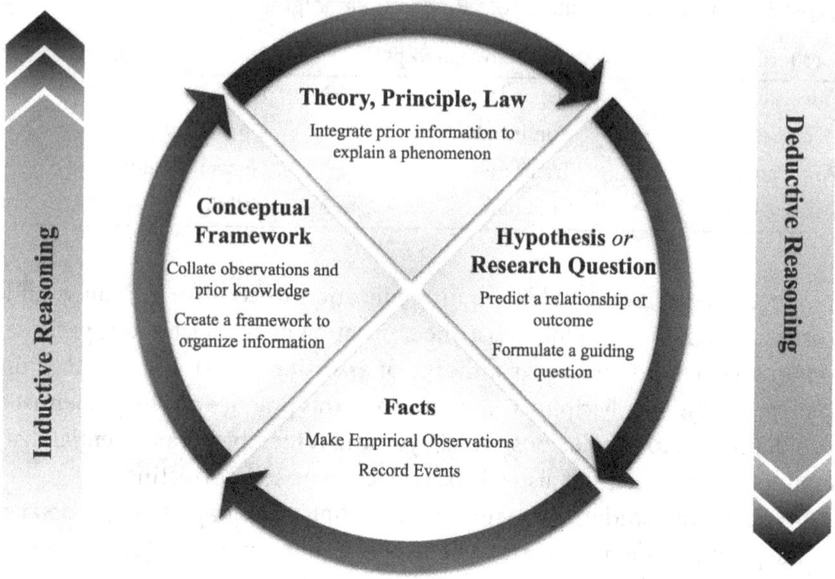

Figure 3.1. A Model of Scientific Thinking. Adapted from Portney & Watkins (2009).

taking daily classes in Pilates, floor-barre, and dance technique the month before returning to work with the company. Some types of research rely more heavily on one approach or the other, but both methods of thinking are used at different stages in most research approaches.

These two forms of reasoning drive the cyclical process of scientific thinking. A key stage for some researchers is the creation of a *theory* based on a conceptual framework, a research question, and knowledge acquired from prior research. Theory-based researchers adopt or devise a theory that integrates *facts* (records of events) revealed in prior research in an effort to explain a phenomenon (Berg & Latin, 2008). They commonly state a *hypothesis* that can be tested to enhance or diminish support for the theory. Other researchers devise a research question based on existing knowledge and design a study to answer their question without forecasting any particular outcome. All researchers conduct studies to reveal new facts and relationships between them, which, when numerous and consistent, may support the statement of a *principle* or *law*. The process is iterative and self-correcting. Principles are often used to guide evidence-based practice. Figure 3.1 depicts the interactive relationships between facts, concepts, hypotheses, and theories, and suggests the influence of inductive and deductive reasoning.

Summary

Research in the dance sciences is aimed at empirically answering questions that allow researchers and practitioners to understand, explain, predict, and influence events that improve dancer health, well-being, and performance. Research concepts provide scientists with a shared language for communicating their approach to their work and their findings. The concepts described in this chapter are offered as a starting repertoire for studying research writing. This basic understanding will grow more detailed and nuanced as your study of empirical research proceeds and your ability to comprehend and use the concepts will become more sophisticated and refined to match the path you choose to follow in your research.

Bon voyage!

Highlights

- Science is a *social* enterprise the product of which is *knowledge*.
- Basic and applied describe a continuum along which scientific research may vary.
- Observation is a fundamental activity of empirical science.
- Methodological integrity must be assessed using standards appropriate for a particular research methodology.

References

Berg, K. E., & Latin, R. W. (2008, 2004). *Essentials of Research Methods in Health, Physical Education, Exercise Science and Recreation* (3rd ed.). Wolters Kluwer.

Kazdin, A. (2011). *Single-Case Research Designs: Methods for Clinical and Applied Settings* (2nd ed.). Oxford University Press.

Koutedakis, Y., Myszkewycz, L., Soulas, D., Papapostolou, V., Sullivan, I., & Sharp, N. C. (1999). The effects of rest and subsequent training on selected physiological parameters in professional female classical dancers. *International Journal of Sports Medicine, 20*(6), 379–383.

Laws, K. (2002). *Physics and the Art of Dance.* Oxford University Press.

Marx, M. H., & Hillix, W. A. (1973). *Systems and Theories in Psychology* (2nd ed.). McGraw-Hill.

Passer, M. W. (2014). *Research Methods: Concepts and Connections.* Worth Publishers.

Payton, O. D. (1994). *Research: The Validation of Clinical Practice.* F. A. Davis.

Portney, L. G., & Watkins, M. P. (2009). *Foundations of Clinical Research: Applications to Practice* (3rd ed.). Pearson Education.

Skinner, B. F. (1953). *Science and Human Behavior.* MacMillan.

Sidman, M. (1960). *Tactics of Scientific Research.* Basic Books.

Additional Resources

Critical Appraisal Skills Programme. *CASP Checklists.* https://casp-uk.net/casp-tools-checklists/

American Psychological Association. (2018). *Journal Article Reporting Standards (JARS).* https://apastyle.apa.org/jars

Qualitative and Quantitative Research: Glossary of Key Terms, accessed 10/14/21. https://writing.colostate.edu/guides/guide.cfm?guideid=90

Kratochwill, T. R., Hitchcock, J., Horner, R. H., Levine, J. R., Odom, S. L., Rindskoph, D. M., & Shadish, W. R. (2010). Single-case designs technical documentation. *What Works Clearinghouse,* (pp. 1–34), https://ies.ed.gov/ncee/wwc/Document/229

Physiotherapy Evidence Database. (1999, June 21). *PEDro Scale.* PEDro. https://www.pedro.org.au/english/downloads/pedro-scale/

II

Creating a Foundation for Research

4

Reviewing Research Literature and Choosing a Topic

LEAH REILLY SHERMAN AND MERRY LYNN MORRIS

Questions to Stimulate Thinking

- What are some tools for brainstorming a topic?
- Is the research process a linear, step-by-step progression?
- What should I look for while reading as I review my gathered research?
- Does my project need to result in a completely new research product?

What does it mean to "create a foundation for research?" In this chapter we discuss important elements to consider and actions to take as you embark on a new research project in the dance sciences. Whether you are an early-career researcher, or you already have substantial research experience, the advice provided here applies to building a foundation for any research project. This chapter focuses upon (1) how to select a research topic, and (2) how to find and review research to justify, inform, and situate your topic. These points are presented as an accessible walkthrough of the process without prescribing a step-by-step, linear recipe for completing research. Importantly, research is both organic and iterative, similar to a choreographic process. Steps are repeated in a cyclical fashion with clarity, depth, and growth arising with each cycle through the phases. Likewise, the research process requires sustained investment in an idea, creative and critical exploration of that idea, and a systematic plan or approach. This chapter offers suggestions for how to engage successfully in a research process. You might also use these recommendations to guide your efforts from your first research endeavors into your professional career.

Selecting a Research Topic

Before starting research, you should have a topic in mind and a sense of the logistics of your inquiry. To begin, reflect upon your experiences in dance. What questions or curiosities have arisen in your dance journey thus far? You might wonder how to improve the execution of a movement, how best to recover from an injury, how to prepare the body for certain types of choreography, or how to cope with the mental stressors of dance. You might have questions about the traditions of dance training practice, emerging trends, instructional methods, technologies in dance, or dancers' nutritional needs. Figure 4.1 illustrates an array of dance research topics.

If you have difficulty brainstorming, consider soliciting possible topics from peers. In conversing with other dancers and researchers, you can expand your knowledge terrain and personal assumptions (Overby et al., 2019). Sharing resources, ideas, and information can maximize research outcomes. Research is often collaborative and is best accomplished through intentional intersection with the research community relevant to your field. Additionally, you might familiarize yourself with trends by skimming dance-related journals or sports science journals (where there may be opportunities to transfer research developments into the dance context). Experiment with free writing on your topic for at least 10 minutes. This is a physical way of mobilizing and organizing your thoughts and ideas early on. Continuously write, without judgment or correction, for a concentrated period. This process can bring insights and questions to the fore and illuminate connections or context for your topic. Return to the free writing process throughout your research to support creativity, clarity, and consolidation of ideas (Overby et al., 2019).

Although any project can seem daunting initially, remember that research is an iterative process; your topic may change as you progress through cycles of investigation, development, and synthesis. While moving through this process, consider three aspects: scholarly importance, logistical conditions, and personal interest. Table 4.1 offers some questions you might ask of each aspect in relation to your topic. Their importance for any given endeavor is at the discretion of the researcher, with guidance from a mentor. Imagine these components working in tandem and having some overlap: the balance you strike is your own.

Figure 4.1. Brainstorming a Topic.

(*A*) Motion capture technology can be used in dance science research to help identify faulty technique and inform how best to execute a movement. Photo: Merry Lynn Morris.

(*B*) A research collaboration might examine teaching of fundamental movement skills for dance. Photo: Footprints Dance Project Society of Alberta.

(*C*) Dance science researchers might investigate technological innovations and their impact on dancer training and performance. Photo: David Shelor.

(*D*) Dance science researchers might explore a variety of different dance forms and genres and the role of community dance in physical and mental health. Photo: Shae Movement North Carolina.

Table 4.1. Research Topic Aspects

Scholarly Importance
Will research on this topic benefit the field?
Has prior research used an approach that may fit my research question?
How will my research add to what is already known about the topic?
Logistical Conditions
Is this research feasible in the time I have allotted?
Are there personnel or funding limitations that will limit my research pursuits?
Do I have access to the equipment, an experienced and interested mentor, and funding I will need to accomplish this project?
Personal Interest
Will this topic keep me interested over the duration of the study?
Is this project consistent with my personal interests?
How will I sustain my motivation as I pursue the project?

In comparison to scholarly relevance and logistic concerns, consideration of "personal" interest may sound like the easiest or least serious of the group. However, enthusiasm for a topic is powerful and will strengthen the research process and the final product you create. If the topic is personally engaging, your commitment level will be higher, your work will be more intentional, and your investment will be evident in the quality of your research, writing, and final product. For early-career researchers, a topic appealing to personal curiosity may sustain your interest more than an assigned topic or one chosen hastily.

There are four points to remember while conducting your research. First, reviewing the literature begins early, continuing as your research grows and your question matures. Productive researchers read early, read often, and read a lot. This is crucial to determining the scholarly relevance of your topic and whether your interest intersects with what is pertinent to the field. The process of finding and reviewing research occurs in the earliest stages of any investigation and continues through the final drafts as you discover new, relevant findings.

Second, research is unpredictable, again mirroring aspects of a choreographic process; a linear trajectory from beginning to end is not frequently the norm. Although there may be unforeseeable shifts occurring during your research, this does not mean it is a haphazard experience. Rather, adopt a systematic approach with consistent intentionality while remaining responsive to new information and challenges. Understand that change happens, and be flexible as you learn more.

Third, the need to develop background around your topic must be approached pragmatically. It is easy to become overwhelmed by the volume of literature you encounter. It is not our responsibility as researchers to review and synthesize every study related to our topic. Realistically, seek a balance between what is important and what is current.

Fourth, dancers new to research might imagine that the most respected scholarship addresses topics from entirely novel and creative perspectives. That expectation can be misleading, however, particularly for those embarking upon this process for the first time. It is essential that you contribute *something* new, not that your entire approach is new. One example lies in *systematic replication,* or the practice of repeating an existing study with one small change (using a new population, setting, methodology, or duration). You might adopt a similar approach, implement analogous protocols, use similar testing instruments, and/or employ the same data analysis methods from an existing study. Although this work is in conversation with published research, by altering just one factor of your study from the original, your findings and final product will differ. This exercise parallels learning to replicate existing dance movements, phrases, and choreographic repertoire—an effective way to learn skills that will be useful as your abilities expand. It is advisable to follow an existing research model when beginning research of your own.

You might choose instead to use *direct replication,* aligning as closely as possible to the original research. As emerging researchers develop a repertoire of such skills, other opportunities for contributing to the dance science knowledge base will become accessible. Good research is built on the work of those who have preceded us. Asking questions and thinking about the issues at hand as you attempt to replicate well-designed and carefully conducted prior research will be instrumental in laying the groundwork for your future efforts. A *direct replication* might be an ideal first effort for a new researcher to develop skills, and a *systematic replication* may better suit a subsequent effort to test what a researcher has learned. Both can make important contributions to the knowledge base in the field (Arnold et al., 2005).

Gathering and Analyzing Research Literature

The remainder of this chapter focuses on two key activities: collecting information and analyzing what you have gathered. In both undertakings, be flexible with your agenda as what you know about your topic grows. Begin with a question rather than an answer—otherwise your research may be shortsighted, simply confirming your preconceptions and drawing conclusions prematurely. Think of these initial stages like a mystery: finding clues and observing patterns with intent to discover significant insights as your conclusion. Rigorously follow the evidence to draw an accurate conclusion to a credible critical inquiry, rather than making assumptions that influence the investigatory nature of your research. The objective is to deepen your understanding of an issue or idea. Openness with your perspective is important; your research will ultimately lead you to more questions.

Collecting Information

Strive to amass a breadth of research on your topic *before* attempting to select a specific research question. Read broadly at first and stay receptive to big concepts. During this exploratory phase, you may refine your ideas; this will remain true later, too, while summarizing what you have found and explaining the implications and opportunities for your topic. Research is an exercise in sharpening your critical thinking, problem-solving, and argument development skills; the process of designing and conducting research is as important as the product you create. Think of this phase as more like rehearsing than the final performance—research activity builds toward creating a research report, article, or presentation.

"Research" is both a noun and a verb, both material and action. It is helpful to conceptualize it both ways as you pursue a project. Table 4.2 lists research materials you will want to consider. Do not limit yourself to only electronic *or* print materials, or only books *or* journals. It is easy to rely on large research databases alone, but in academia many opportunities exist from a variety of resources and disciplines, making the overall quality of your work robust. Approach a large body of research strategically. Seek out literature reviews on your topic, handbooks or primers published on thematic research topics, or even the websites of organizations where key sources and data on a topic exist. These resources may expedite your initial search and will be enriched by subsequent discoveries along the way.

Whether looking for print materials or e-books in a library catalogue or researching articles and abstracts in a database, keep your options open with broad search filters. You might begin with general terms before using discipline-specific jargon. Narrow your results by refining terms but do this after seeing the outcomes from your initial searches. Some filters to consider alone or in combination include title, author, keyword, subject, and call number range. Subject terms and call numbers provide linked results or instances of scholarship connected by related terms and topics. There is no right order or combination of these options, and all are commonly found in library catalogues and electronic databases (Cunningham, 2017). You can

Table 4.2. Types of Research Materials

Types of Research Materials
Print or digital publications
Monographs and edited volumes
Academic journals
Non-scholarly periodicals (magazines, newspapers)
Dissertations and theses
Conference proceedings and abstracts
Open Access (OA) and Open Education Resources (OER)
Multimedia items (images, videos, podcasts)

Table 4.3. Boolean Search Terms

Term	Description
AND	A limiter (returns only results applying to all independent terms searched)
OR	An expander ("OR is MORE"); returns combination of all results for two or more independent terms
NOT	A limiter; excludes specific term from searches of at least two independent terms

combine terms and filters across multiple fields or use the classic Boolean operators offered in Table 4.3.

To search judiciously, record the terms you use, how you combine them, and where you use them (databases, library catalogues, etc.). Also, what does not work is as informative as what does: if a combination does not return the desired results, it might become helpful later as your question changes. While you may start with an aspect of your topic in mind, as your inquiry matures, so might your point of view. If so, reconsider the combinations that were not an exact match in your earlier research. Perhaps this concept is best compared to using improvisation to germinate dance choreography: sometimes you must work through things by re-engaging with prior information or experience and trying something again or in a new way. Keeping notes on your search process will aid efficiency later in refining your research. See *Chapter 7* for a more detailed discussion of search output.

Useful materials can be identified quickly by skimming an abstract or table of contents for key terms. If a resource aligns with your topic, continue by asking: What are the research goals? What methods are employed? And what conclusions were derived from this study? Many articles, specifically in the sciences, follow a similar format: abstract, introduction, method, results, and discussion. Reading the abstract provides an effective initial skimming technique for gleaning a summary; to find more about the methods, outcomes, or other components, go strategically to those portions of the article (Berg & Latin, 2008).

When you find a resource that speaks to your topic, look closely at its reference list. A reference list compiles the sources an author specifically employs in the text—not to be confused with a bibliography, a much longer list of everything accessed during one's research. Several references from a source may be appropriate for your research. In some disciplines "references" is used interchangeably with "citations," but in this chapter we use the former to mean sources included in a reference list. The citation (e.g., author name, date) that is placed within the running text points readers to the corresponding reference. Scanning the reference lists of electronic or print materials will support a more efficient process, potentially leading to resources you did not expect to find. Similarly, a bonus of browsing in a library is unintentionally finding additional titles physically located near those you are purposefully investigating. Research is organic, and sometimes serendipity can pay off! Table 4.4 provides several examples of electronic dance resources, including databases and journals.

While you explore, flag entries appearing in multiple reference lists. Some databases include the number of times a particular reference has been cited. Publications like these may be older, rather than the most current literature, but when you see the same study cited repetitively across the years, take note. These are seminal contributions within a discipline and can be appreciated as the core literature. These are the anchoring voices you seek, and they are the ones with whom current scholars are corresponding and interacting. Indeed, no literature review is complete without acknowledging the contributions foundational to the study itself. One caveat is that medical sciences favor recent publications, like journal articles from the last five years. Depending on the nature of your project, you may need more or fewer historical references versus newer contributions; this balance is best decided considering the objectives of your project and in consultation with your research mentor.

Librarians as Resources

Throughout your research and writing processes, remember that some of the most valuable information is garnered through direct, live interaction with people. Although this section focuses on librarians, faculty or other professionals familiar with your area of inquiry are also assets. Furthermore, you might consider podcasts, TED talks, or other media from reputable people in the field. From your initial research through the final draft, do not hesitate to reach out to your local librarian. For the dance sciences, seek out a librarian who focuses on dance, performing arts, the movement sciences,

Table 4.4. Electronic Resources in the Dance Sciences

Databases
Dance Education Literature and Research Descriptive Index
Dance On-line: Dance Studies Collection
ERIC
International Bibliography of Theatre and Dance with Full Text
International Encyclopedia of Dance
JSTOR
Performing Arts Periodicals Database
PsycINFO
PubMed
SPORTDiscus
Journals
American Journal of Dance Therapy
Arts and Health
Body & Society
Body, Movement and Dance in Psychotherapy
Dance Research Journal
Journal of Applied Arts & Health
Journal of Applied Sport Psychology
Journal of Dance & Somatic Practices
Journal of Dance Education
Journal of Dance Medicine & Science
Journal of Orthopedic & Sports Physical Therapy
Journal of Physical Education
Journal of Strength and Conditioning Research
Medical Problems of Performing Artists
Research in Dance Education

Note: See Appendix C for an expanded list.

sports psychology, or orthopedic medicine. Academic librarians of any specialty offer one-on-one consultations, and as experts in their subject-specific collections they can help you find resources that are not obvious through your independent efforts. Meeting with a librarian can be an effective way to process some of your questions and problems. By exploring your ideas with another person, you practice organizing your thoughts in new ways, and librarians can offer additional insight. There is no shame in asking for help, only opportunity.

Librarians can also advise about other library services. Interlibrary Loan (ILL) and similar multi-library lending services are one such resource. These networks allow researchers to borrow materials from outside their library's collection, opening their research horizons exponentially. Aggregated catalogues can be helpful in scanning the field of publications relating to a specific topic. Browsing WorldCat reveals what materials are available through ILL, for instance. Touting itself as "the world's largest network of library content," this catalogue includes records from libraries internationally, including academic, public, and museum institutions. WorldCat resembles other library catalogues in that it includes multiple fields and filter options for limiting searches by title, author, keyword, and topic. Contact your librarian about requesting these materials; you may need to go directly through WorldCat or your library may have its own form. Your librarian may also introduce library research guides: online publications for specific disciplines, themes, or courses, compiling subject-related resources into one helpful location. Research guides are an excellent place to start any project and to revisit as your research matures.

Citation Management in Research

Proper citation or reference management is a critical research discipline to learn. As you discover resources for your project, record them in a document or by using a citation manager. The latter allows you to save references found in catalogues or databases for review later. You can generally save full-text files with exported resources for quick reference. This kind of efficiency parlays well when you find gaps in your research: you can return to your bibliography and easily identify what you still need. Such knowledge inspires continuous research, helping to inform continued exploration beyond initial resource gathering. Another advantage of tracking your resources is that you will be more efficient in citing these works in the written product, both in-text and in the reference list.

Librarians can be instrumental in strategizing your research workflow, too. This includes using a citation manager or locating the most current information about your discipline's preferred writing style. Your librarian can introduce you to these tools in a consultation or through a library workshop. Table 4.5 compares four major citation manager programs. Ask the librarian which reference/citation managers your institution subscribes to, and try a few to see what you prefer, as each interface varies in its functionality, strengths, and weaknesses. Citation management programs vary in the level of detail they manage. They are helpful in tracking resources, and they can convert reference citations into any format required. The better organized

Table 4.5. Citation Managers

Endnote	https://endnote.com/	Web-based or software Subscription Simple to customize citation styles Work-offline available Compatible with MS Word Attaches pdf files
Mendeley	https://www.mendeley.com/	Software with online account Subscription Cannot customize citation styles Work-offline available Compatible with MS Word, Open Office Attaches pdf files
RefWorks	https://refworks.proquest.com/	Web-based Subscription Simple to customize citation styles Work-offline available Compatible with MS Word, Google Docs Attaches pdf files
Zotero	https://www.zotero.org/	Software with browser add-on Free Difficult to customize citation styles Work-offline available Compatible with MS Word, Google Docs, Open Office Attaches pdf files

your research is from the beginning, the smoother your writing process will be. Staying organized is necessary whether completing a semester-long project or a multi-year dissertation. Additionally, reference managers can automate (although not always perfectly) the process of generating a reference list and the process of formatting citations and references to match the styles used in different journals (see *Chapter 7*).

Analyzing Your Research

Once you have gathered an assortment of applicable resources it is time to analyze them more carefully. In this section we discuss some strategies you might implement. This portion of the research cycle might initially feel intimidating, depending on the depth and breadth of research collected. Start by re-reading, summarizing, and thinking about whether or not your topic is clearly defined and if a particular reference is appropriate. The goal at first is to evaluate what you have, what you know, and what you still need to create the scholarly foundation for your research. In time this cycle of summary and analysis will build contextual knowledge and hone your perspective, enabling your perception of the research question to evolve. Remember that analysis is an effort toward understanding and making meaning, through careful attention to detail and part-to-whole relationships. A valuable mindset is to stay receptive to the material at hand. Immediate judgments of agree/disagree or like/dislike are not productive for deep analysis. As you review materials, notice your thinking habits and work toward re-imagining your perspective or assumptions about a topic in multiple ways. Ask what more you can find, and create an active dialogue with the materials you are reviewing through highlighting compelling insights or details, writing reflective notes and questions in the margins, circling important terms, and defining unfamiliar terms or phrases. This dance-like engagement with your resources can assist in your retention of the material and in deepening your ideas and approach.

Begin with the items that are most likely to provide what you need to get an overview (like an abstract), background context and justification (introduction), general findings (figures and tables), the measurement strategies and the equipment and protocols used, the author's use of theory or methodology, and the author's conclusions. Cull and prioritize your collection, determining what is most or least relevant in illuminating your topic. The ability to narrow your focus to a smaller group of resources provides a more manageable number of sources to evaluate more critically. Try not to get stuck in the details on your first reading. Put the lower priority sources aside but do not banish them—you may return to them as your knowledge evolves. This process of skimming, culling, and prioritizing can be repeated as you prepare your literature review, your research proposal, and final product.

As the second round of evaluation, delve deeply into selected resources by interrogating each article, book chapter, report, or conference proceeding critically. Read broadly at first, then re-read a second and possibly a third time with the intent to discover more details with each visit. This is a skill that can be challenging to master, but it will sharpen with practice. Expect to re-read materials as you conduct your study and write your report, balancing openness to new ideas with staying connected to your research agenda. As you read critically, take notes on the purpose and major findings of each

Table 4.6. Article Review Questions

What was the research purpose? Why is it important?
What did the researchers do? How did they do it?
What were the findings and their implications?
What do you find intriguing or compelling in the existing research?
What questions have been answered? What needs clarification?
What data are being gathered?
Is there an observation protocol you could adapt for your research?
What treatments or training are applied? How are they replicable?
Is the research design appropriate for your project?
How would the elements of the study have to be adapted for your project?

study as well as details that will inform your work. Ask yourself questions as you read. Answering the questions in Table 4.6 will allow you to build context for your research. Step back periodically and consider what themes are being repeated across multiple studies.

These big ideas should be obvious in the abstract, discussion, and conclusions. However, be sure the conclusions offered by the authors are supported by their methodology and the results obtained. If they are not, write your own conclusion for the study and note whether the difference between what you see and what the authors report offers an opportunity for you to fill a gap in the literature by contributing to the developing knowledge about your topic. As you search for a topic that will become the focus of your research, consider patterns that emerge from your review of existing research. Do you see common themes or similar methodological approaches? Do these relate to each other, either congruently or in opposition? How does your point of view converge or diverge from the literature? What or who is not represented in the literature? Ultimately, this reflective exercise is in service to laying the groundwork for the writing you do later.

Summary

In this chapter we propose strategies and tools for beginning your research. There is great power in selecting a topic purposefully, and this can be done in tandem with thoughtful and thorough research practices. We suggest brainstorming as a strategy for identifying a topic speaking to your experiences and observations that will add to the existing body of literature.

We emphasize that research is iterative and non-linear, requiring continuous creative and critical efforts. Remaining responsive to changing conditions and new possibilities is a skill that can be mastered in time. Direct or systematic replication is an effective research strategy, especially for early researchers.

Some of the strategies we discuss for assessing the existing literature include employing multiple databases and types of sources, seeking summative sources such as existing literature reviews, keeping your resources organized, approaching your librarian for research assistance, using citation managers, and being attentive to the frequently cited seminal literature in the field.

As your own body of research progresses, focus on the most relevant sources to grow your understanding of how each item contributes both to your topic and to the broader discourse. Consider which elements of these studies strike you as compelling or unusual, and determine the questions that remain to be answered or information that is yet to be discovered. The work you do to collect, summarize, and analyze the existing research on your topic will provide the raw material for your literature review or annotated bibliography (see *Chapter 7*) and will become key to writing your own research proposal and report (see *Chapter 20*).

Highlights

- Productive researchers "read early, read often, and read a lot."
- Stay organized and efficient by recording your research process—write down which terms you used, where you used them, and whether they worked or not.
- Best practice for new researchers may be to replicate prior research as their first study.
- When reviewing research reports, ask yourself: What questions have the authors answered? What questions have they not answered? What still needs clarification?

Application Exercise

Select a reference management tool and begin building a bibliography for your study. *Chapter 7* offers suggestions for annotating your bibliography.

Editors' Note

We thank Leah Sherman, Merry Lynn Morris, Sarah Kenny, and Jeff Russell for their willingness to shift small portions of content between *Chapters 4 and 7* to allow it to appear in a sequence that is more useful to readers.

References

Arnold, B. L., Gansneder, B. M., & Perrin, D. H. (2005). *Research Methods in Athletic Training*: Chapters 3 & 4—Choosing a topic & reviewing the literature (pp. 25–66). F. A. Davis.

Berg, K. E., & Latin, R. W. (2008). *Research Methods in Health, Physical Education, Exercise Science, and Recreation:* Chapter 3—Reading a research report (3rd ed., pp. 23–34). Lippincott, Williams and Wilkins.

Cunningham, N. (2017). The twenty-first century library. In S-H Lee (ed.), *Scholarly Research for Musicians* (pp. 18–28). Routledge.

Overby, L. Y., Shanahan, J. O., & Young, G. (2019). *Undergraduate Research in Dance: A Guide for Students:* Chapter 3—Choosing and formulating appropriate research questions or project goals (pp. 17–22). Routledge.

Additional Resources

(n.d.). Resource papers. *International Association for Dance Medicine and Science.* Retrieved June 15, 2021, from https://www.iadms.org/page/186

(n.d.). Resource papers. *Journal of Dance Medicine and Science.* Retrieved June 15, 2021, from https://www.iadms.org/page/47

Medical Problems of Performing Artists (MPPA). Retrieved July 11, 2022, from https://www.ingentaconnect.com/content/scimed/mppa

Rosenwasser, D., & Stephen, J. (2011). *Writing Analytically.* Cengage Learning.

Solomon, R., Solomon, J., & Kilcawley, D. (2017). Dance medicine & science bibliography. *International Association for Dance Medicine and Science.* Retrieved June 15, 2021, from https://www.iadms.org/page/29

5

Types of Research

TOM WELSH

Questions to Stimulate Thinking

- Why are there different types of research?
- Is there a hierarchy of research types?
- If so, does the order imply relative value (for research or practice)?
- How do I decide which research approach to use?
- What should I do if my research question is not ready to be answered?

In *Chapter 1* we said that the purpose of research is to answer questions about how the world works, in order to create new knowledge, and to do so systematically. Applied research, you will remember, has the added requirement of demonstrating that discoveries have implications that are clinically or socially significant. In addition to offering insight on how the world works, applied research usually must demonstrate immediate practical benefits for those who will make use of the research findings. Recalling the research concepts discussion from *Chapter 3*, we might say that our purpose is to develop and refine an evidence base that informs teaching, training, clinical practice, and future research in the dance sciences (cf. Portney & Watkins, 2009).

In fields as diverse as the dance sciences, the variety of questions that researchers may want to answer requires a variety of research approaches to address them. Fortunately, there are many well-established approaches that dance science researchers may use to answer questions that interest them. It may be useful for those who specialize in one approach or another to feel that their approach is the most important because of the motivation it provides to pursue their work vigorously. But it is probably more productive for new researchers to view the diversity of research approaches as a collection of tools that can be used to address the different types of research questions they may want to answer at different points in their research careers. Researchers may choose one approach to answer a current question and choose another approach to address a different question as their research intentions evolve in the future.

Various labels are used to describe the types of research used in the collection of domains that make up the dance sciences. Some are based on the type of problem addressed (epidemiology, ethnography, treatment); some are based on the type of data created (qualitative, quantitative); some are based on procedural aspects (cohort, case-control); some are based on perspective (retrospective, prospective); and some are based on the type of comparison made (between-group, within-subject). One classification strategy that can be applied across research domains is to organize the approaches by the functions they perform. In this chapter we look at the types of research that are used in the dance sciences based on their function. We also describe how these different types of research may interact to support a program of research in the dance sciences.

Types of Research

Historic and Analytic Research

Historic and analytic research is designed to report and analyze past events in ways that create insight. There are whole books devoted to conducting research of this type (e.g., Fraleigh & Hanstein, 1999). If your research question fits this purpose, we encourage you to consult this text or another like it for guidance. Some of the functions of historic research can be achieved with description and correlation, which are mentioned in this chapter and described in Part IV of this text.

Descriptive Research

Descriptive research is designed to identify and describe the characteristics of what exists or what is currently taking place in some aspect of the dancer's world. A clear, complete description of a particular aspect of the dancer's world or experience can lead to insights that support, clarify, or contradict intuitions we may have about how the dancer's world works. That information can be used to refine practice in dance training and treatment. For example, if a descriptive study shows that during the three weeks preceding a new season, most dancers engage in less than half the level of physical activity that is normally expected of them during the first week of a new season, company directors may choose to start their seasons differently or they may decide to contract with their dancers to manage their activity during the lead-up to a new season differently.

Descriptive research findings can also suggest new research questions to answer, and they can provide a foundation on which to build other types of research. Staying true to the descriptive purpose can help researchers be sure that the assumptions they use to pursue other types of research are accurate, clear, and complete. Clarity of focus and patience are virtues when conducting descriptive research. Descriptive research can be divided into four sub-categories: observational studies (*Chapter 11*), survey research (*Chapter 12*), qualitative research (*Chapter 13*), and descriptive case studies (*Chapter 15*). Following are brief descriptions of the four approaches to provide an overview of these approaches to research, which can be scaled to fuel a new researcher's developing repertoire.

Observational Studies

Observational studies involve directly observing a phenomenon of interest or analyzing a durable record that is produced by that phenomenon to reveal details and patterns. For example, an observational study might involve counting the number of dancers engaged in movement-based warm-up activities immediately prior to the start of technique classes. Another might involve counting the number of university dancers who see a health care practitioner for a dance-related condition during a particular portion of the season. Descriptive statistics *mean, median, mode, average, range, standard deviation* are commonly used to summarize the observations made in descriptive studies that gather quantitative data. Graphic display is also used to make the relationships among variables conspicuous, particularly in within-subject research. Narrative summaries and emerging themes are commonly used to summarize qualitative data. *Chapter 9* explores ways of summarizing data in order to generate insight.

Surveys, Questionnaires, and Interviews

Surveys, questionnaires, and interviews ask dancers to answer questions or offer ratings to describe or assess events, perceptions, and dimensions of their experience that they are well-positioned to evaluate. A research survey might, for example, ask dancers to report the number of times they exercise at a gym, the number of years they have been dancing more than 10 hours a week, or the number of hours they spend dancing each week. Dancers may also be asked to rate their experience with teachers, choreographers, health care practitioners, or other dancers. Dancers' responses can be gathered in written form using a questionnaire or through interviews with notes made by a researcher or transcripts made from recordings to make the dancers' responses available for analysis.

Descriptive Case Studies, Case Reports, and Case Series

Descriptive case studies—called case reports and case series in medical circles—allow researcher-practitioners to describe a unique or new case they are encountering. These studies may have elements that are quantifiable and elements that are qualitative, depending on the case under consideration and the question the researcher is seeking to answer.

Qualitative Research

Some researchers use the term *qualitative* to identify a particular type of data (Arnold et al., 2005) whereas others use it to identify a distinct approach to research. It is an approach to description that uses intensive observation of individuals (e.g., dancers, teachers, choreographers, clinicians, artistic directors) by a researcher who describes the observations in a narrative summary. Qualitative research is aimed at providing a deeper and more nuanced understanding of a phenomenon of interest. It might involve documenting an injured dancer's changing perceptions during progression from surgery through treatment, rehabilitation, and re-entry into dancing following a debilitating injury. Qualitative research generates data that dancers, teachers, and choreographers find particularly compelling. We have given the approach its own chapter in Part IV.

Integrating Various Descriptive Approaches

The different descriptive research methods (observation, survey, case study, and qualitative research) can be paired in ways that provide additional insight. When qualitative data (emphasizing narrative description) are paired with quantitative data (such as counting the number of times a dancer sees a health care practitioner), the approach may be called *"Mixed Methods"* research. We describe the ways researchers have combined qualitative and quantitative data in Part IV.

Why Start with a Descriptive Study?

Descriptive research can provide insights that are immediately useful to practitioners. For example, lower extremity norms for dancers who are currently performing in professional ballet companies might be used to set goals for dancers studying in preprofessional dancer training programs. Descriptive findings can also be used to build a foundation for other types of research by, for example, documenting a discrepancy between the aerobic endurance required to perform works of a particular type and the baseline endurance in the population from which the performers are being selected. When blazing a trail in a new research area, descriptive research is often an essential first step. If you are proposing research in a new area, you may do yourself a favor by starting with a descriptive study to provide context for your future work in the area. Conducting a descriptive study can teach you how to think about your topic in ways that will help you create other researchable questions. It can also help you select a research approach that will be capable of answering other research questions convincingly.

Is Your Research Question Ready to Be Answered?

We recommend staying open to the possibility that the question you want to answer may not be ready to be answered. Constructing a building starts with creating a supportive foundation and then adding the walls and roof to that strong, supportive foundation. Descriptive research can help you create the background, the experience, and the methodological skills you will need to conduct research that addresses a research question you want to answer. If you are committed to conducting research using an approach other than descriptive in an area that is new to you, it may be wise to choose a topic that already has a substantial descriptive research foundation that you can use to learn about your topic by reading reports by other researchers.

Correlational Research

Correlational research identifies and quantifies relationships between naturally occurring variables and other variables that are not under the researcher's control. It involves asking: "If one variable changes (e.g., fewer performances this season) how does another variable (number of dance injuries experienced) change?" Correlational research might be used to answer questions such as: Is dance experience related to injury risk? Is injury severity related to dropping out of dance? The intention in correlational research is to see whether differences or changes in one variable co-vary with differences or changes in another variable (Huitema, 2011). It is prudent to assume that correlation describes relationships between dependent variables.

Although correlation coefficients can be used to predict an outcome, correlations do not reveal which of the related variables may be the cause and which may be the effect. In fact, corresponding changes in correlated variables may be the result of changes in other, uncontrolled, unmeasured, and possibly yet to be identified variables. To reveal and confirm cause-and-effect relationships, researchers must conduct studies in which the administration of one (independent) variable is controlled by the experimenter in order to evaluate its influence on another (dependent) variable (Barlow & Hersen, 1984).

Intervention Research

Intervention research involves making systematic changes in an independent variable (e.g., treatment or training) and observing the effects on a dependent variable (e.g., performance or capacity) that is important to dancers. Teachers teach dancers to expand their dance abilities; physicians do surgery to repair damaged tissue; physiotherapists, physical therapists (PTs), and athletic trainers treat dancers to re-establish function; and choreographers create and set dances to communicate with or move an audience. Intervention researchers also apply teaching practices, treatment protocols, and training approaches, but they do so with special precision and control and without assuming that an intervention will have a predetermined effect.

Intervention Case Studies

Case reports and case series research is primarily descriptive (see *Chapter 15*). Kazdin (1981) pointed out that the activities performed by clinicians can often be enhanced to turn what clinicians normally do into entry-level intervention research. By measuring outcomes more precisely and applying interventions more systematically, intervention case studies can approach the descriptive intensity of qualitative studies and provide a preliminary quantitative assessment of the effects of a treatment or training intervention on an outcome of interest to dancers. This puts intervention case studies within reach of clinicians, teachers, and other practitioners who cannot afford to compromise their practice in order to engage in other types of research.

Experimental Research

Experimental research is designed to assess and explain causal relationships by carefully controlling the application of an independent variable and carefully observing any corresponding changes in a dependent variable. Researchers might, for example, ask a company director to change the return-to-work protocol for a group of randomly selected company dancers following a lay-off between seasons to see if a more gradual increase in training intensity might reduce the number of overuse injuries that dancers experience as compared with dancers who do not experience the special return-to-work protocol. Or researchers might design a test to see whether teaching dancers an imagery-based cueing technique can improve plié depth and hamstring activation by measuring these variables before and after the initiation of a targeted imagery training program for dancers (Couillandre et al., 2008).

An essential element in experimental research is a convincing demonstration that changes (or differences across conditions) in a relevant *dependent variable (DV)* correspond with changes or differences in a specified *independent variable (IV)*. Experimental designs employ strategies for controlling, and thereby ruling out, the potential influence of extraneous variables in order to isolate and make conspicuous the influence of the treatment or training under evaluation.

There are two main families of experimental research. Between-group experiments involve randomly assigning a sizable number of dancers to one of two or more groups, each of which experiences a different treatment or training condition (IV). A performance or capacity (DV) is usually measured on one or two occasions for each dancer and the average score for the dancers experiencing one condition is compared with the average score for the dancers experiencing the other condition. *Inferential statistical tests* are used to determine the likelihood that any differences observed between the group averages may have occurred due to chance. If the probability of the observed difference appearing by chance is less than a pre-specified threshold (e.g., fewer than 5 chances in 100, $p < .05$), researchers infer that the independent variable probably caused the difference observed across groups.

The other family of experimental research, within-subject experiments, involves arranging for a few dancers to experience all of the training or treatment conditions under evaluation in a controlled sequence. Each dancer's performance or capacity is measured repeatedly over days, weeks, or months. Measurements are displayed graphically (dancer by dancer) to show individual patterns of performance over time and across conditions. Instead of statistical testing, within-subject researchers use precisely timed application of the independent variable(s), replication of the experimental effect, and logic to rule out alternative explanations for any changes observed. If the effect is replicated when and only when the conditions change, researchers infer that the independent variable probably caused the changes in performance across conditions. Best uses for both families of experimentation are described in *Chapters 16 and 17*.

Quasi-Experimental Research

Sometimes there are situations where dance science researchers would like to conduct an experimental analysis but do not have access to sufficient resources, or they do not have sufficient control over local circumstances to make an experimental analysis possible. For example, researchers may not have the luxury of randomly assigning dancers to different conditions, or they may not be able to arrange for dancers in a comparison group to have identical experiences (e.g., same teachers, same rehearsals) except for the independent variable under evaluation. Such limitations make it difficult to rule out the influences of possible confounding variables, a key function of experimental analysis.

Rather than forgo a chance to study the influences of an intervention of interest, researchers facing such constraints might choose to conduct a study that approximates an experimental analysis to see what they can learn under the available conditions. For example, researchers at a multi-level ballet training center may choose to implement a pre-season conditioning program with dancers in one class to see whether the dancers in that class experience fewer overuse injuries than in previous seasons. Any increase or decrease in overuse complaints might be compared with changes observed for dancers in other classes at the training center that do not experience the pre-season conditioning program. Because uncontrolled differences in the dancers' experiences before, during, and after the application of the training program cannot be ruled out as alternative explanations for any difference in overuse injury trajectories, the results of a comparison of this type must be tentative. Future experimental analysis will be needed to draw stronger conclusions, but the work researchers do to conduct a quasi-experimental analysis can inform the design and execution of a subsequent experimental analysis. Quasi-experimental results might also provide interim guidance to practitioners while they await the increased clarity that may ultimately be provided with an experimental analysis.

Systematic Reviews and Meta-Analyses

A relatively new contribution to the researcher's toolkit is the development of systematic reviews and meta-analyses. They are often conducted separately, but together they allow researchers to combine results reported by previous researchers in order to draw general conclusions about the effectiveness of various approaches to addressing issues that are important to dance medicine and science researchers and practitioners. More recently, naturalistic researchers have offered the *meta-synthesis* as a comparable strategy for combining qualitative results across studies. *Chapter 18* explains how to conduct a systematic review and offers resources for learning more about meta-analyses and meta-syntheses.

Relations among Research Types

As you can see, even our simple description of types of research involves a diversity of approaches that can feel overwhelming. However, if you view the approaches as a collection of tools that you may choose from to answer a research question that interests you, their diversity becomes an advantage. You need not use them all, but having a number of approaches to choose from will allow you to address a greater variety of research questions.

The approaches to research described earlier in this chapter can be arranged according to any number of abstract dimensions. We could organize them by the degree of control they are able to achieve over potential confounding variables. We could organize them according to the insight they can provide regarding the conditions under which dancers work. We could organize them by their ability to compare the average performances of large groups of dancers working under one set of conditions to a comparison group working under a different set of conditions. We could arrange them by their ability to reveal patterns of performance over time and across experimental conditions for individual dancers. Whatever framework we might conceive, the researcher's primary concern is to select a research approach allowing the answering of a particular research question. The value of any research approach is determined by its utility for answering a particular research question.

Choosing a Research Approach

Choosing a research approach to match your research question is essential. Chatfield (2000) suggested that it is generally better to fit a research approach to your research question than to try to fit your question to a particular approach. The audience you want your research to speak to and the conventions of the journal where you would like to publish your findings will also influence your choice of a research approach. We include in Part V more guidance for choosing an appropriate avenue for disseminating your research results.

There may, however, be a natural progression through the types of research questions that are ready to be answered and, therefore, the approaches to research that researchers are able to apply as a new area of research is being developed. A progression of research questions may be addressed best by a progression of research approaches. For example, it may be helpful to describe what exists (descriptive research) in a new area of research (such as dance partnering) before investigating which of the currently available approaches to teaching partnering are associated with successful partnering (correlational research). Both of these steps might naturally precede an effort to design and test a new intervention aimed at improving partnering skills for dancers, perhaps first using an intervention case study approach with a pair of partners, and ultimately using a between-group or within-subject experimental approach.

Some researchers follow a line of inquiry that takes them through a series of stages reflecting the sequence of research types described in this chapter so that they can learn about the phenomenon they are studying from a variety of perspectives. Other researchers choose to specialize in a favored approach and limit the questions they ask to those that fit the approach they prefer to use. Either strategy can contribute to the development of the knowledge base in our field. However, new researchers may benefit from the experience they can gain by conducting each of the types of research described. What is learned by doing research of one type will inform work done for another type of research. Another benefit is that experience in conducting a variety of types of research is likely to expand the type of research questions a researcher can choose to answer. Finally, having experience with a variety of research approaches allows researchers to understand the work of their colleagues better and may allow them to engage in collaborative research that will provide a fuller understanding of a phenomenon of mutual interest. This benefit may be especially important in a field as diverse as the dance sciences.

We embrace all types of research that dance science researchers find useful for expanding the evidence base for training and treating dancers. Figure 5.1 offers a way to conceptualize the similarities and differences among the types of research used in exercise science, medicine, education, and the dance sciences. At the beginning of each chapter in Part IV, which includes descriptions of specific research approaches, we identify the types of research questions the approach is designed to address. We outline the requirements for using each research approach as well as the limitations. Previewing these sections in the Part IV chapters will help you identify the approaches that are likely to be a good fit for your research question and the resources to which you have access.

Summary

Various types of research are available to address the different types of research questions that dance science researchers want to answer. The research types previewed in this chapter and described in detail in the Part IV chapters progress from descriptive through experimental research. As

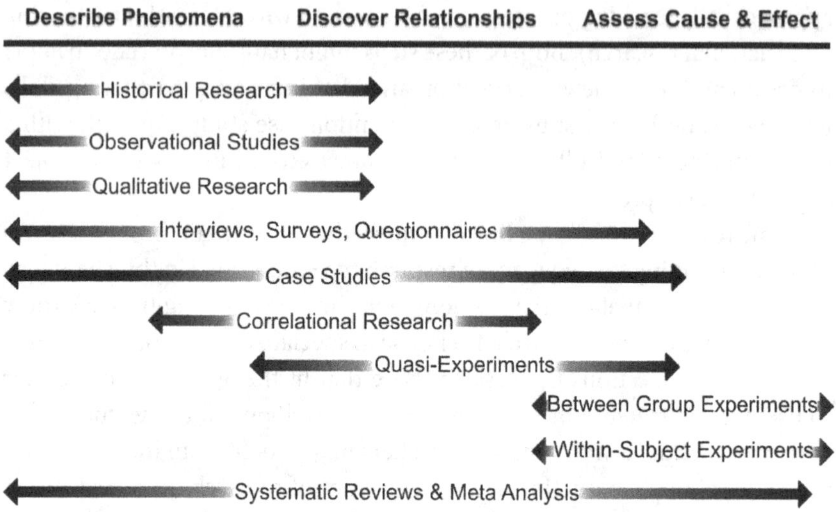

Figure 5.1. Types of Research by Function. Adapted from Portney & Watkins (2009).

the research question you wish to answer becomes clear, you will be able to identify a research approach that is appropriate for answering your question. As you learn more about your topic, your research question may evolve, and you may find that a different type of research is a better fit. The ultimate value of any research approach is determined by its ability to answer the research question you want to answer.

Highlights

- When blazing a trail in a new research area, descriptive research is often an ideal first step.
- The value of any research approach is determined by its utility for answering a particular research question.
- We embrace all types of research that dance science researchers find useful for expanding the evidence base for training and treating dancers.

Application Exercise

Revisit your evolving list of potential research questions and choose one that is currently capturing your interest and that seems feasible with the resources you have available. Explain which approach to research seems to be the best fit for your research question and briefly explain why. Then choose two other research approaches from this chapter and modify your research question to fit each of those approaches. Having more than one question per approach is OK for now. Be sure to write your questions down so that you can see what you are thinking. Then select one question that is appealing and refine that question to make it even FINER—more feasible, more interesting, more novel, more ethical, and more relevant (Hulley et al., 2013).

References

Arnold, B., Gansneder, B., & Perrin, D. (2005). *Research Methods in Athletic Training.* F. A. Davis.

Barlow, D. H., & Hersen, M. (1984). *Single Case Experimental Designs: Strategies for Studying Behavior Change* (2nd ed.). Pergamon.

Chatfield, S. J. (2000, October). Rules of Evidence. In Welsh, T., Chatfield, S., Wilmerding, G., Moore, M., & Hobden, R. *A Closer Look at the Science in Dance Medicine & Science* [Research Forum Presentation]. International Association for Dance Medicine & Science 10th Annual Meeting, Miami FL.

Couillandre, A., Lewton-Brain, P., & Portero, P. (2008). Exploring the effects of kinesiological awareness and mental imagery on movement intention in the performance of demi-plié. *Journal of Dance Medicine and Science, 12*(3), 91–98.

Fraleigh, S. H., & Hanstein, P. (eds.). (1999). *Researching Dance: Evolving Modes of Inquiry.* University of Pittsburgh Press.

Huitema, B. E. (2011). *The Analysis of Covariance and Alternatives: Statistical Methods of Experiments, Quasi-Experiments, and Single-Case Studies* (2nd ed.). Wiley.

Hulley, S. B., Cummings, S. R., Browner, W. S., Grady, D. G., & Newman, T. B. (eds.) (2013). *Designing Clinical Research* (4th ed.). Lippincott Williams & Wilkins.

Kazdin, A. E. (1981). Drawing valid inferences from case studies. *Journal of Consulting and Clinical Psychology, 49*(2), 183–192.

Portney, L. G., & Watkins, M. P. (2009). *Foundations of Clinical Research: Applications to Practice* (3rd ed., pp. 18–23). Pearson Education.

Additional Resources

Payton, O. D. (1994). *Research: The Validation of Clinical Practice* (pp. 46–48). F. A. Davis.

Thomas, J. R., Nelson, J. K., & Silverman, S. J. (2011). *Research Methods in Physical Education* (6th ed., pp. 18–22). Human Kinetics.

6

Research Ethics and Research Ethics Committees

K. MICHAEL ROWLEY AND TOM WELSH

Questions to Stimulate Thinking

- Why is it important for researchers to give special attention to ethical conduct?
- What core values or principles inform behaving ethically?
- Who has a stake in the ethical behavior of dance science researchers?
- What is a Research Ethics Committee and how can researchers engage productively with one?

In this chapter we describe the role of ethics in empirical research with dancers and explain why ethical behavior is of particular concern when conducting research. We briefly review the development of ethical guidelines for research and a few mechanisms that have been developed to promote ethical behavior, including ethical review by an interdisciplinary committee. The guidelines used are centered around core principles for ethical behavior that govern the relationship between researchers and research participants. We devote the heart of the chapter to unpacking this relationship and explaining the role of the local Research Ethics Committee (REC), in helping researchers behave ethically toward their study participants. We use the term *Research Ethics Committee* to make the concept comprehensible across cultures; we intend it to include the diverse collection of labels assigned to this function, such as Institutional Review Board (IRB), Ethics Review Board, Ethics Committee, and Human Subjects Committee.

Ethics, Research, and Interest Groups

The word *ethics* is derived from the Greek term *ethos,* which means character. Ethics is "a system of moral principles" (Stewart & Reider, 2016). Applied ethics uses these moral principles to guide the development of standards specific to a topic or field. Professionals in a field use ethical guidelines to inform their own actions, to evaluate and apply sanctions to the conduct of colleagues, and to establish and maintain the public's trust (Resnik, 2015).

In this chapter we focus on ethical conduct as it relates to *research* with human participants. One national organization in the United States defines *research* as: "A systematic investigation designed to develop or contribute to generalizable knowledge" (Office for Human Research Protections, 2018). Most empirical research with dancers aims to answer questions and create knowledge important to dancers, their teachers, health care providers, and choreographers or company directors.

At least four interest groups have a special stake in the ethical behavior of professionals who conduct research with dancers (see Figure 6.1). The most obvious interest group involves the dancers who comprise the population. Some dancers in this population may become participants and some may not, but all have an interest in the ethical conduct of the researchers. In Table 6.1 we highlight important concepts for researchers to consider as

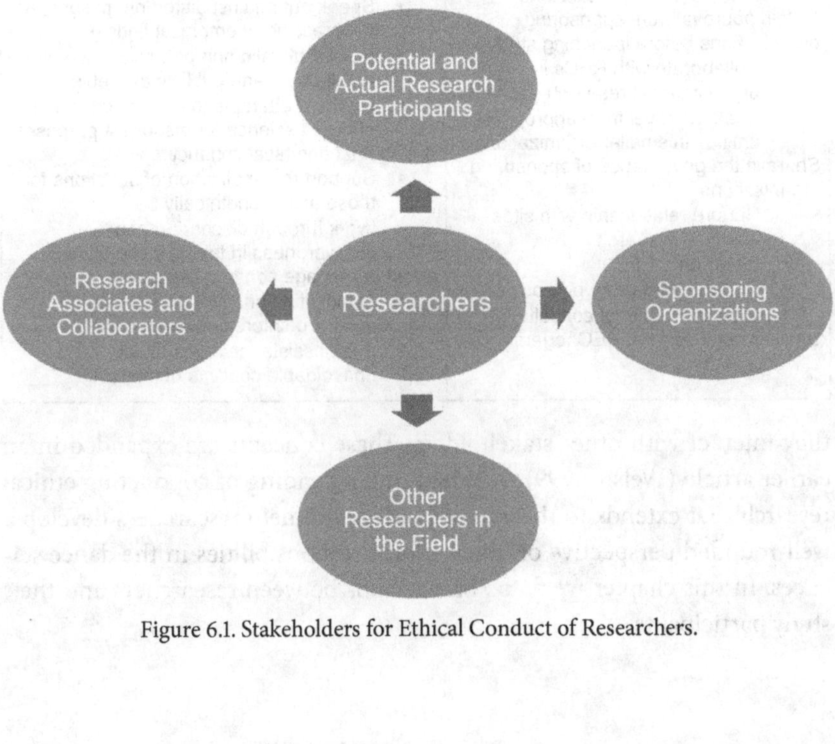

Figure 6.1. Stakeholders for Ethical Conduct of Researchers.

Table 6.1. Ethical Conduct Between Researchers and Other Stakeholders

<table>
<tr><th>Research Associates and Collaborators</th><th>Other Researchers and the Field</th></tr>
<tr><td>Help new, developing researchers acquire knowledge and skills
• Secure the future of our field
• Provide research assistance, fresh perspective, added sense of purpose
Recognize the contributions of others
• Cite the work of colleagues
• Avoid plagiarism
• Offer invitations to co-author
• Help colleagues sustain their research efforts
Avoid exploitation and misuse of influence
• Recognize the special influence mentors have over associates
• Offer fair compensation
• Use others' time respectfully
• Acknowledge contributions by research associates
• Optimize effectiveness of the research team
• Minimize risk to the careers of developing researchers</td><td rowspan="3">Defend free inquiry
• Prevent suppression of research that may be controversial
• Let empirical results speak for the value of all research
Share results openly
• Teach other researchers and practitioners what we learn
• Negotiate limited delays if needed to accommodate proprietary purposes
Present results objectively, fully, accurately
• Minimize chances of misleading audiences
• Avoid misrepresenting purpose, methods, findings, and implications
Submit manuscripts to only one periodical at a time
• Honor the reviewing-publishing agreement with editorial boards
Publish original results only once and identify replications explicitly
• Reduce the risk of confounding the results of future meta-analyses
Evaluate ideas objectively
• Review the work of colleagues fairly
• Support integrity in our field
Prevent distortion, misuse, or suppression of results
• Speak out against distortion, misuse, or suppression of empirical finding
• Insist that valid and potentially useful results be made publicly available
• Confront attempts to use the good will afforded science for dishonest purposes
Act against unethical conduct
• Support the application of sanctions for those acting unethically
• Work through discomfort and awkwardness in taking these actions
Avoid or manage conflicts of interest
• Avoid situations that may create a conflict of interests when possible
• Seek assistance in managing unavoidable conflicts of interest</td></tr>
<tr><th><u>Sponsoring Organizations</u></th></tr>
<tr><td>Minimize risks to sponsors
• Minimize financial, legal, and reputational repercussions
Obtain approval from sponsoring organizations before launching study
• Collaborate with RECs in large, organizational research sponsors
• Seek approval from appropriate entities in smaller organizations
Share in the governance of sponsoring organizations
• Nurture relationship with sites that host research
• Serve on a REC
• Help prepare grant proposals
• Offer professional consultation
Seek assistance from REC regarding risk assessment</td></tr>
</table>

they interact with other stakeholders. These concepts are expanded in an earlier article (Welsh, 1999). A broad understanding of conducting ethical research that extends to these stakeholders will help researchers develop a well-rounded perspective on their ethical responsibilities in the dance sciences. In this chapter we focus on relations between researchers and their study participants.

Principles for the Researcher-Participant Relationship

Core principles of ethical conduct provide a compass to guide researchers in the direction of ethical behavior when multiple stakeholders are at the table. These principles and concepts were developed and updated over the past seventy years in response to ethical violations by researchers around the world. For a more detailed history see Grodin & Annas (1996). The *Declaration of Helsinki* is one such set of guiding principles adopted by the World Medical Association (2013). Countries and scientific organizations have created their own guidelines inspired by and consistent with the *Declaration of Helsinki*. The guidelines vary in how they are organized and applied, but they cover the same basic protections for research participants.

The *Belmont Report*, a foundational document for ethical conduct in research in the United States (National Commission for the Protection of Human Subjects of Biomedical and Behavioral Research, 1979), is used to organize the discussion of the researcher-participant relationship in this chapter. It classifies ethical conduct into three core principles: respect, beneficence, and justice. Being conversant in these principles will help you see your REC's policies as guidelines to interact successfully with your research participants.

Respect

Respecting the individuals who participate in our research as autonomous human beings with aspirations and needs that are every bit as important as our own is essential for researchers. By viewing your REC as a research collaborator, you can improve the chances that your study will be successful and impactful. Ethical researchers protect the welfare and dignity of their participants across all phases of research from recruitment through to study conclusion. Mutual respect can encourage open and honest communication and sincere engagement in the research process. The principle of respect is applied through a few standard research practices, including informed consent, privacy and confidentiality, and debriefing participants after the study.

Informed Consent

Respect for your research participants means asking for their *informed consent* to participate in the research free of coercion or undue influence. To offer meaningful consent, participants must be informed about the research—its purpose, procedures, risks, and benefits—as well as their rights as participants. Acquiring informed consent by describing what is involved in participating and sharing what you hope to learn by conducting the study invites participants to become stakeholders in the project, or partners in the scientific process. Participants will be more likely to comply with study procedures, complete the study, show up for follow-up visits, and be honest in reporting requested information. We provide a generic template for such a document in Appendix A, but local RECs generally have their own requirements or templates. See what is available from your REC before building your own Consent Document.

Components of informed consent include information, comprehension, and voluntariness. Researchers can facilitate comprehension by presenting the information in a clear, organized, and understandable way. Avoiding professional jargon will reduce the chance of confusing or misleading potential participants. Participants should have as much time as they need to review the information about the study before deciding to participate. Some researchers accompany the written description with a brief oral presentation. Providing opportunities to answer questions either during the presentation or individually later can help ensure that the participants understand the essential study features and what will be expected of them as participants.

Participants must decide to participate voluntarily, which means free of coercion or undue influence, and they must be free to withdraw from the study at any time. You may be thinking to yourself, "Of course I would not force anyone to participate in my study!" It is important to be aware that researchers can unintentionally influence participation in a variety of ways. For example, dancers could feel pressure to participate in their teacher's or choreographer's study for fear of biased treatment in the future. It is extremely difficult, for example, to allow dancers to choose freely not to participate (or, later, to withdraw from a study) if they suspect that their course grade or professional reputation may be at risk. Unfortunately, assurances by persons in positions of authority often do not assuage these fears. Researchers who wish to attempt research with their students or with cast members will want to collaborate with their RECs to come up with creative ways to avoid undue influence (e.g., a neutral third party might manage the information and consent processes).

Consent for Research with Minors

A few special populations have a reduced ability to provide informed consent free of coercion or undue influence—children, incarcerated individuals, and individuals with developmental or mental disabilities. Local RECs can guide researchers in identifying these populations based on jurisdiction-specific definitions and implementing specific adaptations to the consent procedures. Especially relevant to dance is research involving children. When research participants are children who cannot legally or functionally provide consent, *parental permission* is acquired with all the same components of informed consent already described. In addition, researchers should acquire *child assent* when children can comprehend basic concepts about participating in the study. A generic Child Assent template can be found in Appendix B. It includes sample language appropriate to be read aloud to children. Just as informed consent turns adult volunteers into engaged research participants, child assent builds engagement, relationship, and respect with child participants.

Privacy

Special steps generally need to be taken to protect the privacy of individual participants when conducting research and when presenting and storing research results. Public presentations should disguise the identities of the individual participants by, for example, assigning a pseudonym, letter, or number to each individual's records. If you wish to include a photograph of a dancer in a research report, de-identify the photo with masking or blurring or gain the participant's permission before adding it to your report.

Confidentiality

Additionally, any personal information gathered in the execution of an experiment should be assumed confidential and should not be shared with anyone beyond the research staff. Observations made and information gathered for research purposes should be used exclusively for the research project. Reveal any limitations on confidentiality that may be mandated by jurisdictional laws, institutional regulations, or departmental procedures when participants are asked to volunteer. If an unanticipated conflict arises during an experiment, researchers should seek the participant's permission before revealing any personal information to anyone other than the research staff.

Debriefing

Some information, such as anticipated and actual results, may need to be withheld during a study to minimize extraneous influences (such as expectancy) on the results. However, researchers should describe and explain the results of their study and answer any questions that participants may have after the study is completed. *Debriefing* completes the information sharing process initiated when gaining participant consent. It helps enhance understanding of the treatment the participants received and any changes that resulted from the treatment. Thoughtful debriefing will increase the dancers' appreciation for research and may influence their willingness to participate in future research projects.

Beneficence

Colloquially, *beneficence,* a second core principle, may mean simply "being nice." In the context of research ethics, however, the term takes on a more specific meaning: maximizing benefit and minimizing harm. Ethical researchers never intentionally do harm to participants, but the risk of unintended harm will always be present. Researchers plan their research to minimize personal risk in the case of adverse events. Ethical researchers also seek to maximize the benefits of their research. Acting in ways that comply with the principle of beneficence will create the best possible experience for research participants and enhance your success as a researcher. The following sections suggest ways of optimizing the risk-benefit ratio.

Minimizing Risks

In a research study, *risk* is the potential for participants to encounter harm. Dance science researchers can minimize risks by, for example, including a reasonable warm-up before physical exertion, scheduling testing at times when dancers are already warm and prepared to dance, and avoiding testing at times when dancers are tired or distracted. Special attention may also be required to protect dancers from assaults on their emotional well-being and dignity. Instructions should be given in a respectful manner.

Another way to minimize risk is to use qualified staff to administer the treatment and testing procedures. Skillful staff are likely to be more efficient and professional in their interactions with research participants. Still another way to minimize the burden on dancers is to use their time efficiently. It may be helpful to schedule participants for testing in small groups, so that none will have to wait long to complete testing. Dancers are a busy group of athlete-artists who will notice sincere efforts to respect their time and will likely reciprocate with a redoubled effort to adhere to the research procedures.

Maximizing Benefits

Direct benefits. The *benefits* of any research study may be direct or indirect. The former provides benefits directly to research participants. In a research study investigating the effects of a ten-week dance conditioning course, for example, a direct benefit to participants may be free conditioning classes for ten weeks. In this case the free classes are guaranteed. Still, it would not be appropriate to state something like "improved flexibility and strength" as a benefit, since this is a potential outcome of the intervention that has yet to be shown. Researchers must remain neutral on anticipated outcomes of the research when informing participants about the study.

When attempting to maximize direct benefits of the study, there can be some complications as well as opportunities. Group comparison research presents a special problem when the primary direct benefit for participation is access to treatment. Participants in the control group do not experience treatment and any of the benefits that may result. One solution is to arrange for control group participants to receive the treatment after the study has ended. Alternatively, researchers can use a within-subject research design in which all participants receive treatment as part of the experiment (Welsh & Chatfield, 1996 ; also Chapter 16, this text). Additional direct benefits might include sharing pre- and post-test physical fitness assessments or personalized exercise recommendations.

Indirect benefits. Indirect benefits do not accrue directly to research participants. Rather, they affect the ways that society, the current state of knowledge, or the dance community may benefit from the research. This may be the case when studying dancers who currently participate in conditioning classes to learn best practices for designing the course you will implement in a future study. These dancers likely receive no direct benefit from you asking them about their ongoing conditioning practices, but the development of a conditioning course that goes on to be supported by research may benefit many future dancers.

Optimizing the Risk-Benefit Ratio

Assessing the risks and benefits of participating in a study usually includes discussion of the *risk-benefit ratio*, which we want to weight in favor of benefit. If there are no direct benefits to participants, risk should be no more than *minimal risk*—a term used to describe risks that are no greater than those encountered in daily life. If risks are greater than minimal but there are no direct benefits, a compelling justification must be articulated to RECs and to the participants in the informed consent process. Before starting your study, you might think about participation from the dancer's point of view. When your study concludes, consider asking the dancers to rate the appropriateness of the training or treatment received and value of the benefits in relation to the time and effort invested (Wolf, 1978).

Justice

Justice, a third core principle, refers to fairness in distribution. Ethical researchers ask whether the burdens of research are being distributed in a manner that is fair and do not fall disproportionately on disadvantaged groups or groups with limited autonomy. The principle of justice often plays out in the selection of research participants, which must be fair and equitable at both the individual and societal levels. Fairness in participant selection at the individual level means participants must be allowed to participate if they wish and if they satisfy the inclusion and exclusion criteria defined by the study. At the societal level, researchers might ask themselves whether participants are being selected based on some factor like availability, manipulability, or access to resources rather than fairness. For example, at first it may seem like a generous idea to test the effects of a new dance training protocol by offering free classes to children in low-income schools. However, if these children will not have access to the protocols after the research has concluded, are these children bearing the risks and burdens of research when the research findings will not benefit them or their peers? Research participants, and those they represent, should generally be the beneficiaries of research in which they are participating.

We want our findings to be relevant and important to the dancers they are intended to serve. Conducting a study without considering the principle of justice can create unintended and potentially unnoticed bias against specific populations, such as racial and ethnic minority cultures, women, children, gender and sexual minority groups, and low-income communities. Thinking carefully in advance about the fairness of both access to and application of your research will help ensure the greatest generality and beneficial impact of your research results.

In conclusion, as researchers design a study, these principles—respect, beneficence, and justice—can serve as a value system to guide ethical decision-making and research practice. Their broad language allows wide application to many aspects of study design but can also make it difficult to understand how they apply to research procedures in a specific and actionable way. Your local REC is empowered to apply the principles in ways that make sense to your community, institution, and research goals. RECs can help researchers balance these diverse interests.

Training in Ethical Research Practice

The Collaborative Institutional Training Initiative (CITI) Program (citiprogram.org) was developed to provide thorough and consistent research ethics and compliance training to researchers around the world. Many RECs require researchers to complete a set of courses offered by CITI or developed by their own institution. The training programs may include Human Subjects Research courses, Conflict of Interest training, or health care information privacy training. Ensuring that all researchers have a fundamental understanding of research ethics improves the reputation of research institutions and opens the doors for funding from both private and public organizations. Your research mentor or REC can point you to the online training site and the modules required at your institution for the type of research you plan to conduct. They can also help you estimate how much time you will need to allow for ethics training so it will not delay your research.

Research Ethics Committees (RECs)

Research Ethics Committees (RECs) are bodies composed of researchers, clinicians, and community members who review and monitor all human research studies with an eye for ethical regulations and principles. Ethical principles alone cannot substitute for a review committee because the evaluation of research projects usually requires judgment. RECs evaluate study applications by using principles outlined in international ethics reports as well as specific regulations outlined by national and local governments. REC approval is often a condition for sponsoring organizations, grant funding, degree progress, and the publication of research findings. Researchers sometimes complain that these committees are too focused on compliance, sanctions, and punishment. We encourage you to view your local REC as a collaborator in the research process. REC members have often collectively reviewed hundreds, perhaps thousands, of research studies, have managed unexpected and adverse events, have learned about diverse fields of study, and have an interest in helping your research succeed.

A primary responsibility of the REC is to protect the interests of the research participants. Questions posed to researchers by the REC may seem technical, obtuse, or as though they are fishing for a "right answer." While it is true that there are some strict requirements, overall your REC wants your research to succeed and will work with you to communicate where the "hard lines" are but also to ensure that the study design remains as you envision it while following ethical guidelines and principles.

RECs will be looking to see that risks to participants are minimized and reasonable compared to benefits, selection of participants is equitable, protections are in place for vulnerable populations, informed consent is acquired and documented, and participant privacy and confidentiality are protected. Local RECs will also have practices and policies in place based on their experience in overseeing research in your local communities and settings. Your REC may not have experience reviewing research in dance medicine and science, but they do have expertise to share what has and has not worked in other fields. You will likely be able to teach your REC the nuances of working with dancers by collaborating in the review process.

Table 6.2. Types of Research Ethics Committee (REC) Review

Type of Review	Description	Estimated Review Time
Full Committee Review	The default method of review: the study application is discussed with a quorum of committee members present and a determination is made.	2 weeks–6 months
Expedited Review	A select few committee and staff members review the study. Research activities with no more than minimal risk often qualify. Inquire with your local REC.	1 week–3 months
Exempt	Often includes research involving simple survey-based research or normal educational practices that are not likely to affect learning adversely. This determination must be made by an REC or staff member and delivered to the researcher in writing.	A few days–2 weeks

Submitting an application for REC review represents an investment of time. The review process often includes multiple review drafts being exchanged between the REC and the researchers as the REC asks for clarifications, justifications, or revisions to the study. REC review can take a few weeks to a few months, depending on the category of the study defined by your local REC (see Table 6.2). Developing aspects of the study design in collaboration with REC members or staff can substantially reduce the number of exchanges required to create an acceptable application. Researchers not familiar with dance medicine and science are encouraged to ask for review or mentorship from established researchers in the field. Doing so will reduce the time needed to get REC approval, and it will optimize the ethical integrity, accessibility, and impact of the study for dancers in a way your local REC may not have the experience to accomplish on your behalf.

A question that may arise is: "When does a researcher's design and development activity cross the threshold into conducting research which requires prior REC approval?" Particularly in applied research, a period of informal observation is needed to understand a problem and its context sufficiently to design a solution that can be tested. The solution and the means for measuring its influence must be pilot tested to ensure that protocols can be applied with precision when the steps intended to produce generalizable knowledge are initiated. In addition, REC study applications often require information that cannot be provided until these preliminary steps have been completed.

With experience and mentorship, researchers learn how to gather the information they need to design and pre-test their protocols without involving any study participants or increasing risk for their institutions. What is learned in this design and development phase informs the preparation of the REC study application, and the research begins after the REC application is reviewed and approved. Some research teams submit two applications: One for the pilot work, making best guesses about what will need to be done, and then submitting a second application for the research study before launching the portion of their work that is explicitly designed to produce generalizable knowledge. New researchers or those facing unfamiliar circumstances should consult their mentor or another experienced researcher for guidance.

Submitting a Study Application

RECs will ask for specific information about your research procedures and documents that will be used during your research. Different RECs have different formatting guidelines—sometimes each element is asked for in a series of questions on web-based software, and other times templates for research protocols are provided on the REC's web page for researchers to follow. It may be difficult at first glance to understand why certain information is requested but remember that RECs are charged with a very specific responsibility of ensuring researchers adhere to ethical principles.

RECs have a vested interest in making sure your study is feasible and will have an impact on the field or on participants' health and well-being. A study with insufficient resources to succeed or with no justifiable impact on the state of knowledge will have no direct or indirect benefits and therefore will not satisfy the principle of beneficence. This is why your REC will ask you about research objectives, resources, and funding. They will need to know who is assisting with the research and if the assistants have completed required trainings. The REC will review all procedures involved in the study including devices, equipment, surveys, and questionnaires. Think ahead to include all research activities your participants will encounter. You may want to leave yourself the option to adapt your procedures. For instance, instead of saying, "Improvement will be assessed with a grand jeté," you might propose "Improvement will be assessed using movements typically performed in the ballet class." Making a broader promise to start will allow you to collect data on a larger number of movements as long as the movements are typical for the class. Discussing your study with diverse experts and your REC committee and staff members will help you propose procedures that are both definite and inclusive.

RECs will need to review details of the study population and how researchers will recruit the participants. This includes reviewing recruitment materials for the study, like fliers, social media announcements, and language to be used in emails and live presentations. They will also review the document participants will be asked to review and sign to indicate their informed consent to participate. RECs want to ensure that the language appropriately reflects the study and is not coercive or misleading. RECs can help researchers develop recruitment strategies and materials that are both ethical and successful.

As part of the study application, researchers must disclose conflicts of interest to their RECs. Broadly, a *conflict of interest* is present when a researcher is involved in more than one organization and serving the interests of one organization may work against the other. For example, a dance teacher developed a new foam roller for intrinsic foot muscles and markets and sells the device. The researcher has an obligation (as a researcher) to acquire unbiased, objective data about the utility of the device. But the researcher may also be motivated (as the seller of the device) to show that the device is effective. These conflicting motivations, or interests, have the potential to compromise the researcher's objectivity and allow bias to influence the study. Researchers often feel confident they can appropriately avoid temptation or prevent bias, but cognitive research about unconscious bias contradicts such a claim. Regardless, even the appearance of a conflict of interest can damage your or your institution's reputation and can undermine the public's trust in the research findings. Conflicts of interest can often be managed with the help of your REC if they cannot be avoided completely.

Summary

The special authority granted to researchers through their role as key actors in the enterprise of science demands ethical conduct. In this chapter we have explored ethical relations between researchers and research participants, and discussed researcher relations with research associates, sponsoring organizations, and the professional community that uses research findings. The principles of respect, beneficence, and justice offer guidance in managing these relations. Governmental regulations, RECs, and professional organizations all have guidelines to help researchers behave ethically toward research participants. Researchers can optimize relations with research participants by recognizing participants as collaborators in the research process who, when treated ethically, can enhance the quality of the work researchers produce.

Highlights

- Ethical researchers protect the welfare and dignity of their participants as partners in the research process.
- Components of informed consent include information, comprehension, and voluntariness.
- RECs have a vested interest in making sure your study is feasible and will have an impact on the field or on participants' health and well-being.
- Researchers can optimize relations with research participants by recognizing participants as collaborators in the research process who, when treated ethically, can enhance the quality of the work researchers produce.

References

Grodin, M. A., & Annas, G. J. (1996). Legacies of Nuremberg: Medical ethics and human rights. *Journal of the American Medical Association, 276,* 1682–1683.

National Commission for the Protection of Human Subjects of Biomedical and Behavioral Research. (1979). *The Belmont Report: Ethical Principles and Guidelines for the Protection of Human Subjects of Research.* US Department of Health and Human Services, Office of Human Research Protections. https://www.hhs.gov/ohrp/regulations-and-policy/belmont-report/read-the-belmont-report/index.html

Office for Human Research Protections. (2018). *Federal Policy for the Protection of Human Subjects ('Common Rule').* US Department of Health and Human Services. https://www.hhs.gov/ohrp/regulations-and-policy/regulations/common-rule/index.html

Resnik, D. B. (2015 December 1). *What Is Ethics in Research & Why Is It Important?* National Institute of Environmental Health Sciences. Retrieved February 5, 2020, from https://www.niehs.nih.gov/research/resources/bioethics/whatis/index.cfm

Stewart, R. J., & Reider, B. (2016). The ethics of sports medicine research. *Clinical Journal of Sports Medicine, 35,* 303–314.

Welsh, T. M. (1999). Ethical standards for experimental research with dancers. *Dance Research Journal 31,* 86–99.

Welsh, T. M., & Chatfield, S. J. (1996). Within-subject research designs for dance medicine and science. *Journal of Dance Medicine and Science, 1,* 16–21.

Wolf, M. M. (1978). Social validity: The case for subjective measurement or how applied behavior analysis is finding its heart. *Journal of Applied Behavior Analysis, 11,* 203–214.

World Medical Association. (2013). *Declaration of Helsinki.* 64th General Assembly of the World Medical Association. https://www.wma.net/policies-post/wma-declaration-of-helsinki-ethical-principles-for-medical-research-involving-human-subjects/

Additional Resources

American Psychological Association. (2002). Ethical principles of psychologists and code of conduct. *American Psychologist, 57,* 1060–1073.

Botham, S. (2018). *Ethics Dance: Promoting Ethical Practices in Dance.* Retrieved April 20, 2020, from http://www.ethicsdance.co.uk/

Collaborative Institutional Training Initiative (CITI) Program. *Research Ethics and Compliance Training.* http://www.citiprogram.org

Council for International Organizations of Medical Sciences (CIOMS). (2016). *International Ethical Guidelines for Health-Related Research Involving Human Subjects.* https://cioms.ch/wp-content/uploads/2017/01/WEB-CIOMS-EthicalGuidelines.pdf

Office for Human Research Protections. (2016). *Human Subject Regulations Decisions Charts.* US Department of Health and Human Services. https://www.hhs.gov/ohrp/regulations-and-policy/decision-charts/index.html

World Health Organization (WHO). (2011). *Standards and Operational Guidance for Ethics Review of Health-Related Research with Human Participants.* https://www.who.int/ethics/publications/9789241502948/en/

7

Writing a Literature Review

SARAH J. KENNY AND JEFFREY A. RUSSELL

Questions to Stimulate Thinking

- How could knowing what others have studied about the topic I am interested in guide the research that I conduct?
- Why would others want to read a review of previous literature that is about the topic of my research?

What Is the Purpose of Writing a Literature Review?

In order to conduct a research project, you must first know what other researchers have done previously that is related to your topic. Writing about others' work has several benefits, the most important of which is educating yourself about the background information that will form a framework for what you want to investigate. In this chapter we focus on how to put into words what you learn from reading other people's research in a way that enables you to be really clear about how your proposed study will relate to what has been done before.

Among many reasons why it is important to write a literature review (or "lit review"), perhaps the most essential and compelling is to explain the context for your study. You must answer the question, "How do I know that this topic needs to be investigated?" After all, why would someone conduct a study that does not need to be carried out? A review of the literature provides you with the rationale for doing your research study because it is based on related work that appears in the existing literature.

Other reasons for writing a literature review support this essential purpose of giving your research a contextual foundation, and these include the following:

1. To describe what is already known about your topic—Providing an overview and analysis of prior research and the current state of knowledge about your topic allows you, and those who read your work, to evaluate whether you are needlessly duplicating effort, making purposeful improvements in prior studies, or properly considering the trials and errors made by others. Writing about what is currently known provides insight into what is not yet known about your research topic.
2. To appraise the existing literature critically—There is a difference between providing a descriptive summary of previous literature and thoroughly evaluating it. To appraise a study critically, you evaluate the extent to which the research was properly conducted so that you can make judgments about how meaningful the contributions of the work are to the existing body of knowledge. Being able to gauge how useful a study is can provide guidance for future investigations. As you write about and critically reflect on the literature that is related to your topic, you can design your project in a way that improves upon what previous researchers have accomplished.
3. To identify gaps in existing knowledge—One way that you can greatly add to the current body of work in dance science is to conduct research that "fills a gap," building on what has been previously reported. When completed, your review of the literature should highlight areas in which additional attention is needed to answer important questions in dance science. No single researcher or research team possesses all the answers; it takes many people—with different approaches, all working over a period of time—to begin to plug the holes in what we know. Your literature review should provide a springboard for you to contribute new understanding of the topic that you have chosen to study.

How to Write a Literature Review

Purpose Statement

Establishing the objective, or purpose, of your review of the literature is step one. Having a clear purpose for what you want to accomplish in your lit review makes the entire process of giving energy to your research much easier. The purpose is the fuel that will drive your literature review forward. Keep it foremost in mind as you write; this will bolster your enthusiasm for your project and provide the basis for progress. In fact, you will find it helpful—as will your reading audience—to state the purpose at the beginning of your review to "set the stage." A dancer would not take the stage unless already warmed up, with the lighting design in place, the stage manager and other staff in position, the music cued, and the stagehand ready to raise the curtain. A dance performance cannot be successful without these preparatory steps that set the stage for a superb effort. Articulating your purpose acts in the same way when writing a literature review, providing the first preparatory step.

Take this chapter, for example. If you go back to the first paragraph, you will see these words: "we focus on how to put into words what you learn from reading other people's research in a way that enables you to be really clear about how your proposed study will relate to what has been done before." This is the purpose we desire to accomplish here; it is the goal we have set for ourselves—to give you the knowledge you need to write a successful literature review. You may want to start your purpose statement like this: "The purpose of this project is . . ." or something similar. The purpose may also be called an aim, objective, or goal; all indicate the direction of your work.

The purpose of a research project develops from that place of curiosity that compels you to pursue research in the first place. The reason why you are embarking on this adventure—and the literature review that is part of it—is because you have a question about something you have observed with your eyes or considered in your mind. Two examples of many research questions you could ask are: "Does six weeks of lower extremity strength training with light resistance increase jumping height in university dancers?" (quantitative research example) or "What are the lived experiences of older adults who participate in a community dance class?" (qualitative research example). Then, as with any question, figuring out the answer is the next step.

When the research approach is from a quantitative perspective, determining a research hypothesis will help you target the information that assists you in answering your question; formulate the content of what will be covered in your review of the literature; and ultimately prepare you to conduct your own research study. Your hypothesis really is a straightforward statement that provides a predicted answer to your research question. Coming up with a working hypothesis is a good strategy for directing your review of the literature. Thinking about our strength training question, a working hypothesis, which is sometimes called an alternate hypothesis, could be as follows: "Six weeks of light strength training will *increase* the jump height of university dancers." It expresses that there will be a difference between one condition (in this case, before weight training) and another (after weight training). Then a good way to start searching and reading the literature would be to focus on previous studies that were designed to improve dancers' jump height (or power). A null hypothesis is one that predicts there will be no difference between one condition (in this case, before weight training) and another (after weight training). For example: "Six weeks of light strength training will *not affect* the jump height of university dancers." Null hypotheses are then tested using statistical analysis of quantitative data to determine if the null hypothesis is to be accepted or rejected. Information about research design, research questions, and hypotheses is covered in detail in other chapters.

When the research approach is from a qualitative perspective, as in our example research question about the lived experiences of older adults who participate in community dance, you want to ensure that when you collect data (e.g., through talking to your research participants, the older adults) you enable all possible experiences to be shared, rather than focusing on a preconceived idea that you have gleaned from the literature. Depending on the qualitative approach you take, you may complete only a partial review of the literature to ensure that the research has not been executed before, and that you are not influenced beforehand by any particular perspective or conclusions about how older adults experience dancing. In such a case, a brief review of the problem or research focus and why it is important, along with determining what we know and what we do not know, are important aspects of a brief review of the literature. You do not come to any preconceived notions or formulate hypotheses with a qualitative approach to research. Qualitative methodologies and their associated research paradigms are addressed in detail in *Chapter 13*.

Writing to Your Audience

So far in this section we have presented systematic techniques to prepare you for writing a literature review. An additional piece of the puzzle that will enable success is to think about your readership. In order to accomplish a high-quality review, you will have at least two audiences. One is yourself; your review will help you assess the state of the knowledge on your topic so that you can mold your own research project appropriately. The other audience is, of course, your readers. For example, if you are a student, your readership will include a professor, a committee of professors, or one or more external reviewers; that is, the people who are charged with supervising your research and assessing your work for academic credit.

Another type of audience includes those who will read your literature review for the benefit of their own research or work (e.g., dancers, dance teachers, health and fitness instructors, scientists, or health care professionals). It is especially important for your reader to understand your perceptions and intentions that formed the foundation of your study. In our digitally driven world, the places where high-quality, publishable writing may end up are quite fascinating to consider. Thus, your review must be clear, accurate, relevant, and structured to meet the needs of as many readers as possible, and certainly the ones who are the most likely consumers of your efforts. Few things are more disheartening in research than a researcher with enthusiasm and exciting ideas who is unable to conquer the challenge of getting thought processes into a written form from which others can benefit.

Once you select the research question and have prepared yourself mentally for writing to your audience(s), it is time to move on to the next step: determining the types of resources that you will use.

Determining Appropriate Resources

Given the ease of online access to a plethora of resources on any given topic, enabling you to collect a number of resources related to your research topic, you must be able to choose which are appropriate resources to include in your written review and which are not. In a buffet at a restaurant, not everything available to you is supportive of healthy living. Similarly, not everything we can access online will support "healthy" research. *Chapter 4* contains many valuable tips for finding appropriate resources, so here we encourage you to focus as much as possible on primary, original research resources rather than secondary, opinion-based publications.

Primary resources are reports of results interpreted by the researchers who themselves conducted the study that is the subject of the report. Using our prior example, if you complete a study investigating the ability of a strength training program to increase jump height in university dancers and publish your results in a journal, this would be considered a primary resource. If a magazine about dance, or your local newspaper, or a blogger, or even an author writing a review of research about dance training for a professional journal incorporates your research results into their work, each of these examples is a secondary resource. The difference is that when you write about your own research, you know about it firsthand. Everyone else who reads your report and writes about what you did has only secondhand information because they did not do the research themselves. This makes all of their writing, no matter how outstanding, secondary. It is always best to use primary resources in your review of the literature because that ensures you are getting your information directly from the original sources, minimizing any possible misinterpretation of research findings along the way.

Dr. Thomas Findley (1989), writing in the *American Journal of Physical*

Medicine and Rehabilitation, offered several excellent techniques for conducting a review of the literature. His goal was to make more bearable what can be a voluminous process of reviewing all the scientific articles about a topic. Basically, in writing a review of the literature you embark on a journey through as many articles about your topic as you can get your hands on. With this, you may have the tendency either to get bogged down in a deep mud-hole of information or to forget where you found crucial tidbits that relate to and bolster your project. To counteract this problem, Dr. Findley recommends using an A, B, C, and X coding system when reading articles. An "A" article is highly relevant and tightly related to the project, a "B" article is somewhat relevant, and a "C" article is possibly relevant. "X" articles may for some reason have seemed potentially relevant when you found them, but they turned out to be unrelated when you examined them more closely. Thus you can save your time by casting those aside.

Once you have your A, B, and C primary research articles, read them with a critical view—making notes in your own words about the quality of the research (e.g., thoroughness of the method, appropriateness of the analysis, confidence in conclusions drawn), what worked for the researchers, what did not work, along with what pieces you can incorporate into your own research—and use your preferred method of highlighting to identify key points in them. These highlights will supply content for a structured outline that you can use as the framework for your writing and as content for an annotated bibliography that will help you keep track of where you find each piece of key information from your reading. When you are finished with this task, you should have a good understanding of how prior investigations integrate with the research you envision, providing a view of the gaps in current knowledge.

Other Practical Techniques for Writing a Literature Review

If you are truly reviewing the literature—every single piece of scientific writing that you can possibly locate about your topic—you could find yourself standing in front of a gigantic mountain of articles, depending on how much your area of interest has been studied previously. The best way to climb a mountain is to take one single step at a time. Of course, depending on your reason for writing a literature review in the first place (i.e., the purpose!), you may indeed feel the need (or be required, as in an academic environment) to head for the summit of that entire mountain of literature. However, for other destinations where you will submit a review (e.g., a grant or scholarship application, a "state of the topic" summary paper, or your own research

Table 7.1. Three Primary Types of Literature Reviews

Narrative Review	A review that focuses on gathering and summarizing existing literature about a topic (not confined purely to research literature) as a means to understand the available knowledge on that topic and assimilate the knowledge into a single document. This type of review does not have to be exhaustive, meaning that there may or may not be an attempt to locate every piece of pertinent literature.
Scoping Review	A review that gathers as much information as possible about a research question—regardless of the type of information—with the intent of outlining what is known about prior research on the topic and creating a framework that informs how one's proposed research will be conducted. A scoping review may or may not be exhaustive, but the aim is usually to be as complete as possible.
Systematic Review	A review that seeks to evaluate in a carefully controlled way all of the research previously conducted about a topic with the objective of identifying the extent to which research evidence exists to support a hypothesis, a method of action, or a clinical technique. A systematic review should be exhaustive so as not to miss important results that have been published previously, and it must follow a prescribed, repeatable methodology (Siddaway, Wood, & Hedges, 2019) to achieve its intended goal.

Note: Based on Grant & Booth, 2009.

article for a journal), scrutinizing a subset or portion of the literature is often adequate. In either case, there are some additional tips beyond what we have already shared that will be beneficial.

Grant and Booth (2009) identified several types of literature reviews, three of which are described in this text (see Table 7.1). In this chapter we focus on how to conduct Narrative and Scoping Reviews, with the primary differences being the level and breadth of resources included in the written review. *Chapter 18* addresses Systematic Reviews in detail.

Annotated Bibliography

As you read a collection of relevant articles from your online search, assemble them into an annotated bibliography. An annotated bibliography is a series of bibliographic references with notes attached to each reference. There are several benefits of this exercise, including keeping your thoughts in order about what you are reading, providing a document that continually refreshes your mind about your topic, and perhaps most important, offering a logical, systematic framework from which to start your own writing process.

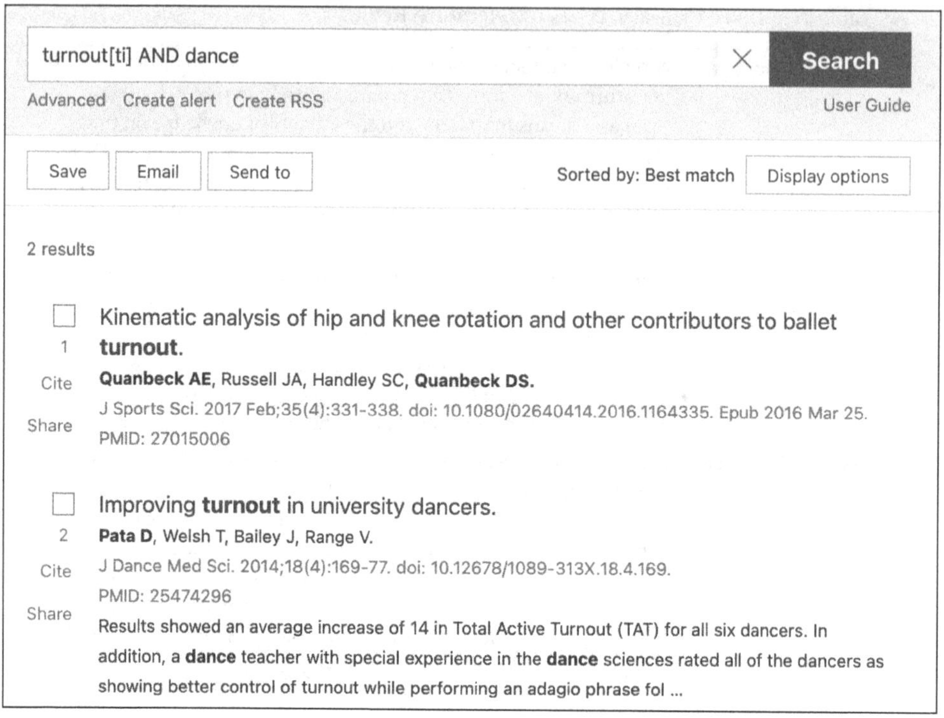

Figure 7.1. Example of Literature Search Output. These are two of the articles that would be returned by the search phrase "turnout[ti] AND dance" using the Pubmed.gov search engine.

Let's look at one simple way of creating an annotated bibliography. Suppose you wanted to study turnout in dancers. A search in PubMed would yield several results, including those seen in Figure 7.1.

After reading each article thoroughly, you would add your personal notes (i.e., annotations) about the key points you find interesting and relevant to your research purpose. Your bibliography might then look something like this:

1. Pata, D., Welsh, T., Bailey, J., and Range, V. (2014). Improving turnout in university dancers. *Journal of Dance Medicine & Science, 18*(4), 169–177.

   Six first-year university dancers participated in a training program to see if turnout improved. "Total Active Turnout" was measured repeatedly—the amount of turnout in degrees while standing on rotating disks. Dance teachers watched videos of the students performing a phrase that required good control of turnout before and after the program and had to judge functional improvement. After training, Total

Active Turnout improved between 9 and 22 degrees in all dancers. Dance teachers were variable in how they rated the dancers' turnout, but generally noticed improvement.

2. Quanbeck, A. E., Russell, J. A., Tonsager, S. C., and Quanbeck, D. S. (2017). Kinematic analysis of hip and knee rotation and other contributors to ballet turnout. *Journal of Sports Sciences, 35*(4), 331–338.

Ten female university dancers' functional active standing turnout was measured in a biomechanics lab using markers and cameras. Recorded more knee rotation and less hip rotation during turnout than prior studies. Mean turnout = 129 degrees measuring the rotation at the feet; 135 degrees measuring with camera system. Hips provided 49 degrees of total turnout. Knees contributed 41 degrees of total turnout.

You can make your annotations as comprehensive, as brief, or as grammatically correct as you wish. Just be sure that you write your notes clearly, so that you can understand what you wrote when you go back to them later. Once you collect a number of articles in your annotated bibliography, you will want to organize the references according to the planned sections of your writing, general topic areas, or some other structure that makes sense to you. Notice that an annotated bibliography developed this way gives you an outline and even some initial phrases and sentences from which you can assemble your overall literature review.

Introduction and Discussion as Targets

Another helpful tidbit is to write with your Introduction and Discussion sections of your research paper in mind. When you do, the literature review will fit seamlessly into the rest of your document, whether it is a journal article, thesis, or funding proposal. An Introduction sets the stage for your research. A Discussion is an interpretation and explanation of how your results align with the current state of the literature. Thus, when writing your review, try to anticipate which of the facts, data, and interpretations you have discovered in your reading will be most suitable for your Introduction and Discussion sections. Do some outlining or create some notes accordingly so that you do not forget how all the relevant pieces weave together into a cohesive presentation of your research.

Citing Resources

From an ethical standpoint (see *Chapter* 6), it is crucial to give credit for original ideas and unique ways of constructing sentences to the person who wrote them in the first place. Including references as you write your sentences acknowledges the specific sources from where the information was originally derived. In addition to providing recognition of prior authors' efforts, proper citations allow your readers to find the sources you used, so they can read further on their own.

It is important to note that there are multiple strategies to format your references, both within the text (as you are writing your sentences) and as a list at the end of your research paper or report. Where you intend to submit your literature review will define how you structure the citations, usually according to specific formatting requirements or a citation style guide. A citation style guide is a set of standardized rules for publication. Each academic institution, peer-reviewed journal, and professional organization adopts its own specifications for citing and formatting references, as outlined in their style guide. These guidelines are instrumental for organizing your research, reviewing the literature, and identifying patterns within scholarship because they put authors, dates, topics, and sources in context.

For example, the editors of this book provided chapter authors with guidelines for how to arrange the references (i.e., American Psychological Association style, or APA). Another style commonly used in the dance sciences is the American Medical Association style (AMA). While both are used in our field, be sure to consult the style guide required for the venue where you plan to submit your research manuscript. Many citation managers allow users to build filters to match unusual specifications. This can save time and avoid tedium from manually reformatting your references when submitting a manuscript to a journal or organization. If you have questions about a particular citation style, your librarian can help.

Below are some samples of different formatting schemes for one of the journal articles that we referred to earlier. Notice how the information content (author names, publication year, article title, journal title, volume and issue number of the journal, page numbers of the article) is essentially the same; the pieces are just arranged a bit differently in each instance. Once again, always consult the recipient of your writing—whether it is a professor, department, university, funding agency, journal, or other place—for guidance about how to assemble your references properly, as there are many, many details involved and many, many variations for constructing your citations.

American Psychological Association (APA) Format

Quanbeck, A. E., Russell, J. A., Tonsager, S. C., & Quanbeck, D. S. (2017). Kinematic analysis of hip and knee rotation and other contributors to ballet turnout. *Journal of Sports Sciences, 35*(4), 331–338.

Chicago Manual of Style Author-Date Format

Quanbeck, Amy E., Jeffrey A. Russell, Sarah C. Tonsager, and Deborah S. Quanbeck. 2017. "Kinematic Analysis of Hip and Knee Rotation and Other Contributors to Ballet Turnout." *Journal of Sports Sciences* 35, no. 4: 331–338.

American Medical Association (AMA) Format

Quanbeck AE, Russell JA, Tonsager SC, Quanbeck DS. Kinematic analysis of hip and knee rotation and other contributors to ballet turnout. *J Sports Sci.* 2017;35(4):331–338.

Typically, it is desirable to avoid using direct quotes in your writing unless a particular author you discover presents words in such an innovative and compelling way that you feel you have no choice but to reproduce them in your own work. In this case, use quotation marks around the quote. If the quote is longer than three lines, it should be indented from the main paragraphs of the document rather than set off by quotation marks. Note that for direct quotes you must provide the page number(s) where your reader can locate the exact quotation in the original source. Moreover, it is a lesson of utmost importance that just writing an author's words in a different way—that is, paraphrasing or restating them "in your own words"—does not release you from the responsibility to indicate properly the source of the information, phrase, or idea.

Managing all the references you consult is easily done through online software tools introduced in *Chapters 4 and 9.* Although providing product reviews and explaining their similarities and differences is not within the scope of this chapter, basically the Corporation for Digital Scholarship's *Zotero* (Corporation for Digital Scholarship, no date), Mendeley Ltd.'s *Mendeley* (Mendeley Ltd., 2021), and an online version of Clarivate's *EndNote Basic* (Clarivate, 2022) are reference management software programs that are available without charge to students and faculty on many university campuses. *Zotero* is an open-source product, while the other two are proprietary. *EndNote* also has a program with more features than the free version that is marketed by Clarivate; it has academic pricing for students and faculty, as well as institutional licensing. *Refworks* is owned by ProQuest LLC (2022), a division of Clarivate. It is software typically licensed by a university or other institution so that students and employees can access it.

Regardless of which one you use, all online reference management software products store the necessary components of your references (e.g., authors, article titles, name, volume and issue of journals, book titles, page numbers, etc.). They can catalogue the references you find from websites, literature searches in digital databases, and articles in PDF format. Bibliographic details compiled by the referencing software program should be checked for accuracy. For example, ensure that each entry is complete with information placed correctly and with proper punctuation. This will enable precision in the citation, regardless of the citation style used.

Once you have typed a statement in your literature review that requires citation (for example, you might write something like this that needs to be cited: According to Quanbeck et al., the rotation of the knees contributes an average of 41° to total turnout.), the referencing software allows you to place a citation marker at the correct spot in the written text. Thus, when you are finished writing, you will have a series of precisely formatted citations in the text, all properly keyed to your reference list that the software will have assembled according to your pre-selected configuration. We encourage you to investigate each of these programs and to discuss the options with your librarian, professors, and colleagues so that you can make an informed decision about which software to use.

Summary

Writing a strong and comprehensive literature review is an important ability to master for any researcher, regardless of the research paradigm. The process will help you understand what others have done before you and will yield a foundation on which to build your proposal for research. It can be a fascinating journey through the history of your topic of interest while simultaneously being a seemingly cumbersome task. Reading one article at a time, then writing one sentence at a time, will help keep the task manageable. Ensuring that you first have a purpose for your research efforts will assist you in stepping out on the right foot. Asking good research questions will provide a framework from which to build your review. Incorporating tips and tricks that make the process of reviewing and writing about the literature more effective will help you enjoy the experience of producing a final document that is of a high standard.

Highlights

- Literature reviews answer the question, "How do I know that this topic needs to be investigated?"
- Your purpose statement is the fuel that will drive your literature review forward.

- Focus as much as possible on primary, original research over secondary, opinion-based publications.
- When reading all the articles, use a coding system (e.g., an "A" article = highly relevant; "B" = somewhat relevant, and "C" = possibly relevant).
- A citation management software program is essential to stay organized and efficient.

Editors' Note

We thank Leah Sherman, Merry Lynn Morris, Sarah Kenny, and Jeff Russell for their willingness to shift small portions of content between *Chapters 4 and 7* to allow it to appear in a sequence that is more useful to readers.

References

Clarivate. (2022). EndNote: Accelerate your research. *EndNote.* Retrieved March 18, 2022, from https://endnote.com/

Corporation for Digital Scholarship. (n.d.). Your personal research assistant: Zotero is a free, easy-to-use tool to help you collect, organize, cite, and share research. *Zotero.* Retrieved March 18, 2022, from https://www.zotero.org/

Findley, T. W. (1989). Research in physical medicine and rehabilitation. II: The conceptual review of the literature or how to read more articles than you ever want to see in your entire life. *American Journal of Physical Medicine and Rehabilitation, 68*(2), 97–102.

Grant, M. J., & Booth, A. (2009). A typology of reviews: An analysis of 14 review types and associated methodologies. *Health Information and Libraries Journal, 26*(2), 91–108.

Mendeley Ltd. (2021). I "M" discovery. *Mendeley Ltd.* Retrieved March 18, 2022, from https://www.mendeley.com/

ProQuest LLC. (2022). RefWorks®. *RefWorks.* Retrieved March 18, 2022, from https://about.proquest.com/en/products-services/refworks/

Siddaway, A.P., Wood, A.M., & Hedges, L.V. (2019). How to do a systematic review: a best practice guide for conducting and reporting narrative reviews, meta-analyses, and meta-syntheses. *Annual Review of Psychology, 70,* 747–770.

III

Designing and Conducting Research

8

Participants, Setting, and Variables

TOM WELSH, JATIN P. AMBEGAONKAR,
AND LYNDA MAINWARING

Questions to Stimulate Thinking

- How do researchers select participants for their studies?
- When would researchers want to conduct dance research in a laboratory? . . . in the field?
- What makes an observation or measurement valid?
- What is the difference between measurement reliability and procedural fidelity?
- What is the value of pilot testing?

In Part III we offer an overview of the steps you will take as you plan and conduct empirical research in the dance sciences. These steps are often referred to as research methods. We describe them in a manner that acknowledges the diversity of research approaches used in our multidisciplinary field to help you recognize the common intentions of the research approaches detailed in Part IV.

The steps involved in planning and conducting empirical research commonly include:

- Selecting participants and securing a setting,
- Designing an observation (DV) protocol,
- Designing an intervention (IV) protocol (some approaches),
- Managing, analyzing, and presenting data,
- Organizing the elements to optimize methodological integrity.

Some steps will not apply to some types of research. For example, qualitative, descriptive, and correlational studies do not specify independent variables or use interventions. We have organized the steps into a progression, but it is common for researchers to move from one step to another and back again. Even if a step cannot be completed until later (e.g., analyzing data), you must think through in advance how each step will be performed to ensure, for example, that the data you are gathering will satisfy the requirements of your analytical tools. We encourage you to begin building the components of your study as you read these chapters. As with building a choreography, research requires patience, time to reflect, and a willingness to rework what you have built as you gain insight into what you are aiming to achieve.

In this chapter we address the research activities commonly described in the first four sections of the *Methods* in a research report: Selecting participants, choosing a setting, making observations, and (when relevant) applying treatment or training. Each step is informed by the research question or hypothesis. The discipline dancers acquire with dance training serves them well in addressing these highly disciplined research activities.

Selecting Participants

Identify a Target Population

An important early step in designing your research methods is identifying the *population* you want to study. A *population* refers to the "members of a clearly defined set or class of people, objects or events that are the focus of an investigation" (Payton, 1994, p. 34). A population might be, for example, dancers at a certain stage of their training, such as first-year studio dancers or professional dancers; teachers in dance training programs, or athletic trainers who work with dancers. Inanimate entities such as different brands of Irish Step Dance shoes or dance floors may also comprise a population. A population can also be comprised of groups of individuals, such as competition dance teams or dance trainng programs. In these cases the collection of individuals (team, training program) become the unit of analysis. Having a target population in mind as you design your study will help you make decisions about how to proceed.

Purposeful Selection

How participants are selected for a study depends on the question researchers want to answer, and the research approach they choose to address it. For example, a participant for a Case Report is selected because he or she presents with a condition the researcher finds worth describing (see *Chapter 15*). Researchers engaged in qualitative and other types of descriptive research frequently select participants who possess special characteristics or face a special challenge, such as being hypermobile or recovering from a debilitating injury. Researchers interested in learning how individuals respond to a newly developed treatment or training protocol may select participants based on their potential to improve a certain dance ability or the opportunity to address a challenge they face. Choosing study participants in this way is called *purposeful* or *purposive sampling* (Kowalski et. al., 2018).

Non-Probability Sampling

Researchers who do not select study participants at random from their target populations are using a *non-probability sampling* approach, some of which are extensions of the purposeful selection approaches described earlier. The simplest approach is a *sample of convenience*, which involves inviting participants who satisfy a study's inclusion criteria (see later description) to volunteer to participate in a study. Convenience sampling is the most common approach to selecting participants for studies involving human beings (Evans & Rooney, 2018; Portney & Watkins, 2009; Barlow et al., 2009).

Researchers who are interested in learning how a particular condition manifests in a number of individuals may use a *snowball sampling* approach, in which each participant is encouraged to recommend or recruit others who present with the same condition to volunteer for the study. Researchers who want to ensure that their sample represents various segments of their population might use a *quota sampling* approach in which they select, perhaps at random, a representative proportion of each segment of their populations (e.g., 30% 1st year dancers, 25% 2nd year dancers, etc.). Specific applications of these and other sampling strategies are described in the Part IV chapters (also see Payton, 1994; Portney & Watkins, 2009). Figure 8.1 depicts some commonly used sampling strategies.

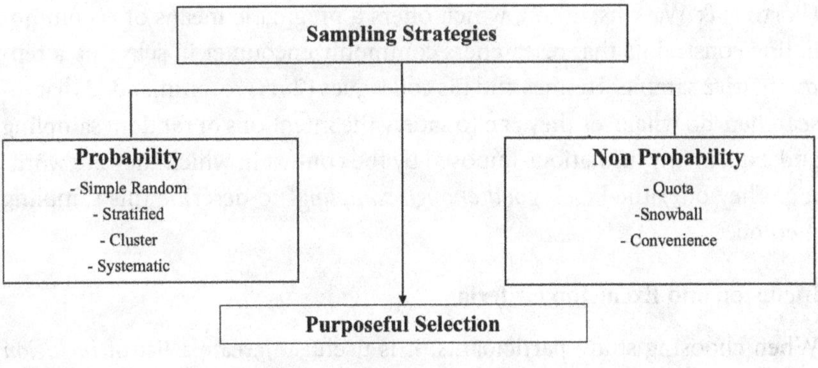

Figure 8.1. Participant Sampling Strategies.

Probability Sampling

If you plan to use inferential statistics to generalize your findings to a population, you will want to select a representative sample of participants from the population targeted by your research question. A perfectly representative sample would include every individual in the population, but of course that approach is rarely practical, especially if a population is large and geographically dispersed. The practical alternative is to select a subset of the population, a sample of individuals to represent the population (Payton, 1994). The ideal approach to ensuring that a sample accurately reflects the population is to select randomly a sizable number of participants directly from the entire population. For a sample to be random, each individual in the target population must have an equal opportunity to be selected.

When conducting survey, interview, and other types of research in which the primary contact with participants can occur through telecommunications or correspondence, it may be possible to select from the whole target population. For example, a survey can be sent to a randomly selected sample of all university dance faculty in a particular country. Online interviews can be conducted with a random sample of physiotherapists who are on-staff at the 10 most prominent dance companies in Europe. Researchers might use a *systematic sampling* strategy by, for example, inviting every tenth physician on a registry of dance medicine practitioners in Australia to take part in a study. Other sampling strategies that are commonly used in survey research are described in *Chapter 12*.

When research requires in-person observations or direct interactions with study participants, it is more common to select participants randomly from a segment of the population to which researchers have reasonable access to perform assessments or apply interventions. This strategy employs a *sampling frame* (Kowalski et al., 2018), also called an *accessible population* (Portney & Watkins, 2009), which offers a pragmatic means of accommodating constraints that researchers commonly encounter in selecting a representative sample. Thomas and his colleagues (2015) recommended that researchers do whatever they can to satisfy the intentions of random sampling and explain any limitations imposed by the context in which they are working. They offer the label *"good enough sampling"* to describe this sampling method.

Inclusion and Exclusion Criteria

When choosing study participants, it is useful to create a list of *inclusion criteria* that specify who from the pool of available participants is eligible to participate. Inclusion criteria describe characteristics that a potential participant must have to be considered for a study. For example, if your target population for a Conditioning for Limbo program is young, healthy, beginning Limbo dancers, you might specify that participants must be between the ages of 10 and 15 and be able to pass a brief general fitness test. You might also specify that participants have to be available to participate in hour-long training sessions held at <name and location of facility> two afternoons per week for 10 weeks beginning <date> and ending <date>.

It is also useful to specify *exclusion criteria.* For example, you may want to exclude participants with an injury that may compromise their performance of the skills or abilities you will be observing and exclude dancers who have an injury that may be aggravated by the training, treatment, or testing you plan to use in your study. Exclusion criteria can also be used to disqualify a participant's data from the analysis of the results if someone is unable to complete a sufficient proportion (e.g., 85%) of the scheduled training or treatment sessions. Specifying any data-excluding criteria before your study begins will head off the impression that you may be deleting data selectively to make your results look better. Using explicit participant inclusion and exclusion criteria will also improve the homogeneity and the representative ness of your sample and facilitate data analysis. To optimize their utility, inclusion and exclusion criteria should be more than simple opposites; they should be complementary.

Sample Size

Deciding how many participants to recruit for your study will be influenced by the methodology you use and by pragmatic considerations such as access to participants and the setting where you will conduct your research. Case reports, some qualitative studies, and some within-subject experiments may be conducted with a single individual or a small number of participants. Group comparison studies commonly employ a sizable number of participants to support statistical analysis. When conducting hypothesis-based research, typically with multiple participants, groups, or times, you will want to conduct a *power analysis* to determine how many participants to recruit. Power is explained in *Chapter 9.*

Applying Research Findings

Researchers who conduct their study with a large enough sample of participants who have been selected at random from a target or accessible population will be able to use statistical inference to forecast the applicability of their findings beyond the sample of participants who actually took part in the study. The concepts of generalizability and external validity are involved (see *Chapter 10).* Describing *participant demographic information* such as age, sex, level of dance experience, aspirations (e.g., preprofessional, recreational), and other characteristics germane to the topic under study can also help practitioners estimate the potential relevance of the research findings to those they normally train, treat, or manage and the contexts in which they work. The ultimate test of the applicability of research findings, regardless of how participants are selected or described, is whether similar results can be achieved with other participants, in other settings, or with other researchers or practitioners administering the treatment or training, a process called replication (Barlow et al., 2009; Kazdin, 2011; Sidman, 1960). We should remember, however, that for some types of research, general applicability is less important than gaining a deeper understanding of a phenomenon of interest. Again, the research question guides the methods used.

Choosing a Research Setting

Another important decision you will need to make early in the research process is where to conduct your study. The place where researchers make observations and deliver treatment or training is called a research setting. One dimension along which settings can vary is lab(oratory) and field. Working in a lab (e.g., exercise physiology, motion capture) will normally allow greater control of experimental variables. Working in a field setting (e.g., dance studio, clinic) may enhance the practical applicability of the procedures under evaluation (see Table 8.1). Lab and field can be viewed as directions on a continuum. Thoughtful researchers seek a balance that matches their research priorities. In a jump training study, for example, researchers may choose to make strength, endurance, and power measurements in an exercise physiology lab and to conduct jump training and technical skills assessments in a dance studio.

Some researchers modify a lab setting to make it more like a dance studio, for example by importing a portable limbo bar or ballet barre, installing a Marley floor surface, or using a musical recording instead of a metronome to standardize tempo during testing. Other researchers create dance-specific assessments of strength, alignment, and endurance that can be performed in a dance studio (Deckert et al., 2007; Wyon et al., 2003). Yet another choice is to arrange a collaboration among colleagues with access to complementary resources. For example, a dance trainer and teacher might collaborate with an exercise physiologist to conduct a study that is made possible by their combined skill sets (e.g., Couillandre et al. 2008).

Table 8.1. Typical Characteristic of Laboratory and Field Settings

Factor	Laboratory	Field
Environment	Artificial	Real-life
Ability to control variables	Higher	Lower
Potential influence by uncontrolled variables	Lower	Higher
Similarity to the target setting	Lower	Higher

Making Observations

Virtually all empirical research with human participants involves observation or assessment of a performance, capacity, tendency, perception, or experience. Recording observations may be the primary focus in some types of research. In qualitative studies and case reports, researchers describe what they observe, sometimes in great detail. In observational descriptive studies, researchers generally quantify what they observe by reporting frequency, duration, or proportion, for example, or by measuring a physical property or capacity. Records of observations become data, and collecting good data is fundamental to effective empirical research (Sidman, 1960). For consistency in perspective, we approach the topics in the rest of this chapter mostly from a quantitative research perspective but offer important caveats for other research approaches. *Chapter 13* is devoted to the unique aspects of qualitative research methods and other research approaches are similarly treated in the Part IV chapters.

When reading the literature, you may encounter a variety of labels that refer to the concept of making and reporting observations: outcome variable, assessment, outcome measure, test result, or dependent variable (DV). Regardless of which label is used, effective assessments in quantitative research generally possess three qualities; they are *valid, reliable, and responsive* (see Welsh, 2003). We describe each quality separately and then consider how they interact.

Assessment Validity

A valid assessment is one that assesses what it purports to assess. It is relevant to the question you are trying to answer or the hypothesis you want to test. Review the purpose statement you wrote while reading *Chapter 7.* What capacity, performance, perception, or experience will you need to assess to answer your research question? You may, for example, want to know whether dancers find a particular treatment or training approach mentally stressful.

You might use a mental stress scale or gather self-reported qualitative information about participants' perceived stress.

In the dance sciences, determining how to assess an outcome that is relevant to your research question may be an important prerequisite to conducting your research. Two common approaches to evaluating the relevance of a measurement approach to the phenomenon under study include logical (or face) validity and construct validity. Logical validity involves asking key stakeholders (e.g., dancers, teachers, clinicians, choreographers) whether your measurement approach offers a reasonable representation of the ability or capacity you are studying. For example, you might provide experienced ballet teachers with front view images of dancers at the ending position of développé a la seconde (lifting one leg, knee first, high and to the side) and ask whether an increase in the angle between the standing leg medial malleolus, the gesture leg ASIS, and the gesture leg medial malleolus represents an increase in développé height (Morgan, 2016). The assessment can be conducted informally by simply asking several teachers the question and, perhaps, offering a few photographs to illustrate the measurement (logical validity). To determine construct validity, you might convene a panel of experts (for the example above, perhaps nationally recognized ballet teachers) to make ratings of whether a measurement approach adequately represents the capacity, ability, or phenomenon under study. Baumgartner and colleagues (2016, pp. 114–134) offer a comprehensive discussion of measurement validity. You will want to be sure your outcome variable adequately assesses a capability related to your research question.

Assessment Reliability

In addition to assessing an outcome in a manner that is relevant to your research purpose, your assessment approach will need to generate consistent or *reliable* results. If you were to use a stretchy tape measure to measure a dancer's standing height and you pulled the tape a little harder on some days and a little less hard on others, you would get different measurements, not because the dancer's height had changed but because the measurement tool and the way it was being used were not consistent across measurements. It is common for researchers who work in laboratory settings to calibrate their instruments at the beginning of each measurement session by checking their measurement device against a constant standard. An example would be putting a known 1kg weight on a digital scale and using the fine-tuning adjustment to ensure that the scale is showing exactly 1kg before making any new measurements.

It is also common for researchers in a variety of settings to define operationally the steps involved to ensure measurement consistency. Specifying the actions to be taken by the observer (or tester) is a first step toward promoting consistency. To create a reliable assessment protocol, imagine what you would want a trusted colleague to do on a day when you were unavailable to make the assessments yourself and write it out, step-by-step. Operationally defining your protocol will permit you to use more than one observer. It will also allow you to describe clearly how assessments were made when you prepare your research report, and it will support replication of your study by other researchers and practitioners in the future.

It is common for observers to be trained to perform assessments with pilot participants before collecting any data that will be used in a study. In some fields, measurement reliability is assessed periodically throughout a study, and observers undergo retraining if reliability deteriorates, a phenomenon known as *observer drift* (Bailey & Burch, 2018). Several approaches to assessing the reliability of research observations are used in the dance sciences.

One approach to assessing measurement reliability is to see if a single observer can make consistent, repeated assessments of the same capacity, skill, or event. A clinician might use a goniometer (a long-armed protractor) to measure external hip rotation for 5 to 10 dancers in series, twice through, possibly in random order and, perhaps, without the clinician being able to see the scale on the goniometer. As each measurement is made, an assistant records the score and resets the goniometer before handing it back to the clinician to make the next measurement. Comparing the clinician's first measurement with the same clinician's second measurement for each dancer would yield a degree of agreement by correlating the collection of test and retest scores (see *Chapter 14*). An assessment of this type is referred to as *intra-observer, intra-tester, or intra-rater* reliability because the paired measurements are made by the same observer. The term *measurement stability* is something used to describe this type of measurement reliability (Berg & Latin, 2008).

Another approach is to see whether two or more independent observers can obtain the same measurements using the same protocol independently. Assessing consistency in this way is referred to as *inter-observer reliability*. It assesses agreement between observers, and the term *objectivity* is sometimes used to describe it (Berg & Latin, 2008).

One technique for quantifying the reliability of the paired assessments is to calculate the Intraclass Correlation Coefficient (ICC), which is explained in *Chapter 14*. Another technique is to calculate Inter-Observer Agreement (IOA) by pre-specifying a degree of difference that will be considered an agreement (e.g., a difference less than 0.5 degrees), counting the number of times the paired observations agree, and calculating the percentage of all observations on which the two observers agree (Bailey & Burch, 2018). Portney and Watkins (2009) describe a variety of other approaches to assessing measurement reliability and validity.

Assessment Responsiveness

A third quality of competent assessment that is essential in intervention research is the *responsiveness* of the dependent variable (DV) to changes or differences in the independent variable (IV). For an assessment protocol to be useful for intervention research, it must be responsive (or sensitive) enough to reveal changes or differences in the influence of the training or treatment (IV). Conducting an experimental analysis offers the ultimate test of responsiveness, but you can increase the chances that your DV will be responsive to your IV by choosing an assessment protocol that has been shown to be responsive in a previous study or during pilot testing.

Assessment Trade-Offs

There are some trade-offs that new researchers need to consider when building a study. First, as creative individuals, dancers may be inclined to invent their own measurement protocol. We suggest that you first see if you can find a previously published protocol that you can fit to your study. If you can find an existing protocol, you will be able to devote more of your time to working on other aspects of your study and you will be able to compare your results with those obtained in other studies that used the same measurement approach. If you cannot find an existing assessment tool that will let you answer your research question, you will need to develop your own and validate its use. Rowe and Mahar (2006) explain how to conduct a measurement validity study.

Second, optimizing one of the three measurement qualities—validity, reliability, and responsiveness—may compromise one of the others. For example, a highly valid assessment may prove hard to measure reliably or a highly reliable assessment may not be sufficiently responsive to your intervention. As you choose a measurement system, be sure to balance the influence on all three aspects of your measurement protocol.

Specifying Independent Variables

Some research approaches are aimed at identifying naturally occurring or already present variables that occur together. For example, if dancers in a particular training program are progressing more rapidly than dancers in another training program, researchers may ask themselves: What are the teachers doing that is unique or special? How do the training practices differ from the norm? What is different about how dancers are selected? What special ancillary services are available? The possibilities, of course, are limitless. The aim is to identify variables that may be related to or co-vary with important outcomes variables. To assess the influence of one variable on another, researchers need to specify the second variable as clearly as they have specified their outcome variable. What will you and your research associates need to do to get a reliable assessment of other variables? Correlation research (*Chapter 14*) allows researchers to study relationships between variables. One of the variables that co-varies with a beneficial outcome might become an independent variable (IV) in a subsequent study.

Interventions

Intervention case studies and experiments are designed to assess the effects of an intervention. They do so by systematically applying an independent variable (IV) and observing its influence on a specified outcome (DV). An intervention might be applying a treatment or training program or making a change in a company policy or practice. For example, if researchers suspected that dancers are more likely to improve their dance abilities while taking a class from technique teachers who use more positive reinforcement than aversive control in their classes, the researchers might operationalize a teaching approach that emphasizes positive reinforcement and compare dancer improvement under that condition with improvement under a condition that emphasizes critiquing errors.

Whether an intervention is adapted from the work of other researchers or expert practitioners, researchers must operationalize how the intervention will be applied as they conduct their study. Operationally defining an intervention is similar to operationally defining a measurement protocol. It involves creating a list of steps or operations that those who will implement the intervention are to follow. To conduct an intervention study, particularly an experimental analysis, the treatment or training protocol needs to be specified as precisely as your measurement protocol.

Procedural Fidelity

Researchers in some fields have adopted the practice of assessing the fidelity with which their intervention procedures are applied during a study (Peterson, Homer, & Wonderlich, 1982). *Procedural fidelity* allows researchers to be sure they are actually testing the variable(s) they intend to test. A useful initial step is to turn the operational definition of your independent variable into a checklist that those who will apply the intervention can use to guide their efforts, much as airline pilots follow a checklist in preparing to take off and land. A next step would be to have a research associate observe several randomly selected sessions in each experimental condition using a comparable checklist to record compliance with the intervention protocol. Comparing the paired records can provide insight into the fidelity with which your intended intervention is actually being administered.

Pilot Testing

Just as dances have to be rehearsed to give them the best chance of achieving the choreographer's intentions, experimental procedures have to be rehearsed to ensure that they will have the best chance of generating data that will clearly answer your research question. As in dance, engaging intensively with your process (e.g., observation, intervention, and data management) will be necessary to determine what you need to do and how you need to do it.

Consider using the normal process of creating and rehearsing a new dance as a model for pilot testing your research. In the early stages give yourself the freedom to try a variety of approaches before settling on exactly how you will administer your observation protocol and, if relevant, your intervention. As you discover elements that are working, operationalize them and evaluate the consistency with which they are being applied. See if your research team can administer your procedures consistently by assessing inter-observer agreement and procedural fidelity with a pilot participant who is not in your potential participant pool. The results will indicate where your protocols need refinement.

The final stage of pilot testing is like a dress rehearsal; make it as much as possible like the conditions that will be in place when your study participants are present. You might use your procedural fidelity checklist to decide when your study is ready to be *performed.* To borrow an over-used but functional saying, "success is what happens when preparation meets opportunity."

Prepare thoroughly so that you will be ready to succeed when your study is launched.

Managing Bias and Expectancy

In all research the aim is to minimize, acknowledge, or mitigate risk of bias. In naturalistic studies, researchers declare their biases, prejudices, and expectations to be transparent, and they build a reflexive description of such positions into the research report. They observe and describe phenomena and acknowledge that research has a subjective objectivity and is positioned within a particular theoretical framework. This approach differs from experimental analyses in which variables are manipulated systematically to reveal their influence on an outcome variable of interest. In experimental studies, the influence of experimenter bias is mitigated by operationalizing the experimental protocols and by checking to be sure they are being followed as the experiment is conducted and the data are analyzed.

Researchers also must take special precautions to reduce the influence of *expectancy, demand characteristics* and *reactivity* to minimize their influence on the participant's performance as potential confounding factors. If study participants are sensitized to the anticipated effects of an intervention, that knowledge may change their performance independent of the effects of the variables under study. If it is possible to prevent the participants from knowing whether they are receiving a potentially beneficial treatment *(that is, blinding),* that influence can be reduced as a possible confounding factor. If those applying the treatment and the experimenters can also be prevented from knowing who is receiving the treatment *(double-blinding),* the potential influence may be reduced even further. Just experiencing the special treatment required for experimentation can cause participants to react differently than they might otherwise (reactivity). Thoughtful organization of the components of a study (research design, see *Chapter 10*) can mitigate such confounding. If such influences cannot be eliminated entirely, their possible influence on the results should be acknowledged in the research report.

Summary

Selecting the participants, setting, and variables for your study will involve some disciplined and time-consuming work, just as building a new dance does. Operational definitions, reliability assessments, and procedural fidelity checks are tools that support these activities. Pilot testing sessions are comparable to rehearsals, program runs, and dress rehearsals. They will prepare you and your collaborators to perform your roles with precision to optimize your chances of generating data that will contribute meaningfully to the field.

Highlights

- Your research question and methodological approach will guide your approach to selecting participants and setting.
- A measurement system must be valid, reliable, and responsive to be useful for intervention research.
- Pilot testing research protocols before launching your study is as important as rehearsing choreography before the opening curtain.

References

Bailey, J., & Burch, M. (2018). *Research Methods in Applied Behavior Analysis* (2nd ed.) Taylor & Francis.

Barlow, D. H., Nock, M. K., & Hersen, M. (2009). *Single Case Experimental Designs* (2nd ed.). Pearson Education.

Baumgartner, T. A., Jackson, A. S., Mahar, M. T., & Rowe, D. A. (2016). *Measurement for Evaluation in Kinesiology* (9th ed.). Jones & Bartlett Learning.

Berg, K. E., & Latin, R. W. (2008). *Essentials of Research Methods in Health, Physical Education, Exercise Science and Recreation.* Wolters Kluwer.

Couillandre, A., Lewton-Brain, P., & Portero, P. (2008). Exploring the effects of kinesiological awareness and mental imagery on movement intention in the performance of demi-plié. *Journal of Dance Medicine & Science, 12*(3), 91–98.

Deckert, J. L., Berry, S. M., & Welsh, T. M. (2007). Analysis of pelvic alignment in university ballet majors. *Journal of Dance Medicine & Science, 11,* 110–117.

Evans, A., & Rooney, B. J. (2018). *Methods in Psychological Research* (3rd ed.). Sage.

Kazdin, A. E. (2011). *Single-Case Research Designs: Methods for Clinical and Applied Settings* (2nd ed.). Oxford University Press.

Kowalski, K. C., McHugh, T-L. F., Sabiston, C. M., & Ferguson, L. J. (2018). *Research Methods in Kinesiology.* Oxford University Press.

Morgan, K. (2016). *Increasing Développé Height in University Dancers.* Honors thesis, Florida State University.

Payton, O. D. (1994). *Research: The Validation of Clinical Practice.* F. A. Davis.

Peterson, L., Homer, A., & Wonderlich, S. (1982). The integrity of independent variables in behavior analysis. *Journal of Applied Behavior Analysis, 15,* 477–492.

Portney, L. G., & Watkins, M. P. (2009). *Foundations of Clinical Research: Applications to Practice* (3rd ed.). Pearson Education.

Rowe, D. A., & Mahar, M. T. (2006). Validity. In T. M. Wood & W. Zhu (eds.), *Measurement Theory and Practice in Kinesiology*, pp. 9–26. Human Kinetics.

Sidman, M. (1960). *Tactics of Scientific Research*. Basic Books.

Thomas, J. R., Nelson, J. K., & Silverman, S. J. (2015). *Research Methods in Physical Activity* (6th ed.). Human Kinetics.

Welsh, T. M. (2003). A primer on measuring dancer capacities. *Journal of Dance Medicine & Science, 7*(1), 5–9.

Wyon, M., Redding, E., Abt, G., Head, A., & Sharp, N. C. (2003). Development, reliability, and validity of a multistage dance specific aerobic fitness test (DAFT). *Journal of Dance Medicine & Science, 7*(3), 80–84.

9

Managing, Analyzing, and Presenting Data

JATIN P. AMBEGAONKAR

Questions to Stimulate Thinking

- What are the steps in the research data cycle?
- How can researchers ensure data integrity and security?
- What types of data do researchers use in dance science?
- What are common analysis techniques available to dance science researchers?
- How can you present research results to make meaningful interpretations of data?

Working with good data is a fundamental requirement for effective empirical research (Sidman, 1960). Data are most useful if they are reliable, relevant, and have as little unidentified bias as possible. In this chapter you will learn how to manage, analyze, and present your data to make meaningful interpretations. In the most basic sense, a datum (singular) or data (plural) form the raw material(s) to answer the research question(s). Without good data, your findings may be compromised or uninterpretable. Thus it is essential that researchers employ good data management practices.

Steps in the Research Data Cycle

Conducting pilot testing of the study procedures, instrument, and interview questions and training all members of the research team are all essential steps for good data collection. Being prepared and having contingency plans for how to address anticipated and unanticipated problems that may arise is

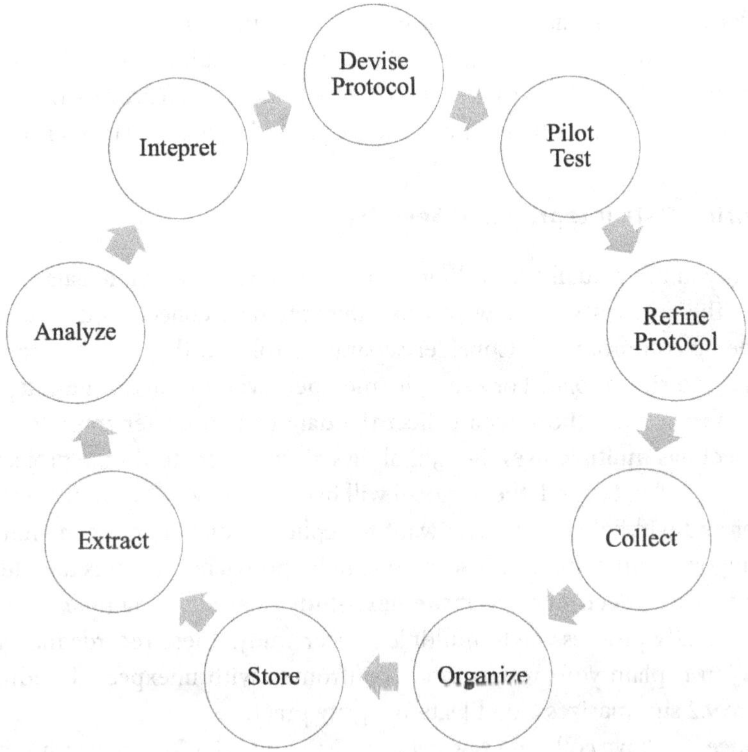

Figure 9.1. Steps in the Research Data Cycle.

essential. For example, consider how will you address absences from testing. What will you do if essential equipment malfunctions? (Examples might be calibration settings on your video, or a malfunctioning audio recorder.) How will you treat raw data compromised by unanticipated conditions? (For instance, when asking questions about a performance anxiety during dance, an unexpected event unrelated to the study creates anxiety.) In research terms, rather than postmortem analyses—that is, analyses of what happened after the fact—you can do a "*pre-mortem*" analysis. Put simply, in this analysis you go step-by-step over your protocol (see Figure 9.1) and try to consider problems and issues. Having considered these issues before they occur will prepare you to respond to complications, even if you will not be able to anticipate all possibilities, and it can reduce chances of unanticipated issues compromising your data collection. When pilot testing, decide the type and form of records needed for the study. Then share this protocol with all members, so that everyone understands their roles and responsibilities.

For example, when examining if a dancer's hip strength changed after taking part in a Pilates program, have documents (e.g., informed consent forms, instruction sheets, data recording sheets) pre-printed and ready as a packet before the dancer comes to the studio. Set up all equipment, so that when the dancer arrives for testing, you are ready to work. This process will come naturally to dancers who have experience preparing for a performance.

Ensuring Data Integrity and Security

When you are actually recording data, create a way to review data in real time. This process will allow you to adjust the data collection if your data are being compromised. Consider separating roles in the research team to reduce the risk of bias. For example, members who evaluate data integrity are different from those who collect the data or administer training. This protocol has multiple uses. First, it allows all members to do the project in the same order. Second, the protocol will help as you write up your research paper and will help others who want to replicate your methods in another setting or population. Third, some research approaches use this data feedback to guide decisions about the next study phase; for example, a single dancer study progressing to multiple dancer study. These records may also help you explain your results when confronted with unexpected findings. Figure 9.2 summarizes Good Data Integrity practices.

Once you have collected raw data, store them well. Most Research Ethics Committees (RECs) require that research data be stored in a secure location. Of course, keeping data secure is also in your best interests. Always keep your master data records safe and secure. Make multiple back-up copies of your raw data in hard-copy and electronic forms, and store them in multiple locations, and send to collaborators for added safeguarding. You are investing a great deal of effort in collecting data. How much is it worth to protect that investment?

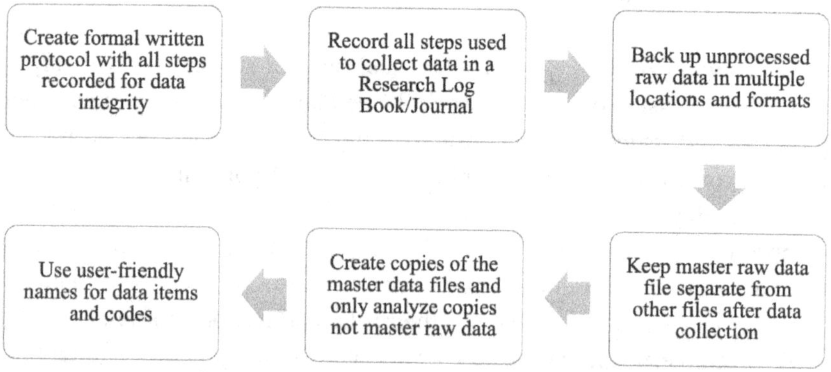

Figure 9.2. Summary of Good Data Integrity Practices.

Data Types in Dance Sciences

Dance science is multidisciplinary. Researchers use multiple data types to answer questions, including quantitative data (e.g., numbers) or qualitative data (e.g., narrative summaries, text) or data in visual form (e.g., graphs). Often quantitative researchers observe phenomena, develop research questions and hypotheses, design and conduct studies, answer the questions, and strengthen the confidence in the results by replicating the studies. Qualitative researchers generally observe and analyze information in natural settings to explore phenomena, understand issues, and answer questions. Qualitative researchers are able to provide rich descriptions and explanation of social phenomena (Creswell & Poth, 2017). See Table 9.1 for general quantitative data types and *Chapter 13* for qualitative data types.

Common Data Analysis Techniques

The data analysis techniques you use depend on your research questions. Think about each data point you have collected. Then ask: How will these data help answer my questions? This way, you will be able see that each question will have its own data analysis technique and may or may not require statistical testing. Then think of ways of presenting data so that readers can understand your results.

For example, with quantitative data, you may need to provide readers *descriptive statistics* that describe your actual data points and how your data are distributed in numerical and graphical displays. Table 9.2 describes some of the measures of central tendency and dispersion for quantitative data to allow for a sense of the overall data distribution. Similarly, when considering qualitative data, tables with the themes and associated quotes that emerged from analyses can be useful in some approaches. Also described in later chapters are additional details about *Inferential Statistics,* which infer the probability that two or more datasets differ by more than what can be predicted by chance (e.g., t-tests), or can predict other values based on data (e.g., regressions).

Descriptive Analyses

Before conducting formal data analyses, start with a graphical data display to see how your data look (e.g., frequency histograms, box and whiskers plots, violin plots). This first level *descriptive analysis* conducted before statistical testing can help identify possible data recording errors, indicate data dispersion and outliers, or show you if data are multi-modal (e.g., more than one

Table 9.1. Quantitative Data Types

Type of Data	Properties	Data Analyses Functions	Measure of Center	Interpretations	Dance Science Examples
Nominal	Non-Parametric	Can count frequencies	Mode	Can say one is different from another	Female and Males
	Distinct Categories	Cannot make a mean			Left and Right Leg
	Order of categories has no meaning; i.e., one category is not greater or lesser than the other				
	Cannot combine to make a mean				
Ordinal	Non-Parametric	Can count frequencies	Median	Can say one is different from another	Ankle sprain: 1st degree, 2nd degree, 3rd degree
	Distinct Categories	Cannot make a mean		Can rank	1st place, 2nd place, 3rd place
	Ranked in specific order—ascending or descending			Can say one is higher or lower, greater or smaller than another	Education: High school, Bachelors, Masters, Doctoral
	Gaps or intervals between are not necessarily equal				
Interval	Parametric	Can count	Mean	Can say one is different from another	Temperature in dance studio
	Distinct Categories	Can rank		Can rank	Time
	Specific order	Can add, subtract, and calculate means		Can say one is higher/lower, greater/smaller than another	
	Intervals between categories fixed, i.e., equal units			Can say one is "X" units greater or smaller than the other	

Ratio	Parametric	Can count	Mean	Can say one is different from another	Height
	Like interval in addition also has a true zero point	Continuous		Can rank	Weight
		Can rank		Can say one is higher/lower, greater/smaller than another	
		Can add, subtract, and calculate means		Can say one is "X" units greater or smaller than the other	
		Can multiply, divide, and other mathematical functions		Can say one is "X" times greater or smaller than the other	
				Zero point indicates a total absence of whatever ability or property is being measured	

Table 9.2. Measures of Central Tendency and Dispersion for Quantitative Data

Measure	Properties and Interpretation
Mean	Arithmetic mean
	Sum of all measurements divided by number of observations in dataset
Mode	Most frequent value in dataset
	Used with nominal data
Median	Middle value that separates the higher half from the lower half of dataset
Standard Deviation	Average deviation of data around sample mean
	How much variation exists in dataset?
	How far do the values within the sample differ from sample mean?
Standard Error of Mean	Mathematical estimate of sampling error
	How far sample mean of data is likely to be from true population mean?
Confidence Intervals	Intervals around a measure (e.g., mean) that has a specified chance (e.g., 95%) of containing the true measure (e.g., population mean)
	Typically, 2 values specified at (±) 95%

number where there are multiple data points). Use this information to clean data (e.g., recode, decide how to address outliers), so that your subsequent data analyses will be easier and meaningful.

Then examine the measures of central tendency (mean) and dispersion (standard deviation) and calculate Confidence Intervals (CI). The CI is a range of values that is likely include the true population mean value with some confidence and is expressed as a percentage (%). You can calculate a CI for any confidence level you want, but the most commonly used value is ±95%. This 95% CI is a range (upper and lower) where you can be 95% certain that the values contain the true mean of the population. See Table 9.2 for commonly used central tendency and dispersion measures for quantitative data.

When interpreting data, you may need to make inferences from statistical analyses of raw data. Inferential statistical tests can be divided into two categories: *parametric and non-parametric tests.* For example, if you want to know how many push-ups a dancer can perform on average, you would have to contact all dancers all over the world (i.e., the population) and ask them to do push-ups. However, realistically this may not be possible. So you collect data from a representative subset (a sample) of student and professional dancers. Then you would take values like means and standard deviations from this smaller sample and use these values to estimate push-ups numbers for the entire population (parameters).

These estimates are based on the assumption that the underlying population from which the sample was taken is representative of the entire dancer population. In other words, you assume that in your sample, some dancers did fewer push-ups, some did more push-ups, and others were somewhere in the middle (i.e., in a bell-shaped curve, where more scores are near the center, and the frequency diminishes as scores move further from the center). Alternatively, if you only chose elite professional dancers, their values would likely be at the higher end, for example, and the bell would curve to the right side. Thus your estimates would not be normally distributed and would not be representative of your population. In this case then, you would have to use non-parametric tests. These tests do not rely on assumptions about the shape or parameters of the underlying population distribution.

There is not one single rule to help choose between using parametric and non-parametric data analyses. However, generally if you have smaller sample sizes (e.g., less than 30 participants), it is prudent to examine your sample distribution to ensure that it is normally distributed before conducting parametric analyses. If you find that the data are higher on one end, then consider using equivalent non-parametric analyses. Sometimes there is no single way to decide whether to use a parametric or non-parametric procedure, and you may need to consult a statistical expert.

As mentioned, it is important to check how many participants you need to answer your question (i.e., what sample size you require to be confident in your results). To check this, conduct a *power analysis.* This provides a way to estimate the sample size required to obtain a statistically significant difference based on the anticipated difference(s). Conducting a power analysis will help ensure that your study is more likely to yield statistically significant results. A power analysis is most useful when performed before actual data collection (also known as *a priori analysis*).

Theory-based research is often concerned with testing *hypotheses,* or well-informed guesses, about how the world should work, based on an existing, evolving, or proposed theory. As described in previous chapters, in quantitative research, you will define a null hypothesis (H_0), where you posit that no differences exist, and an alternative hypothesis (H_1), where you posit that differences do exist.

Often, based on your research question complexity, describing raw data, central tendency measures (means), or how data are distributed (standard deviations) may be just the starting point to answering your question. For example, you are examining if Pilates training improves dancers' ability at

Table 9.3. Common Quantitative Data Analyses Types, Tests, and Results

Type of Analyses/Tests	Meanings and Examples of Statistics and Results
Descriptive	Frequencies, percentages, ranges (Quartiles—1/4th)
Central Tendency	Mean, mode, median, standard deviation, Confidence Intervals
Effect Sizes	How large are the changes as compared to the central tendency and deviations from the mean?
Reliability	
Inter-item	Cronbach's alpha
Inter-rater	Cohen's kappa
Non-Parametric Comparative	
2 Groups	Chi-Square
More than 2 Groups	Wilcoxon rank sum test/Mann-Whitney U test, Kruskall Wallis test, Wilcoxon signed-rank test
Parametric Comparative	
Simple transformations	*Z* test
2 groups or 2 measurement times	t-tests (paired, unpaired), 1-way Analyses of Variance (ANOVA)
More than 2 groups, or 2 measurement times or combinations of more than 2 groups and/or measurement times	ANOVAs
More than 2 groups with a single covariate (i.e., outside variable that can influence your independent and/or dependent variables)	Analyses of Co-Variance (ANCOVA)
Combinations of more than 2 groups and/or measurement times with multiple covariates	Multiple Analyses of Co-Variance (MANCOVA)
Relationships and Predictive Models	
2 continuous measures	Pearson product-moment correlations
2 non-parametric ranked data measures	Spearman rank-order correlations
1 dichotomous and 1 continuous measure	Point bi-serial correlations
Prediction of one measure by another measure	Simple regression
Prediction of one measure by a combination of measures	Multiple regression
Prediction of a measure by nominal or ordinal measure(s)	Logistic regression
Relationships and Predictions of multiple variables	Structural equation models, Hierarchical models

Type of Analyses/Tests	Meanings and Examples of Statistics and Results
Epidemiological	
Prevalence	Proportion of population who has a condition at a specific time
Incidence	Occurrence of new cases in a population during a specific period of time
Incidence Proportion	Proportion of the sample who get the condition within a given period of time as a percentage of the sample at risk
Incidence Rate	Occurrence of new incidents during a certain time period
Clinical Incidence	Average number of incidents per participant during a certain time period
Risk Ratio	How much higher/lower the risk of the rate of a variable is relative to another variable
Rate Ratio	The measure of change per unit of time
Odds Ratio	Estimate of odds of exposure among participants who had the incident to the odds of exposure among participants who did not have the incident

push-ups. Say one dancer did 21 push-ups before (pre) versus 22 after (post) training (change of 1); a second did 18 push-ups before and 22 after training (change of 4); and a third did 15 push-ups before and 22 after training (change of 7). In this case, by simply looking at raw numbers (number of push-ups), you cannot say whether a change of 1, or 4, or 7 was "enough" for you to conclude that Pilates improved push-up ability.

So here, you will need to use *Inferential Statistics* (e.g., a paired t-test comparing the pre- versus post-training values) to determine how much change is "enough" to infer that Pilates training improved strength and then estimate the probability that the intervention (i.e., taking part in the program) was responsible for changes in the number of push-ups completed.

Later, you may also want to examine whether the amount of training (e.g., every additional 30 minutes of training per week) is related to the amount of improvement (e.g., 1 push-up increase). To evaluate these relationships, you would use *Predictive Statistics,* such as correlation and regression analyses to examine relationships among variables and estimate whether changes in one variable can predict changes in the other. For an even broader examination of injury or performance changes over longer time periods you would use epidemiological statistical analyses (see Knowles et al., 2006). Table 9.3 describes some common data analyses types, tests, and results that you can use to answer research questions. See Keppel & Wilkens (2004) for additional information.

Probability

When conducting statistical tests, you need to understand the concept of "statistical significance." This concept is commonly assessed with a term called the probability-value (p-value). While the p-value can be a useful statistical measure, we need to be careful not to misunderstand or misinterpret what the tests assess and what the results can actually tell us. To understand this issue, let us consider the basic concept behind the p-value.

Put simply, the p-value is the probability that a particular outcome may have occurred by chance. This value was introduced by Ronald Fisher as an informal way to assess whether evidence was worthy of a second look, and not as a definitive test (Fisher, 1973). Fisher suggested that researchers should first set up a null hypothesis (H_0, assumption of no difference) that they want to disprove: for example, there are no differences between two groups (amateur and professional dancers' jump height). Next, researchers should assume that this null hypothesis is true (no differences between the groups) and calculate the chances of getting this null result or even more extreme results (that there are differences between groups). Then the researchers compare these values to what is actually observed. This observed probability after conducting the experiment is the p-value.

The smaller this p-value, the greater the likelihood that the null hypothesis is false—that is, that there are actual differences between the variables. In standard scientific practice, a p-value that is less than or equal to .05 is considered a statistically significant difference. So here, you accept the risk that the observations do not accurately represent the population in 5 out of 100 times the experiment might be run. Clearly thus, there are chances for error in interpretations. What are these errors in interpretations, and how do they work?

Type I and Type II Errors

When conducting hypothesis testing, a Type I error is when there is *no* actual difference (i.e., not significant), but you commit an error by concluding that there is a difference—or an error of commission. Alternatively, a Type II error is when there is an actual difference (significant, according to the statistical test), but you omit making this conclusion and conclude that there is no difference; that is an error of omission. In statistical terms, the probability of a Type I error is Alpha (α), while the probability of a Type II error is Beta (β). Power (described earlier in the chapter) is the probability of *not* making a Type II error and is mathematically noted as $1-\beta$. Table 9.4 describes different possibilities of Hypothesis Testing Outcomes and Errors.

Table 9.4. Hypotheses Testing Outcomes and Errors

		Null Hypothesis (H_0) is True	
Decision	Do not reject	Correct inference (true negative)	Type II error (false negative)
	Reject	Type I error (false positive)	Correct inference (true positive)

Recall from our earlier discussion on power and sample size that you want to increase your study's power to increase the chances that it will reveal a statistically significant difference. Several factors can affect power, including the significance level (α) that you choose, how much Type I error you accept before beginning your study, your sample size (the larger the sample, the less risk of error), the variability in your measured variable (how much your data deviate from the mean), and magnitude of the effect of the variable (that is, how much difference are you looking for to make a meaningful interpretation).

There is a trade-off here, and you will need to consider the relative value of increased power and the work needed (more participants, more data points) to get the increased power. Generally, researchers consider power of 0.80—that is, that the mean sample size is sufficient to yield a p-value that satisfies the stated goal (e.g., $p < 0.05$)—80% of the time and above as an adequately powered study. Several free online statistical programs can calculate these sample sizes if you have means and variance from prior pilot data.

Traditionally the Type I error rate or the probability (p-value) of making an error is kept at or below 5% (i.e., 0.05). You are also trying to control for Type II error and aim for the power of your results to be at or above 80%—that is, the chance of a false negative is 20%. In the words of Daniel Lakens, the goal of controlling the errors is not to make a fool out of yourself too often (Lakens, 2019). So by setting the α at 0.05, you are assuming that you will not make a mistake more than 5% of the time in the long run.

However, it is important to understand that these 5% and 20% error rates are somewhat arbitrary. You can set the value depending on your research question. If you can justify different error rates, change them. For example, the margin of error is tighter in pharmaceutical research, where false negatives may cause mortality. Thus, in such research these numbers are often 1% ($p < 0.001$); that is, chances of making a Type I error are 1/100. Of course, this is again a balancing act, as when you adopt a more rigorous p-value, your chances of making a Type II error increases. Then you will need to balance this risk by increasing your sample size—which may require more resources and time to complete the study. Researchers also use the term "trends" to describe these differences that achieve a p-value just above the predetermined break point (e.g., 0.06). While some researchers use this practice, to ensure research rigor we encourage you to select a predetermined p-value and discuss your study results using this break point.

Given these notes, it is easier to understand why using only p-values is being increasingly criticized. Recently some scientific journals have even begun discouraging using p-values, and instead asking authors to report means, standard deviations, and CI's (see explanation of CI's in Table 9.2). While a detailed description is beyond the scope of this chapter, these values do not depend only on the somewhat arbitrarily predetermined p-value, and rely rather on understanding of the dispersion of the data.

Clinimetric Properties of Outcome Measures

In many instances, dispersion descriptions and statistical significance may not be enough by themselves to interpret data meaningfully. For example, consider a group of dancers taking part in twice weekly, 1-hour-long strength training program for 12 weeks. Upon data analyses, you find that the dancers' vertical jump improved from 20.2 cm to 20.4 cm after taking part in the training. With a large enough sample, this change may be "statistically" significant. However, are the results actually meaningful and of value to the dancers? Is the time invested in achieving this 0.2 cm improvement worth something to the choreographer in that this added height can be used to create better choreography? There is no simple answer to these questions.

In these cases you will need to conduct clinimetric analyses. Clinimetric analyses examine whether statistical changes actually mean anything meaningful. One strategy is for researchers to use clinimetric analyses to translate the outcome measure (e.g., in this case jump height) into a meaningful value for stakeholders (e.g., dancers) who may benefit from the research. Another approach is to use a social validity assessment as described in *Chapter 10* on validity. Figure 9.3 describes some clinimetric properties of outcome measures and provides a visual depiction of how these are generally related in increasing order of difference from the original mean. Additional details can be found in other references (Riemann et al., 2018).

How to Analyze Data

As you can see, conducting data analyses can get complicated. While you can conduct the analyses by hand or using simple datasheets, the process can become tedious and complicated, depending on your research design. Thus, consider using specialized statistical software. Commercial software

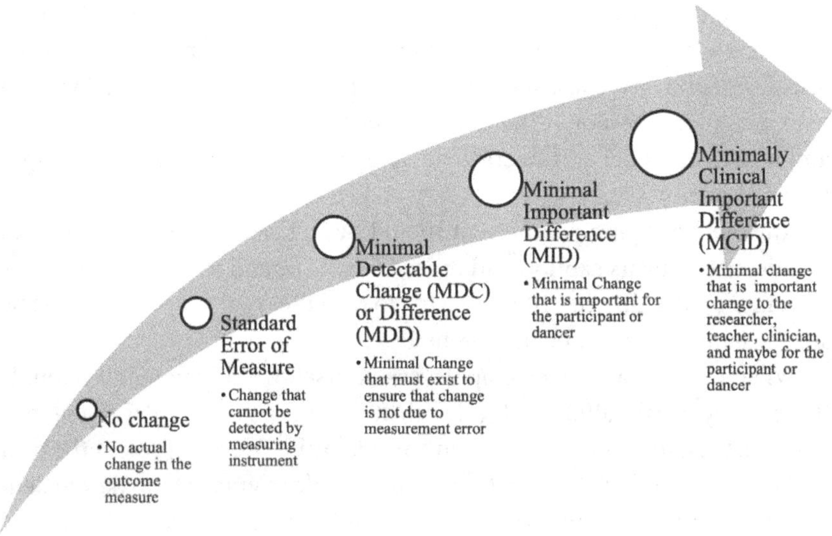

Figure 9.3. Clinimetric Properties of Outcome Measures.

programs exist for quantitative (e.g., SPSS®) and for qualitative data analyses (e.g., NVIVO®).

Multiple free software programs and programming languages are becoming available to researchers (e.g., JASP, https://jasp-stats.org/, Jamovi https://www.jamovi.org/, and R https://www.r-project.org/). These programs offer a user-friendly experience for both researchers and non-statisticians. These programs are also open-source, which means that the designers share the way they coded the programs and how they analyzed raw data so that others can use and improve the data analysis techniques via peer-to-peer sharing and review. These programs also offer ever improving visual and graphical display options. While they do have a learning curve, they offer free tutorials to help conduct a specific data analysis technique based on a question.

How to Present Data to Make Meaningful Interpretations

After you conduct the data analyses, you should think about how you want to organize and present your data. The information should tell a "story" and be smooth, like a well-choreographed dance performance. It should lead readers logically through the steps you took to answer your research questions. Thus, the order in which you present your data and results is important. The data organization and presentation should be systematic and consistent and in the order that you asked your questions.

Data can be presented in text form or can be summarized graphically to provide a visual display of relationships, patterns, or trends. Often, *visual displays* (tables, graphics, and figures) can provide a more nuanced view of your data than text alone. The following sections describe specifics of tables and figures and offer tips to create meaningful and attractive ways to organize and present data.

Any table or figure you present should be sufficiently clear, well labeled, and described by its caption and indexed by its legend to be understood by your intended audience without having to read the results section; it must be able to stand alone and be interpretable.

When creating a table or graphic, ask yourself the following question: Is this graphic worth 1,000 words (Hertel, 2018) or even 100 words? If your answer is not an unequivocal yes, then consider other ways to present the data in the text or combine information from a couple of graphs to make efficient use of your reader's time and effort. See additional references at the end of the chapter that provide good details on how to create graphics.

Tables

Tables are extremely effective tools for making comparisons and observations across data (Boers, 2018). Tables have four basic purposes (Wainer, 1997). They have an explanatory function, a communication function, a storage function, and an aesthetic function. When creating tables, consider what actually goes into them. Some details may vary, but all tables have a few universal elements (Figure 9.4):

- **Title**: Should be free-standing and may include population, dependent and independent variables, type of information delivered, and units of measurement
- **Row and Column Headings**: Variables and categories
- **Body**: Information (data, numbers, text, pictures, images)
- **Symbols**: Depicting items to emphasize in table body
- **Notations and Footnotes**: Descriptions of symbols or added footnotes

See other tips when creating tables:

- Do not add too much information
- Use white space well; keep enough space so that readers can see data clearly
- Do not be redundant; if data exist in text, do not repeat in table and vice versa

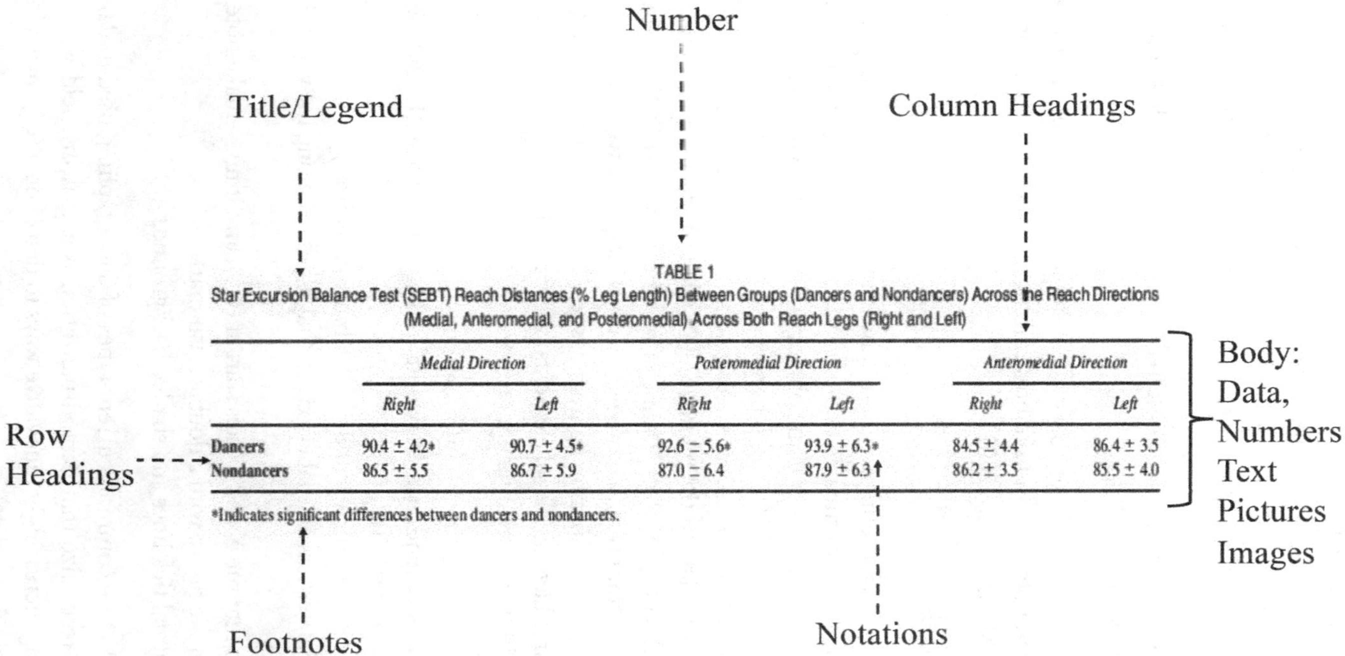

TABLE 1

Star Excursion Balance Test (SEBT) Reach Distances (% Leg Length) Between Groups (Dancers and Nondancers) Across the Reach Directions (Medial, Anteromedial, and Posteromedial) Across Both Reach Legs (Right and Left)

	Medial Direction		*Posteromedial Direction*		*Anteromedial Direction*	
	Right	*Left*	*Right*	*Left*	*Right*	*Left*
Dancers	90.4 ± 4.2*	90.7 ± 4.5*	92.6 ± 5.6*	93.9 ± 6.3*	84.5 ± 4.4	86.4 ± 3.5
Nondancers	86.5 ± 5.5	86.7 ± 5.9	87.0 ± 6.4	87.9 ± 6.3	86.2 ± 3.5	85.5 ± 4.0

*Indicates significant differences between dancers and nondancers.

Figure 9.4. Common Components of a Table.

- Organize rows and columns in the same order as in text (e.g., dancers first, non-dancers second)
- If you want readers to compare data points, put these data in columns so that readers look at the information from top to bottom rather than putting these data in rows where readers have to look from side to side
- Numbers: Whole numbers preferred; if decimals are needed, round to 1–2 places (maximum)

For additional details about how to create tables to report specific statistical analyses and notations and symbols commonly used in science publications, refer to Morgan et al. (2002).

Graphical Displays

While tables are often useful to show numbers, sometimes a table may not be enough or may not be optimal to depict relationships or comparisons across data. Here, you may consider using photographs, line drawings, charts, and graphs (Lane & Sándor, 2009). Photographs are useful to show research protocols, instruments, or positions when conducting research and must be attributed as appropriate to the photographer. Flowcharts are useful to show the research design processes. Data can also be presented in the form of graphs. In the sections that follow, see general principles of when and how to use some of the most common figure types (e.g., plots, graphs, charts). Like tables, all figures have universal elements (Figure 9.5).

See some tips when creating figures:

- Do not pack in too much information
- Graphs should tend to be horizontal rather than vertical (landscape rather than portrait)
- Use error bars to represent data variability graphically
- Rotate the vertical axis label axis so that it lies flat along the vertical axes
- Confirm that the X and Y axes start at (0,0), and if not—make note of this in the figure with a double slash mark
- Use minimal tick lines and shaded backgrounds

Following are some common current types of data graphing used in dance sciences, but recognize that data visualization is a growing field with researchers using increasingly innovative ways to depict data (e.g., heat plots, radar plots, raincloud plots).

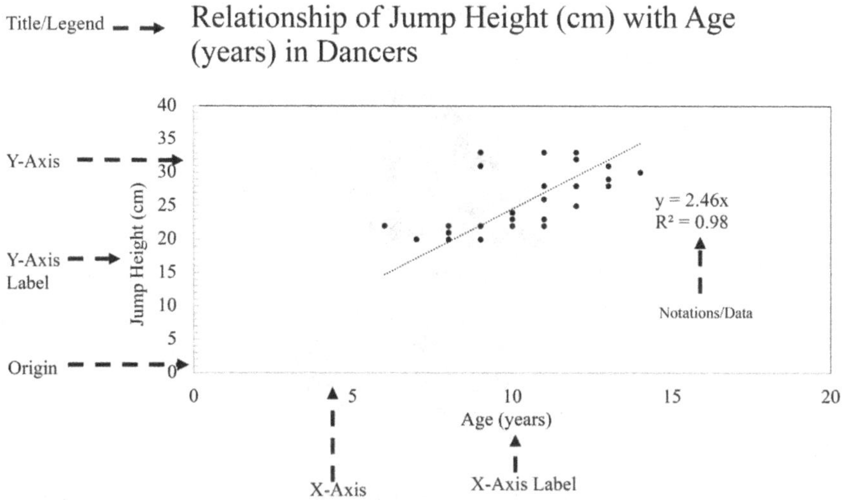

Figure 9.5. Common Components of a Figure.

Plots: When analyzing data, one of the first analyses is to check data distribution using plots. Plots provide information about the normality and dispersion of the data and allow you to identify any outliers. Box plots, violin plots, histograms or "raincloud" plots (which combines box plots, raw jittered data that adds some visualization using data points to scatterplots to see better the information contained in the data, and a split-half violin) are especially useful to describe and depict the spread of data and possible outliers.

Pie charts are useful for displaying the proportion of scores falling into particular categories, such as the number of dancers visiting a particular injury clinic (Figure 9.6). However, these charts are primarily descriptive in nature and so should not be extensively used.

Bar graphs with error bars are used when you are comparing a single variable across several measurements (e.g., groups, times) (Figure 9.7). Stacked bar graphs can also show proportions. However, recently the use of extensive bar graphs has come under criticism (Martinez, 2015; Weissgerber et al., 2015). One of the major problems of bar graphs and line graphs that depict means only is that different distributions of data (e.g., symmetrical, bimodal, outliers, unequal samples) may still result in the same bar graph. To address this limitation, authors suggest the use of other plots (see Hertel, 2018).

Line graphs are used to display patterns over time. They are commonly used in intensive case studies and within-subject experiments. However, like

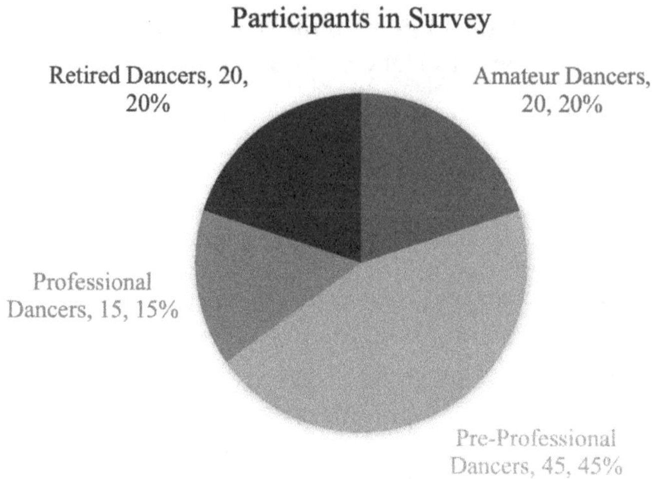

Figure 9.6. Pie Chart Example.

bar graphs, they have similar limitations of inability to describe data dispersions (Figure 9.8). Single-case experiments and other time-series approaches to research rely on the graphic data display to evaluate the degree of experimental control exerted by one variable on another variable.

XY scatterplots are plots of X and Y coordinates showing a score on two variables. Plotting data this way can show whether the two variables show a "relationship"—that is: do they change in value together in a consistent way?

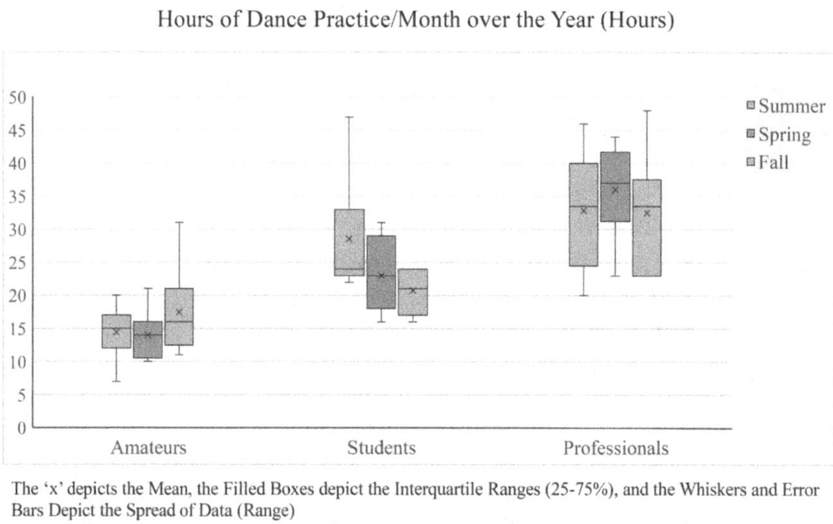

Figure 9.7. Bar Graph with Box and Whiskers. Example depicts hours of dance practice per month over the year and across dancer levels.

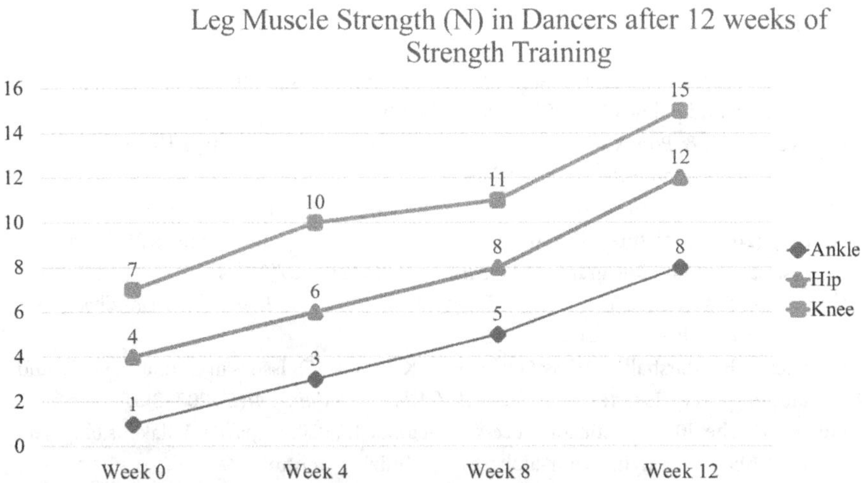

Figure 9.8. Line Graph Example. Strength in dancers' right leg muscles at three joints after 12 weeks of strength training.

If relationships exist, additional information about the strength of these relationships (R-values) and the ability of one variable to affect the magnitude of change in another variable (r^2) can also be depicted in the scatterplot with the "line of best fit" showing this relationship (Figure 9.5).

There are multiple other ways to depict data graphically, and while detailed explanations of these display types are outside the scope of this chapter, you can review the references for additional information. You may also consider working with statisticians to devise appropriate data analysis methods. This assistance can be valuable as long as the researchers' questions remain at the center of study design. Statistics facilitate the analysis of your data on the way to answering your research question.

Summary

Overall, in this chapter you have learned how to work with data—the actual pieces of information that you use to answer your research question(s). In this information age, with social media and other quick means to disseminate information, it is becoming increasingly important to depict your data in a user-friendly manner. Always remember, at the end of the analyses, you as a researcher must fully comprehend your study results and analyses, so that you can assume full responsibility for the findings. Understanding how to collect, analyze, and interpret data logically will allow you to understand and present your findings to others.

References

Boers, M. (2018). Graphics and statistics for cardiology: Designing effective tables for presentation and publication. *Heart, 104,* 192–200.

Creswell, J. W., & Poth C. N. (2017). *Qualitative Inquiry and Research Design: Choosing Among Five Approaches* (4th ed.) Sage.

Fisher R. (1973). *Statistical Methods and Scientific Inference* (3rd ed.). Hafner Press.

Hertel, J. (2018). A picture tells 1000 words (but most results graphs do not): 21 alternatives to simple bar and line graphs. *Clinical Sports Medicine, 37*(3), 441–462.

Keppel, G., & Wickens, T. D. (2004) *Design and Analysis: A Researcher's Handbook* (4th ed.). Pearson Prentice Hall.

Knowles, S. B., Marshall, S. W., & Guskiewicz, K. M. (2006). Issues in estimating risks and rates in sports injury research. *Journal of Athletic Training, 41*(2), 207–215.

Lukens, D. The 20% statistician. Accessed October 1, 2020. http://daniellakens.blogspot.com/2019/05/justifying-your-alpha-by-minimizing-or.html

Lane, D. M., & Sándor, A. (2009). Designing better graphs by including distributional information and integrating words, numbers, and images. *Psychological Methods, 14*(3), 239–257.

Martinez, E. Z. (2015). Description of continuous data using bar graphs: A misleading approach. *Revista da Sociedade Brasileira de Medicina Tropical, 48*(4), 494–497.

Riemann, B. L., & Lininger, M. R. (2018). Statistical primer for athletic trainers: The essentials of understanding measures of reliability and minimal important change. *Journal of Athletic Training, 53*(1), 98–103.

Sidman, M. (1960). *Tactics of Scientific Research.* Basic Books.

Wainer, H. (1997). Improving tabular displays, with NAEP tables as examples and inspirations. *Journal of Educational and Behavioral Statistics, 22*(1), 1–30.

Weissgerber, T. L., Milic, N. M., Win ham, S. J., & Garovic, V. D. (2015). Beyond bar and line graphs: Time for a new data presentation paradigm. *Public Library of Science Biology, 13*(4), e1002128.

Additional Resources

Cohen, J. (1965). Some statistical issues in psychology research. In B. B. Wolman (ed.), *Handbook of Clinical Psychology* (pp. 95–121). McGraw-Hill.

Morgan, S., Reichert, M. T., & Harrison T. R. (2002). *From Numbers to Words: Reporting Statistical Results for the Social Sciences.* Allyn & Bacon.

Tufte, E. R. (2001). *The Visual Display of Quantitative Information* (2nd ed.). Graphics Press.

10

Methodological Integrity and Research Design

LYNDA MAINWARING, TOM WELSH,
AND JATIN P. AMBEGAONKAR

Questions to Stimulate Thinking

- How do we know when research findings are credible?
- What procedures can help you determine whether your intervention had an effect?
- What is methodological integrity?
- How can methodological integrity be compromised?

As you may recall from the opening chapters, the ultimate purpose of research is to contribute to our understanding of how the world works. In *Chapter 5* we described a continuum of types of research conducted in the dance sciences. Like a well-choreographed dance, research comprises many components designed in a particular way through a process that involves many decisions and refinements.

In the dance sciences we are building knowledge about dance and dancers: we want to ensure that the knowledge gained is based on appropriately conducted research and that we can be confident in the interpretation of results and the conclusions drawn. How do we know that the research results are credible and that we can have confidence in them? How do we know that the results of one study represented in the research report might reflect what occurs in a dance studio or a dance school or a company in a different country?

These questions pertain to methodological integrity, which we define as the demonstration of procedural rigor and quality in research. We achieve this with research procedures that are not inconsistent with the underlying assumptions associated with the paradigm of inquiry, and the use of strategies to ensure transparency, validity, and the appropriate acquisition of data. The focus of this chapter is on methodological integrity as it relates to the concept of validity; that is, the extent to which the research examines what it purports to examine, is relevant, and represents the studied phenomenon appropriately. We describe the importance of validity; the way threats to validity influence research outcomes, inferences, or conclusions; and how the research design that you choose influences your research results.

Research Design

Research design is the overall structure or framework for the research. It sets the stage for how the data are collected, generated, or analyzed and provides a framework for understanding the kinds of threats that might challenge the methodological integrity and, thus the interpretation of research findings.

Campbell and Stanley (1963) reviewed how specific types of research designs address internal and external validity within the quantitative research paradigm. Similarly, naturalistic inquiry has various strategies to ensure credibility, or validity, of findings in the myriad qualitative research approaches. Designing your research in ways to ensure that you can have confidence in your findings, regardless of the research paradigm, is an essential part of the scientific enterprise.

Quantitative Research and Methodological Integrity

Methodological integrity includes determining that the research findings are valid—that they can be attributed to the experimental conditions and that they can be generalized beyond the study's context, participants, and the historical, social, and cultural time period. Ascertaining that the research findings are valid and have not been compromised is an important consideration in the research process.

The concept of validity has two aspects: one related to measurement (see *Chapters 8* and *12*) and the other related to research design and data analysis. We describe the latter here.

Internal Validity

When designing research, scientists focus on two types of validity, internal and external. *Internal validity* refers to the extent to which an experiment demonstrates that the application of the independent variable (IV) accounts for a difference or change in the dependent variable (DV) (Kazdin, 2011). It is the ability to rule out rival hypotheses for the cause-and-effect relationship(s) examined. Experimental studies are designed to isolate the influences of one variable on another. Controlling as many sources of influence as possible on the variables under study enhances a study's internal validity by isolating the influence of the IV, which increases confidence that what you are manipulating (e.g., training frequency) is the probable cause of any changes observed in the DV (e.g., fitness measured by maximum oxygen uptake).

Methodological approaches to experimental research vary, but they have a shared intention: the key function of experimental research is to isolate the influence of the IV on the DV to provide logically compelling evidence that changes in the IV consistently correspond with changes in the DV. Experimental, and quasi-experimental designs, and designs without an equivalent control group or condition (often called pre-experimental designs) offer different degrees of internal validity.

Convincing experimental analyses are easier to conduct in a laboratory setting where the researcher can precisely control the independent variable as well as *extraneous variables* (also called *confounding variables)* that may influence the interpretation of results. When conducting between-group experiments, for example, there is usually an experimental group and a control group to which participants are randomly assigned. This research design, often referred to as a randomized control trial (RCT), generally has strong internal validity (i.e., a high probability of revealing a cause-and-effect relationship) if all aspects of the research are tightly controlled and if the sample size is sufficiently large.

Two frequently used RCT designs are the pre-test/post-test design and the post-test only design. These designs explore the changes in a dependent variable before and after an intervention, and between a treatment group and a no-treatment group, respectively. In *Chapter 16*, we explain a different strategy used to isolate the influence of an IV on a DV in within-subject experiments.

External Validity

External validity is the degree to which research findings are applicable beyond your study (Arnold et al., 2005, p. 94). Whereas internal validity is concerned with the integrity of the relationship between the independent and dependent variables, external validity is concerned with whether the findings are likely to be applicable to other persons, settings, and times different from those used in the experiment (Portney & Watkins, 2009).

When you tightly control the research process in a laboratory, you may limit the influence of extraneous variables on the dependent variable and reduce the opportunity for outside influences to interfere with changes in the dependent variable. Optimizing internal validity by tightly controlling the environments in which dancers work during a study may not adequately forecast the applicability of the findings to the natural environments in which the intervention will be applied in practice. Optimizing internal validity may compromise external validity.

Threats to Validity

Internal and external validity of a study can be compromised in numerous ways. Thinking about the types of confounding or extraneous variables that might compromise the interpretation of your research results is critical as you design your research. It can be overwhelming to discover that all the work you did to collect your data has been for naught because you did not think through the design sufficiently to address the potential threats to validity of your research findings. Thoughtful planning and execution can help you avoid such a disappointment.

Threats to internal validity might be imposed by factors that we have organized into three categories: (1) Participant Characteristics, (2) Research Procedures, and (3) Intervention or Treatment Factors. The categories are not mutually exclusive. Table 10.1 identifies common threats to internal validity and provides examples and strategies for managing each threat. Part IV describes strategies for addressing threats to validity more fully.

Table 10.2 identifies threats to external validity and describes examples and strategies for management. A common thread in several of the threats to external validity is that research participants might be influenced by the special features of participating in a study. As described in *Chapter 8*, researchers must take special precautions to reduce the influence of *expectancy* or *demand characteristics* to minimize their influence as potential confounding factors on participant performance.

Practical Significance of Research

With dance science research having an applied focus, we expect research findings to have practical value for key stakeholders. We also want to know that the findings are grounded in sound research methods that lead to valid and reliable results. To ensure that a change in the DV is a result of the IV, researchers may conduct statistical tests and make inferences or draw conclusions about the findings based on the test results. Recall from the discussion

Table 10.1. Threats to Internal Validity and Possible Management Strategies

Type	Description	Example	Possible Management Strategies
Related to Participants			
History	A change in events or context at the time of study that might compete with the IV to influence the DV.	A change in curricula or impact of COVID-19 on performance opportunities from 2020 to 2022.	Ensure data are collected in the same historical period when comparing groups and isolate application of the IV.
Maturation	Natural changes over time or with experience.	Growth spurts that may affect performance.	Consider possible influences of maturation.
	Any growth, learning or maturation that changes the participants (Ps) but is not part of the study may affect the DV.		Limit study periods and be aware of natural developmental timelines. Use of sizeable control group with comparison of pre/post change scores.
Attrition	Ps loss because of failure to complete the study. This changes the sample or group composition.	Several dancers in a comparison group drop out of a study because they are injured, which may make the groups no longer equivalent on a relevant dimension.	Determine reasons Ps drop out and address during the study.
			Report sample size changes due to attrition, describe reasons, and explain implications for the study.
Selection Bias	Ps across study conditions (groups) differ on characteristics that influence the DV from the beginning of the study.	In a study examining the effect of upper body strength training intervention versus a control (no training) on ability to lift partners, hip hop dancers may have stronger upper bodies than tap dancers.	Use control groups.

(*continued*)

Table 10.1—*Continued*

Type	Description	Example	Possible Management Strategies
			Randomly assign Ps to intervention and control groups, using a procedure that distributes the two types of dancers equally across groups.
Selection-Maturation Interaction	Something internal to members of one of the groups affects measurements of that group but not the other.	Dancers in one group of a two-group study are one year older and the hormonal changes in that group affect the intervention.[a]	Use control groups.
			Randomly assign participants to intervention and control groups.
Related to Procedures			
Instrumentation	Changes in instrument calibration or subjective assessments can influence the measurement and influence the DV.	Grip strength measured by dynamometer with ink on the gauge so that the researcher needs to estimate the exact score on each repetition.	Calibrate instruments regularly.
		Each estimation has a margin of error that affects the scores differentially.	Monitor each test.
			Record testing.
			Train assessors.
			Have the same person read scores of all participants during testing.
Statistical Regression Toward the Mean	A statistical phenomenon whereby participants who exhibit unusually high or low first-test scores are likely to be closer to the mean on retesting.	Dancers who score very low on a performance test at the beginning of the year are likely to score higher on the same test at a later time during the year. The higher score could be a result of the tendency for scores to regress toward mean scores on retest.	Use clear inclusion and exclusion criteria.

Type	Description	Example	Possible Management Strategies
			Use larger sample.
			Eliminate extreme scores (outliers) using pre-determined criteria (e.g., outside 3 SDs of the group mean).
			Require repeated, stable baseline scores before applying the IV.
Testing (or Practice) Effect	The influence that initial testing has on subsequent testing. This applies to instances where the pretest is not part of the IV. In tests of physical or psychological performance it is common for scores to improve during the first few trials as the task being tested is learned and fine-tuned.	A dancer completes a pre-injury assessment. During the season the dancer is concussed and the same assessment is re-administered. The score is better than the initial test, which is likely due to a practice effect. The researcher may wrongly interpret the score as no cognitive dysfunction post- concussion.	Administer a series of pre-tests to account for learning or practice effects.
			Use different forms of the test.
Related to Intervention			
Expectation Effects	Researchers may have an expectation about the outcome of a study that might (1) unintentionally influence Ps to respond in a particular way consistent with the research hypothesis, and (2) influence the interpretation of results.	A researcher uses standardized scripts for imagery and performance conditions. The researcher reads the instructions slowly with emphasis on certain words to the imagery group in line with the researcher's beliefs about the efficacy of imagery. The researcher's beliefs influenced an unintentional change in the instructions between groups, thereby influencing the outcome of the study.	Train researchers on test administration and preparation of standardized communications such as scripts and testing procedures.
			Blind Ps to study outcomes.

(continued)

Table 10.1—*Continued*

Type	Description	Example	Possible Management Strategies
			Blind both the Ps and the experimenters to the study treatments (double-blinded).
Diffusion of treatment	Ps in a control group are familiar with the treatment applied to the experimental group. Ps could be influenced by knowledge of the treatment and decide to add a similar treatment to their own lifestyle. There is contamination across groups.	Dancers in experimental group who are administered relaxation and core strength training discuss their experience with control group Ps. Dancers in the control group begin core strengthening exercises. The no treatment group is compromised.	Blind Ps to the experimental conditions.
			Minimize or stop contact across groups or explicitly inform Ps not to share any study-related information with anyone.

Note: a. Dancers taking part in intense physical training assigned to one group in a two-group study of weight changes during training. The intense training affected hormones and delayed menarche in more than half of the dancers in this group. The hormonal changes in this group delay pubescent changes in the hormonal milieu that would typically result in fat gain. Researchers mistakenly conclude that training was the only cause of body fat reduction, when in fact, the changes were also influenced by maturation and selection.

on p-value in *Chapter 9* that a measure of statistical significance is designed to tell us how likely it is that our results may have occurred by chance. It suggests that a difference is probably real, but it does not measure the size of the effect nor whether the difference is important (Arnold et al., 2005, p. 194).

Several dimensions beyond statistical significance bear on the practical importance of applied research findings: magnitude of change, consistency of the effect across individuals, affordability, and cost vs. benefit. A single estimate test statistic alone is insufficient for understanding what your data mean. The utility and misinterpretations of probability testing, and the underuse of other means of assessing change have been examined recently in the literature. The American Statistical Association released a statement that includes six principles for the proper use and interpretation of p-values

Table 10.2. Threats to External Validity and Possible Management Strategies

Type	Description	Example	Possible Management Strategies
Reactivity to Experimental Conditions	Behavior change attributed to knowledge of observation. (Sometimes called the Hawthorne effect.)	Dancers cognizant they are observed during warm up in the studio compared to dancers being observed through a one-way mirror.	Use control group.
			Blind Ps to study purpose.
			Use unobtrusive observation techniques.
Multiple Treatment Interference	Exposure to more than one treatment may affect responses differently in those who experience only one treatment. Multiple treatments could interfere with each other in ways that reduce the effectiveness from what would be observed if a single treatment was applied.	Dancers who take part in both a somatic cueing imaging intervention and a core strengthening program intervention may achieve different alignment control than dancers who only engage in a somatic cueing imaging intervention or a core strengthening program.	Randomize treatments.
			Counterbalance treatment order.
			Have each participant experience only one treatment.
Reactive Effects of Pre-Testing	Using tests that influence performance of the measured variable (e.g., by sensitizing dancers to normally unconscious aspects of their performance), but those who take part in the training without testing are not influenced.	Dancers whose alignment is assessed prior to participating in an intervention that improves self-confidence might perform differently from dancers whose alignment is not tested.	Use control group.
Interactions			
Selection	If study Ps have special characteristics not common in the target population, the ability to generalize the findings is compromised.	Dancers with hypermobility recruited to a study that examines hip rotation during turnout.	Randomize Ps selection.

(*continued*)

Table 10.2—*Continued*

Type	Description	Example	Possible Management Strategies
			Describe selection process clearly (inclusion and exclusion criteria). Describe Ps characteristics relevant to generalization clearly.
Setting and Treatment	Highly controlled environments, such as labs, may not yield the same outcomes as real-world settings.	Assessing dancers' preperformance anxiety in a lab under simulated conditions.	Create environments similar to real world.
History and Treatment	Conducting a study during a particular historical context so that findings cannot be generalized to past or future situations.	Testing dancers' attitudes toward economic challenges during the COVID-19 pandemic.	Replicate studies over time to ensure that results are not time or context dependent.
Limited Population Tested	Using an unrepresentative sample to represent a population.	Use of college student dancers to represent all dancers.	Diversify sampling or generalize only to studied population.
Gender, Race and Cultural Biases	Limited or no diverse representation of gender, race, and culture.	When a study recruits only male or female dancers or includes only one race or culture.	Diversify sampling or generalize only to studied population.

(Wasserstein & Lazar, 2016). Andrade (2019) concluded the following: "All findings should be interpreted in the context of the study design, including the nature of the sample, the sample size, the reliability and validity of the instruments used, and the rigor with which the study was conducted" (p. 214).

Turning to the issue of meaningful interpretation of research results, there are three types of validity that assess the practical value of treatment, training, practice, and policy change in applied research: clinical validity, social validity, and ecological validity. They are not mutually exclusive, but for clarity, we discuss each separately.

Clinical Validity

When working with dancers, there may be critical decisions to make based on screening assessments. For example, in a clinical research study that examines whether concussions have occurred in a sample of dancers, it would be critical to know that the tests used to assess concussion can differentiate between those who meet the criteria for concussion diagnosis from those who do not. *Clinical validity* refers to the predictive value of a screening test for a particular clinical outcome (e.g., the likelihood that a dancer has a concussion). Clinical validity is determined by assessing the sensitivity and specificity with which a particular test identifies persons with a defined symptom profile on an administered test.

Sensitivity is defined as the probability of correctly identifying, from persons with a known condition, those who do have the condition (i.e., identification of *true positives*), and at the same time, not classifying other persons as being without the condition when, in reality, they do have the condition (i.e., avoiding *false negatives*). Specificity is the proportion of people without a condition, who are correctly classified by a screening test as not having the condition (*true negatives*), and at the same time, not classifying persons as having the condition when they do not (i.e., avoiding *false positives*). Sensitivity and specificity address the accuracy of a screening test relative to a reference standard (Trevethan, 2017).

Clinical Significance

Clinical significance refers to whether the effects of a treatment or training protocol are large enough to be practically important to the recipients and other stakeholders. A simple approach to quantifying the influence of treatment is to report the magnitude or percentage difference in performance across conditions (Berg & Latin, 2008). Another simple approach is to calculate a *standardized response mean (SRM),* also called an *efficiency index.* It is the ratio of change from pre-test to post-test divided by the standard deviation of the change score (Portney & Watkins, 2009).

Cohen (1988) offered a slightly more involved approach to assessing magnitude of effect called *effect size* or Cohen's *d:*

$$\text{effect size} = \frac{\text{Difference in Scores across Conditions}}{\text{SD of Scores in the Control or Baseline Condition}}$$

where SD is standard deviation. Cohen's approach to comparing the magnitude of the effect takes variability of the baseline performance into account. He offers a rough, abstract valuation (.20 small effect; .50 moderate effect; .80 large effect).

Another approach to assessing the practical importance of applied research findings is calculating *Explained Variance* (r^2), where r^2 is the square of the correlation coefficient (Berg & Latin, 2008; Payton, 1994). Explained variance (also called *coefficient of determination*) indicates the proportion of variance in Y scores that is accounted for by the variance in X scores. For example, if the correlation between dancers' experience with a new fitness training program and the degree of improvement on fitness tests is .60 ($r = .60$), the explained variance would be .36 ($r^2 = .36$). This means that thirty-six percent (36%) of the performance improvement can be attributed to the new training program while other, presumably unspecified, variables are responsible for 64% of the change.

Other approaches to judging clinical significance involve reference norms, demonstrating the elimination of dysfunctional habits or exceeding initial capacity limitations, and showing meaningful improvements in diagnostic assessments from before to following treatment or training. Portney and Watkins (2009, pp. 18–23), Kazdin (2011), and *Chapter 9* offer extended discussions on assessing clinical change. Of course, the actual importance of any assessment of the magnitude of effect will depend on the context in which it appears (Arnold et al., 2005). Meaningful evaluation of the importance of any observed effect may ultimately reside in the hands of the practitioners who may be interested in adopting a new approach to treatment or training, or in the hands of administrators who may be considering a shift in policy or practice for an institution they manage.

Social Validity

The construct social validity *(or social significance)* identified in the field of behavior analysis in the 1970s aimed to address social and cultural issues that had relevance for behavior change interventions. Strategies for assessing the practical value and acceptability of research findings were developed (Wolf, 1978). They entail asking participants at the end of a study and before debriefing to rate the following:

1. The importance of the performance being studied
2. The degree of change they experienced during the study
3. The acceptability of the procedures used
4. The benefit-to-cost ratio from the participants' perspective

Social validity has been reported in multiple studies over the past 40 years with refinements and elaborations offered (e.g., Carter, 2010). Baer and colleagues (1987) emphasized that the "hallmark of any applied discipline ought to be effectiveness" (p. 322) and that social validity should be a second measure of intervention effectiveness. Social validity might be a useful predictor of both adoption by end-users and the likelihood that a protocol evaluated in a research study will be sustained in practice after the study concludes

(Schwartz & Baer, 1991; Welsh et at., 1994). Dance science researchers may wish to incorporate social validity measures to involve stakeholder perspective as another component for evaluating their research findings.

Ecological Validity

Ecological Validity is an extension of external validity that emphasizes how likely it is that findings from a particular study will generalize to the "real world." If the parameters of a study include features that are not present in real-world settings, findings may not generalize to clinical, teaching, or performing contexts. Such findings would lack ecological validity. The ultimate determinant of ecological validity is the ability to replicate, or at least approximate, findings when practitioners attempt to apply the evaluated treatment, training, and policy practices in their real-world contexts (Barlow et al., 1984, 2009; Kazdin, 2011). Applied researchers may help practitioners make a reasonable forecast by carefully identifying the demographic characteristics of study participants and teachers or clinicians, by operationalizing the procedures used to apply the training or treatment, and by describing relevant aspects of the context in which the study was conducted.

Qualitative Research and Methodological Integrity

Strategies for methodological integrity are integrated into the qualitative research process. The research design, for example, in qualitative research can be modified as the research progresses. The *emergent design* provides an opportunity to add other perspectives to the unfolding picture represented in the findings. *Triangulation* is a strategy that involves using multiple data sources to increase credibility and validity. Qualitative researchers may keep a record of procedures applied and decisions made with an *audit trail.* They also identify their biases, relevant background, familiarity with the topic, and interest in the research by engaging in a process called *reflexivity.* Explicitly declaring these insights in the research report provides readers with context for interpreting the results and the conclusions offered.

Currently there are debates in the literature about how to assess and ensure the quality of qualitative methods. Three different approaches can be identified: (1) application of positivist validity criteria to qualitative research, (2) adherence to *parallel* (Patton, 2002) but separate criteria for establishing trustworthiness, and (3) evaluation of the quality of each qualitative study on its own merits without reference to standardized criteria. With respect to the second approach, *trustworthiness* in qualitative research can be determined by criteria originally developed by Guba and Lincoln (1985): those criteria remain the gold standard for many researchers and include procedures to ascertain credibility, dependability, confirmability, transferability, and coherence/authenticity in the research. They are not the same as the criteria in the quantitative paradigm (validity, reliability, and objectivity), but they have similar properties and aims. Irrespective of the paradigm of inquiry or the methodological approach, rigorous, transparent, and coherent research (i.e., consistency among research methods, assumptions, and reporting) ought to be pursued in any research endeavour.

Each research methodology uses different strategies to make valid or credible inferences. Trying to apply the assessment tools designed for one methodological approach to a methodology that has different underlying philosophical assumptions would be like judging a flamenco festival performance with rules designed to judge a hip-hop competition or judging an ice-skating performance with the point system created for gymnastics competitions. As you evaluate the studies you read, keep in mind that the process of research requires careful consideration of the phenomenon studied, the research participants, the research design, the procedures for generating and analyzing data, and methodological integrity strategies to promote confidence in the research and the findings. These considerations are part of the research enterprise whether the study collects qualitative or quantitative data and whether it relies on comparing group means and variability or individual performance across isolated conditions. Quality assessment tools are described in context, in the Part IV chapters, and details for strategies to establish trustworthiness in qualitative research specifically are described in *Chapter 13*.

Summary

There is no perfect design that protects against all threats to *validity, trustworthiness,* or *authenticity* of research findings. Every design has its strengths and limitations. Whether engaged in experimentation in the classic sense, clinical practice and its idiosyncratic design approaches, epidemiology and its concern for the occurrence, cause and control of health disorders, or qualitative research and its pursuit of authentic and genuine representation of the human experience in context, researchers are always balancing the ideal design with real-world concerns for feasibility and economy. Extra care in the research process helps defend against threats to design weaknesses. Research in dance medicine and science is still in its infancy, but we can build and expand upon the foundational knowledge made available to us from previous research.

Highlights

- Rigorous, carefully designed research is critical in all research paradigms.
- Standards and conventions for ensuring your research is credible and valid are available for various approaches to empirical research.

References

Andrade, C. (2019). The p value and statistical significance: Misunderstandings, explanations, challenges, and alternatives. *Indian Journal of Psychological Medicine 41*(3): 210–215.

Arnold, B., Gansneder, B., & Perrin, D. (2005). *Research Methods in Athletic Training.* F. A. Davis.

Baer, D. M., Wolf, M. M., & Risley, T. R. (1987). Some still-current dimensions of applied behavior analysis. *Journal of Applied Behavior Analysis 20*(4), 313–327.

Barlow, D. H., & Hersen, M. (1984). *Single Case Experimental Designs. Strategies for Studying Behavior Change* (2nd ed.). Pergamon.

Barlow, D. H., Nock, M. K., & Hersen, M. (eds.). (2009). *Single Case Experimental Designs: Strategies for Studying Behavior Change.* (3rd ed.). Pearson.

Berg, K. E., & Latin, R. W. (2008). *Essentials of Research Methods in Health, Physical Education, Exercise Science, and Recreation.* Wolters Kluwer.

Campbell, D. T., & Stanley, J. C. (1963). *Experimental and Quasi-Experimental Designs for Research.* Rand McNally.

Carter, S. L. (2010). *The Social Validity Manual: A Guide to Subjective Evaluation of Behavior Interventions.* Elsevier.

Cohen, J. (1988). *Statistical Power Analysis for the Behavioral Sciences* (2nd ed.). Lawrence Erlbaum Associates.

Guba, E. G., & Lincoln, Y. S. (1985). *Naturalistic Inquiry.* Sage.

Kazdin, A. E. (2011). *Single-Case Research Designs: Methods for Clinical and Applied Settings* (2nd ed.). Oxford University Press.

Patton, M. Q. (2002). *Qualitative Research & Evaluation Methods* (3rd ed.). Sage.

Payton, O. D. (1994). *Research: The Validation of Clinical Practice.* F. A. Davis.

Portney, L. G., & Watkins, M. P. (2009). *Foundations of Clinical Research: Applications to Practice* (3rd ed., pp. 18–23). Pearson Education.

Schwartz, I. S., & Baer, D. M. (1991). Social validity assessments: Is current practice state of the art? *Journal of Applied Behavior Analysis 24*(2), 189–204.

Trevethan, R. (2017). Sensitivity, specificity and predictive values: Foundations, pliabilities, and pitfalls in research and practice. *Frontiers in Public Health, 5,* 349–357.

Wasserstein, R. L., & Lazar, N. A. (2016). The ASA's statement on p-values: Context, process, and purpose. *American Statistician 70*(2), 129–133.

Welsh, T. M., Miller, L. K., & Altus, D. E. (1994). Programming for survival: A meeting system surviving 8 years later. *Journal of Applied Behavior Analysis* 27, 423–433.

Wolf, M. M. (1978). Social validity: The case for subjective measurement. *Journal of Applied Behavior Analysis 11,* 203–214.

IV

Choosing a Research Methodology

11

Descriptive Research Methods

JENNIFER DECKERT AND MARGARET WILSON

Questions to Stimulate Thinking

- Where have you seen descriptive research being used in the dance sciences?
- What types of questions in dance science lend themselves to descriptive research?
- How do I describe what dancers experience in conditioning, training, and performance?
- What steps do I need to take to begin a descriptive study in dance sciences?

This chapter introduces descriptive research that is designed to identify and describe the characteristics of what exists or what is currently taking place in some aspect of the dancer's world. As noted in the opening chapters of this book, describing an aspect of the demands placed on a dancer, or the dancer's experience, can lead to insights that support, clarify, or contradict intuitions we may have about dancers' experiences with dance training or performance. Describing and understanding what dancers are experiencing in conditioning, treatment, training, and performance may help researchers, dancers, and teachers discern how to enhance performance. This type of research can lay a foundation for research questions that might be addressed through other methods. With accurate, clear, and complete description and analysis, descriptive research is often a first step in a new program of research.

Descriptive research encompasses a number of research methods. Some of these methods have their own chapter in this section (Part IV) of the text (survey, qualitative, and correlation research; case studies and small-N studies; within-subject and between-group experiments; systematic literature reviews; and dance injury epidemiology). And each form uses different tools, including surveys, questionnaires, descriptive data, and direct observations. This chapter briefly describes common characteristics in descriptive research methods and describes in more detail how observational research studies are conducted and reported.

Essential Features of Descriptive Research

"Observation is an everyday activity. . . . What sets analytical observation apart from everyday experience is planning and control" (Chatfield, 2000). Observation is an important part of descriptive research that provides information about a population, characteristic, or trait. Descriptive research is a non-experimental approach that describes things "the way they are." Descriptive research may include collecting information through review of records (logbooks or questionnaires), surveys, interviews, or guided observation. It can be used to describe meaningful characteristics of a distinct group or to develop normative information. In most cases it is simply a measure or record of events that would have taken place anyway (Berg & Latin, 2008, p. 243). In descriptive research there is no manipulation of an independent variable to assess a causal relationship. The aim is to identify and describe variables as they exist in a specific environment. Because descriptive research is observational, it can look at a specific group in a variety of settings, such as in a dance studio, dance clinic, or performance group or dance company.

Descriptive research is exploratory; its results can be used to inform a subsequent study (Berg & Latin, 2008). Descriptive research can describe the characteristics of individuals or groups, and it can document existing phenomena or describe how variables change over time (Portney & Watkins, 2009). Descriptive research is based on criteria-driven observation and can answer many of the following questions:

- Who (how many and which people, what are their characteristics)?
- What (what are the actions, expectations, behavior, and exceptions to the behavior)?
- Where (where do interactions take place)?
- When (what is the timing and frequency of the activity; what time of day, or year, or training/performance timeline)?
- How (how do we see different interactions and reactions)?

Descriptive research can use qualitative or quantitative data. Descriptive research is often conducted in the form of a case study—looking at one participant in a given situation—which Payton (1994) describes as *nominal* descriptive research. *Example:* What is the response of a dancer to a specific training program, or what are some of the distinctive characteristics we see as a result of participation in this program? Or, if a dancer is rehabilitating from an injury, a descriptive researcher might ask what are expected functional outcomes of a specific treatment the dancer is receiving? In either case, the researcher would want to have carefully designed questions or criteria developed to obtain objective answers. Nominal research is used to learn about the topic of interest "before the development of specific and answerable questions" (Payton, 1994, p. 91).

Descriptive research also can focus on a group. This is often referred to as *normative* descriptive research, which "describes average or typical characteristics of a given sample" (Payton, 1994, p. 94). Developing a set of guiding questions or criteria is vital in descriptive research to help identify variables and perhaps lead to the development of additional research questions.

In all forms of research, the researcher must always ask "who am I?" in relation to the phenomenon studied (Sparkes & Smith, 2014, pp. 100–104). Identifying your perspective on the criteria and topic is important for reducing the influence of personal bias (seeing what you expect to see, rather than what is really there).

Research Questions Appropriate for Descriptive Research

As you progress in the research process you may find that there are more questions than answers. Descriptive research provides an opportunity to explore some of those questions, and often serves as a precursor to subsequent studies. Descriptive research can provide context for a population, question, or condition to be researched.

Example: A first-time researcher begins observing a group of dance students and notices that ankle injuries seem to have a high rate of incidence. Upon further investigation, she learns that there has been little documentation of the injuries and no previous research about the prevalence of ankle injuries. Questions begin to swirl in the researcher's head about ways to prevent injuries, but first a better understanding is needed of who, what, where, when, and how. This is where descriptive research shines. The researcher begins compiling questions: What is the percentage of dancers in this population with ankle injuries in the past year? What was the specific diagnosis?

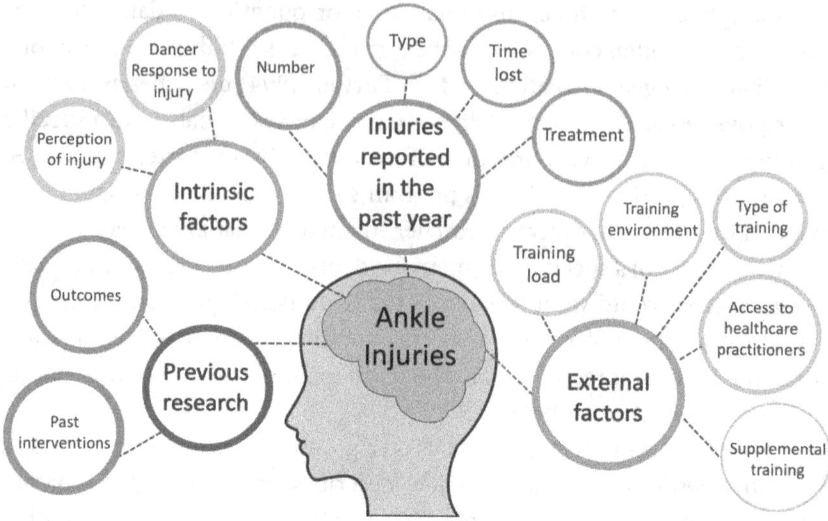

Figure 11.1. Mind Map of Possible Research Topics.

What type of treatment was received? What are the dancer's perceptions of injury? How did this perception change over time? What types of research have already been conducted on ankle injuries in similar populations? What does the rehabilitation process involve? How many hours of training do the dancers' receive? Do they train and rehearse on a daily basis? Does this workload change throughout their training period? Do the dancers have access to a physical therapist or athletic trainer/therapist for early treatment? Do they participate in conditioning or cross training activities, and if so what type?

The list of questions on even one topic can feel overwhelming initially, but by choosing those most relevant to your area of interest, you are able to gather large amounts of information and gain clear insights into a specific population. Descriptive research works to refine assumptions and provide context to guide future questions. See Figure 11.1 for a picture of the types of questions a researcher might ask and how they relate back to the initial inquiry.

Descriptive Research Examples

(1) Dancer Compliance

A researcher wants to look at the participants' motivation to engage with the conditioning program. Before deciding whether to measure physical changes, a descriptive researcher could provide a survey or conduct an interview with the participants on perceived benefits of the program. Using a survey, based on a Likert scale, the researcher could periodically have the participants evaluate their overall feeling about the conditioning program—are they getting stronger, are they feeling more comfortable with the exercises? Do they notice a benefit from the conditioning program in their technique classes? Are there specific exercises in the conditioning program that are most beneficial to their dance classes? The researcher could then follow up on these questions in a structured or semi-structured interview. This information could provide insight into how dancers view the interaction between their conditioning work and dancing while also laying the groundwork for much more specific questions or criterion measures for a follow-up study.

(2) Dancer Response to Injury

A first-time researcher who had experienced several major injuries over the course of her dance career had also experienced a variety of emotional responses while recovering. In speaking with her peers, she began to see trends in injury response and was interested in how other dancers, who had also experienced injuries, perceived the effect of the injury. The researcher created a survey to collect data on dancer response to injury. The survey asked dancers about their perceptions of the injury over time, specifically their response immediately after the injury, one month after, six months after, and then one year, as well as their emotional state at each of these points in time.

(3) Direct Observation

A researcher is interested in the number of corrections students receive in dance class. Documentation of observations might include:

- a tally of the type of correction (technical, artistic, timing, accuracy, etc.)
- comparison of the numbers of corrections given in different styles of dance
- identifying priorities in providing and receiving feedback

(4) Documenting Data

Stein et al. (2014) used a case-series design to document dancer concussions over a five-year period in a large teaching hospital. They documented gender and age, dance style, causes for the concussion—as well as the time of presentation (when symptoms occurred), duration of symptoms, and severity of symptoms. From this study they recommended future studies to evaluate dancers' recognition of concussion symptoms and care-seeking behaviors as well as proposing guidelines for a safe return to dance.

Types of Descriptive Research Methods

Case Study and Case Series

Two common variations in observational research are the case study and case series. In a case study the researcher focuses on one participant or one group to see how they react to a specific stimulus or protocol. For example, when interviewing a dancer recovering from an injury, questions could focus on if the dancer knew of available support systems and class modification options to develop a log for the participant to use to track the return to full dancing.

Conducting multiple case studies on the same problem or topic can be considered a case series. Case series can also involve collecting previously reported information to reveal a pattern not previously defined (Mainwaring et al., 2015). In both cases the participant pool is limited, so the researcher can focus on honing observational techniques. Beginning with observation is not just for novice researchers but also for experienced researchers moving into a new area of research. Conducting observational research does not require an intervention; it is simply a way to see what is going on in each situation. This is a useful starting point in all research settings. *Chapter 15* reviews case studies and case series.

Developmental Research

Developmental research is a description of typical individuals at various stages of life or the description of existing groups (Miller, 2017). These individuals could be a cross-sectional sampling of various segments of a population you are observing using a survey or repeated observations over time. For example, you can create a survey for dancers based on their participation in Pilates, which can be given just to a sample selected from the population of all dancers taking the course, or as a pre-class, mid-way and final survey that helps them rank their experiences in that class.

Normative Research

Normative studies are often directed toward a specific age group, gender, occupation, culture, or disability and describe typical or standard values for characteristics of a given population. Norms are usually expressed as an average, or mean, within a range of acceptable values. Norms can also represent standardized scores that allow interpretation of responses with reference to an arbitrary "normal" value (Portney & Watkins, 2009, p. 304).

Literature Reviews

Descriptive research can also be a focused review of literature. Nemececk and Chatfield (2011) undertook a descriptive analysis of the types of research being conducted on dance teaching and technique from 1966 to 2005. They summarized the information into four areas, thus providing educators with a range of information. This publication described the state of current research, but it also made a call for more research and re-examination of current research that would add to the existing body of literature. Similarly, Durfee et al. (2004) reviewed conference research presentations for the International Association for Dance Medicine & Science annual meetings from 1992 to 2002. They looked at conference abstracts during this time to describe the changes in types of presentations since the inception of the organization. *Chapter 7* covered how to write a literature review, and *Chapter 18* describes systematic reviews of the literature.

Longitudinal Studies

Often a researcher needs to track and document changes that occur over time. This type of study is referred to as *longitudinal* because the same individuals are tested throughout the study. Whereas personal characteristics remain constant, differences observed over time can be interpreted as developmental change (Ployhart & Vandenburg, 2010, p. 98). *Chapter 16* describes studies that might follow a small number of dancers over time or studies that use dancers as their own controls by following them longitudinally.

Epidemiological Questions

An epidemiological study looks at patterns and causes of health and disease conditions in a defined population. Descriptive research in clinical settings can look at the incidence, prevalence, distribution, and possible control of factors relating to health in a dance population. To address clinical questions, research might track dancers' scores on wellness screenings each year of participation in a dance training program. *Chapter 19* reviews dance injury epidemiology.

Tools Used in Descriptive Research

Surveys or Questionnaires

Surveys are often used as a source of information about a specific group that describes their characteristics, experiences, observations, or opinions. Surveys might also provide information on training experience, previous injury, or risk factors for injury. Studies that use surveys are focused on a particular issue but may also include open-ended responses from the participants. This provides more context for interpretation of association among the given variables (Portney & Watkins, 2009, p. 314). For example, a survey might ask a participant to list supplemental training protocols they engage in, with an open-ended question at the end of the questionnaire. Open-ended questions give participants an opportunity to provide more context. Questionnaires and surveys can include ranking options, ratings, and agree or disagree and open-ended questions. *Chapter 12* addresses survey research and discusses in more detail the types of questions one may use.

Rubric for Observation

A rubric or descriptive table can be used to streamline observations. With specific descriptors, creating a rubric allows the researcher to create discerning observations. A rubric offers the researcher ways to create detail in the observations. For example, a movement or response could be rated as (1) occurs all the time, (2) occurs most of the time, or (3) occurs some of the time.

Interview

Interviews are frequently used as follow-up for observation to confirm or correct the researchers' findings. Interviews are discussed in *Chapter 12* as part of survey research and in *Chapter 13* as part of qualitative methods.

Ranking

A descriptive researcher could have a specific list of items being sought in observations and tally the number of times a specific action is seen. Payton calls this *ordinal measurement,* and it can be quantitative (numerical ranking) or qualitative. A Likert Scale is a useful tool for ranking (Payton, 1994, p. 59).

Types of Data Used in Descriptive Research

Qualitative data (*Chapter 13*) often includes observational or descriptive methods to provide thick, rich description that can provide context for theory generation (Sparkes & Smith, 2014). *Quantitative data* involves statistical, mathematical, or computational measurements of observable phenomena (Payton, 1994). A researcher could develop a logbook or worksheet for tracking attendance in a supplemental conditioning activity. Additionally, the researcher could provide a survey or questionnaire for the participant in the conditioning activity to rate their investment and the benefit from the session. These data could be tallied over time and summarized at the end of a semester, training season, or defined measurement period. *Descriptive statistics* are frequently used to summarize qualitative and quantitative data. In descriptive research one can report central tendency (mode, median, or mean) and standard deviation to summarize the frequency of responses or measures. In this way a researcher can identify answers to questions pertaining to a group (Payton, 1994, p. 101). Information on data management and analysis can be found in *Chapter 9*.

Ethical Considerations and Bias

All research raises ethical considerations, which must be managed carefully by researchers, including descriptive researchers. It is important to consider the ethical nature of what you are observing and how you are collecting the data. You should always obtain consent from the participants involved or the parents of a young dancer you are observing or interacting with. *Chapter 6* provides guidance on addressing the ethical issues involved in conducting research with dancers.

One requirement that is present in all research but may be overlooked in descriptive research is avoiding bias. Researchers must always be aware of the potential for personal or professional bias to influence the ways they implement their researcher protocols and their interpretation of data. This is true even when the body of data is speaking for itself, as is the case in descriptive research. It is important to consider your own bias when asking questions, selecting participants, making observations, or collating data. Do you dismiss a specific type of response because it is inconsistent with how you believe the participants should respond? Have you biased your research question based on your past experiences? Are you looking for something so specific that you fail to recognize additional factors that may be affecting the outcome? These are all questions to consider as you approach your descriptive research project. We all have biases; the goal is to observe a specific phenomenon without allowing our biases to influence our observations, analyses, and the interpretation of our results and expectations of outcome.

One final consideration is that the participants may change their behavior because they are being observed. Known in research as the *Hawthorne Effect*, it is an important limitation in any observational study and should always be considered when designing a method of data collection and determining how the researchers will interact with their study participants.

Strengths of Descriptive Research

Descriptive research is an ideal approach for new researchers and those seeking more information prior to conducting additional studies in a new area of research. The variety of descriptive research approaches available to dance science researchers provides options for researchers with a reduced budget, limited supplies, or access only to small samples.

Additionally, it allows the researcher space and time to refine measurement protocols, variables, populations, and assumptions surrounding their research question. Conducting a descriptive study helps researchers understand the context in which they may be conducting future research that builds on what they learn through description. Under-researched topics or populations are well served by descriptive research methods, and while it may be tempting to dive in and immediately fix a perceived problem, careful and thoughtful researchers seek first to understand the problem and the population fully. By conducting a descriptive research study, you will be contributing to the foundational knowledge in the field and providing important insights to inform current and future research.

It is important to develop and define terms of measurement for your study. In order to learn more about the population you are studying, you must carefully hone the research question you want to address. In some studies it is important to have only one question in focus, but as descriptive research is about generating information to describe a population or condition, you may want to generate a series of questions you are interested in knowing more about. As you are conducting a review of the research literature available on your topic (dancer's responses to injury, for example), you will learn about questions that have been generated by other studies, which will provide guidance or justification for your study.

Limitations of Descriptive Research

As with all types of research there are several limitations and requirements to consider with descriptive research studies. However, types of descriptive research vary, so each approach will provide its own list of considerations, which are more clearly outlined in upcoming chapters. We offer an overview of these considerations here.

The first basic requirement is that you must have access to dancers, clinicians, and other participants that you want to describe, and for research that requires extended observation it is important to consider consistency of observation and access to the population. Descriptive research may also require access to medical records, past survey responses, injury reports or personal logs/journals. Collecting this type of data may require additional approvals and it is important to consider the amount of access you will need prior to initiating your research.

Next it is important to ask the right question and be clear to report only on the phenomenon under question. Since the primary objective of descriptive research is to gather information, it is imperative that your question is *specific* to the population and topic being studied. In order to achieve this goal, the method of data collection must contain *clearly defined parameters*—this includes specific survey questions, valid measurement protocols, consistent data collection, and explicit directions for research assistants and participants. The use of previously validated and standardized measurement tools and surveys is one way to help ensure valid data collection. These include tools like a goniometer to measure joint angles, Rate of Perceived Exertion scale, Likert scale for survey responses, or specifically validated surveys such as the EAT-26 (Garner et al., 1982) to assess disordered eating behavior.

Summary

Descriptive research:

- Serves as a great first step in any research endeavor.
- Can be combined with other protocols or serve as a stepping stone for other research protocols.
- Demands a systematic approach that is ideal for first-time researchers as they learn to look and evaluate a question.
- Provides a detailed picture of a population and useful data for reformulating and posing new questions.

Highlights

- Descriptive research provides an excellent point of entry for new areas of research.
- Beginning researchers will gain important methodological skills by conducting a descriptive study as their first research effort.
- Descriptive research lets you look at things "as they are"; can focus on one person, or a group, or just focus on review of literature; and can help you define your research question.

Application Exercise

Think about a question you have been wanting to research. What made you interested in that question? Perhaps it was something you have observed in your classes, rehearsals, or choreography. Can you develop a systematic, replicable approach to observing a key aspect of the phenomenon you are interested in studying? Begin with a list of questions that might shape your research question or define the phenomenon of interest. Choose one question or topic and list some possible descriptive words for summarizing what you observe. Notice which of the questions or objectives occur most frequently—but don't discard the others. Descriptive research can respond to a broad range of experiences. Narrow your focus on these criteria so that you can remain objective in your observation. Consider research tools such as surveys, questionnaires, or structured interview questions to help you refine your method as you look at a population or condition.

Suppose you are interested in how Pilates training benefits dancers in your studio. Design a list of specific concepts you (as a researcher) would expect that they learn and apply from this supplemental activity. What are some of the core principles from Pilates that are important to dancers? (For example, breath, centering, control, or precision.) Come up with a check-sheet that allows you to observe and record instances in which the participants in the Pilates class are utilizing these core principles. Next, come up with 3–4 survey questions you could provide to the participants in the Pilates class to rate their experience in the class. Finally, develop a questionnaire that allows for simple answers (yes, no, improving, not improving, etc.) and open-ended questions (e.g., "How are you linking Pilates training with your dance training? Please provide specific examples"). As you collate your observations and summarize the data from the surveys and questionnaire, what questions have you answered, and what new questions have emerged? Keep asking the questions, and you are on your way.

References

Berg, K. E., & Latin, R. W. (2008). *Essentials of Research Methods in Health, Physical Education, Exercise Science and Recreation.* Wolters Kluwer.

Chatfield, S. J. (2000). Rules of evidence. In Welsh, T., Chatfield, S. J., Wilmerding, G., Moore, M., & Hobden, R. *A Closer Look at the Science in Dance Medicine and Science,* International Association for Dance Medicine & Science 10th Annual Meeting, Miami FL.

Durfee, N. F., Welsh, T. M., & Dunn, J. (2004). Trends in IADMS conference presentations. *Journal of Dance Medicine & Science, 8*(3), 82–88.

Garner D. M., Olmsted M. P., Bohr Y., & Garfinkel, P. E. (1982). The eating attitudes test: Psychometric features and clinical correlates. *Psychological Medicine, 12*(4), 871–878.

Mainwaring, L., Keeler, C., Chou, E., Brown, D., Hiller, C., Welsh, T., Shah, S., Russell, J., Harwood, M., Couillandre, A., Air, M., Angioi, M., & Ambegaonkar, J. (2015, October 9). *How to ask questions: Introduction to survey research* [Research Forum Lecture]. International Association for Dance Medicine & Science 25th Annual Meeting, Pittsburgh PA.

Miller, S. (2017). *Developmental Research Methods* (5th ed.) University of Florida Press.

Nemececk, S. M., & Chatfield, S. J. (2011). Teaching and technique in dance medicine and science: A descriptive study with implications for dance educators. *Journal of Dance Education, 7*(4), 109–117.

Payton, O. D. (1994). *Research: The Validation of Clinical Practice* (3rd ed.). F. A. Davis.

Ployhart, R. E., & Vandenburg, R. J. (2010) Longitudinal research: The theory, design, and analysis of change. *Journal of Management 36*(1), 94.

Portney, L. G., & Watkins, M. P. (2009). *Foundations of Clinical Research: Applications to Practice* (3rd ed.). Pearson Education.

Sparkes, A. C., & Smith, B. (2014). *Qualitative Research Methods in Sport, Exercise and Health: From Process to Product.* Routledge.

Stein, C. J., Kinney, S. A., McCrystal, T, Carew, E. A., Bottino, N. M., Meehan III, W. P., & Micheli, L. J. (2014). Dance-related concussion: A case series. *Journal of Dance Medicine and Science 23*(1), 26–33.

Additional Resources

Milner, S. C., Gray, A., & Bussey, M. (2019). A retrospective study investigating injury incidence and factors associated with injury among belly dancers. *Journal of Dance Medicine and Science, 23*(1), 26–33.

Rodriguez, M., Bolia, I. K., Philippon, M. D., Briggs, K. K., & Philippon, M. J. (2019). Hip screening of a professional ballet company using ultrasound-assisted physical examination–diagnosing the at-risk hip. *Journal of Dance Medicine and Science 23*(2), 51–57.

12

Surveys, Questionnaires, and Interviews

EMMA REDDING AND LILIANA S. ARAÚJO

Survey research enables gathering a lot of data across a large sample of participants administered at one time or multiple time points. In this chapter we describe essential features of survey research, types of research questions suited to survey research, and sampling techniques. Two main methods for conducting survey research (questionnaires and interviews) are discussed. We draw attention to important considerations in devising questions and interpreting responses, and we comment briefly on analytical techniques. We conclude with a discussion of the benefits and limitations of survey research.

Questions to Stimulate Thinking

- What are the main features of a scientific survey?
- How might you design survey research to ask dancers about the types of imagery used in class?
- If you wanted to investigate injury prevalence among tap dancers in your geographic area, how would you select your sample?
- What are the differences between questionnaires and interviews?

What Is a Survey?

The choice of methodological approach in any research should be driven by the research question and the goals of the research. Surveys are essentially about approaching a sample selected from a population, asking a set of questions, and interpreting the answers in order to understand more about that population. The extent to which you meet this goal is determined by the representativeness of your sample, the type of questions and how you ask the questions, and the procedures followed to conduct the survey. Surveys are often used to gather national information on public opinion, as a component of market research designed to understand consumer preferences, or to gather information about socioeconomic, contextual, and behavioral characteristics of specific target populations in specific situations. In dance research, surveys have been used to learn about the prevalence of injuries (e.g., Jeffries et al., 2020), to learn about dancers' attitudes toward supplementary fitness training, and to understand dancers' use of mental imagery (e.g., Nordin & Cumming, 2006a, b).

Types of Surveys

When using surveys, researchers attempt to measure objective facts, such as age or employment income, and subjective states, such as attitudes toward a particular method of training or behaviors regarding healthy eating choices. There are different types of surveys including cross-sectional, longitudinal, and correlational or explanatory surveys.

A *cross-sectional survey* is a "snap-shot picture" of the current population or situation (e.g., the Fit to Dance surveys—Brinson & Dick, 1996, Laws, 2005).

A *longitudinal survey* is durational in nature (across time) and attempts to measure a potential change (e.g., personality in dancers—Bakker, 1991).

A *correlational (or explanatory) survey* provides for comparisons (e.g., dancers' perceptions of the physical, cognitive, affective, and social benefits of partnered dancing—Lakes et al., 2016).

Conducting Surveys

There are two main ways in which surveys are conducted: (1) questionnaires, and (2) interviews. Questionnaires usually comprise a standard set of questions (or *items*), which follow a predetermined order. Two advantages of questionnaires are that they can reach large groups of people simultaneously, thereby providing fast and comparable results, and they are less labor-intensive than interviews.

Self-administered surveys have the advantage of allowing respondents to complete them in their own time and privately. Because there is an element of anonymity in self-administered surveys, they may encourage more open and honest responses, which is particularly helpful when collecting sensitive information (Tourangeau & Smith, 1998). Nevertheless, self-administered surveys depend on people's goodwill and motivation to complete them and require a certain level of reading and writing skills. In this sense, self-administered questionnaires may encourage those who are particularly interested in the research inquiry over those who are not.

Conversely, interviews may be the preferred choice for some people as there is no reading or writing involved. The advantages of interviews are that questions can be tailored to the context and individual (if appropriate to the research goals), and the rapport built between interviewer and respondent can elicit more meaningful and honest responses.

Questionnaires are often sent to a sample of people, either in person or by post, email, or online survey programs (e.g., Survey Monkey). Interviews typically take place in person or on the telephone; more recently, interviews are conducted using online meeting and conference platforms such as Skype or Zoom. There are some clear advantages with computer-assisted surveys, including:

- Instantaneous results
- Data can be extrapolated immediately for analysis
- Questions can be skipped if not relevant to the respondent
- Errors and inconsistencies can be identified easily
- Greater access to people who live far away

Different methods of collecting survey data have been compared (e.g., Cannel et al., 1987; Fowler, 2014). Findings indicate that research into the age at which dancers begin their formal training is suited to a questionnaire, whereas research into dancers' experiences of their training may be better suited to an interview. The method of data collection may affect survey costs, sample size, choice of questions, and response rate.

Sampling for Surveys

Sampling is a procedure to create a small subset of the population that is considered to be representative. The first step is to determine the population from which to draw the sample. For instance, you may be interested in finding out more about professional tap dancers aged between 18 and 35 years living in London, England. The sample will be a smaller group of tap dancers who are typical of the larger population.

The aim in sampling is to ensure that the sample fairly represents the population. Researchers can obtain information about the behaviors and attitudes of a population by asking a smaller sample of people. However, errors can be made if careful measures are not taken to ensure that the sample is representative. When selecting a sample, researchers should consider the sample size, sample frame and design, and response rate. Researchers should also document sampling methods so that they can check the reliability of the sampling method and can guide replication in future studies.

Sample Size

One of the most common questions asked in survey research is how big the sample should be. When determining the sample size, it is important to take a systematic and unbiased approach so that data collected from a sample can be accurately extrapolated to the population, reducing the margin for uncertainty and increasing the representativeness of the data. Sample size can depend on access to the population, resources available, and the logistics underpinning the research as well as the types of questions asked.

While some researchers believe that the larger the sample, the more representative of the population it will be, a large sample will not automatically produce representative results if the sampling process is biased or poorly designed. Researchers should conduct a power analysis to identify the recommended sample size. Some propose that a specific fraction of the population (2% or 5%) is necessary for results to be valid and meaningful. Error rate and response bias are important considerations in determining sample size. For example, to be confident in the results, researchers recommend a sample of at least 1,000 participants for large national surveys, a minimum of 100 participants for smaller studies, and 30 for even smaller-scale studies. In fact, a small sample may be more appropriate if the questions asked are open-ended and/or the research objectives are to gather rich and deep understanding of specifically targeted individuals or a homogeneous group.

In reality, establishing an appropriate sampling strategy is as important as balancing the resources available to dedicate to recruitment of participants and ensuring a good response rate. Therefore, more than the size of the sample, the sampling strategy should determine ways to reduce the error rate and the response bias.

Sampling Frame and Design

A sample frame is the group of people who could potentially be selected to complete the survey. The researcher should confirm that those within the sample frame share the characteristics of the population of interest, to reduce biasing the sample results. For example, in the tap dancer project described earlier, the sample frame should include all professional tap dancers living in London between the ages of 18 and 35 years. The sampling design is an outline of how the sample will be created and who could potentially be selected (the sample frame). The design is an explanation of the procedures to be undertaken to obtain a representative sample.

Types of Sampling

Although some research methodologies use non-random sampling procedures (see *Chapter 8*), especially in qualitative studies, most survey research uses one of several random sampling procedures.

Simple random sampling is a technique whereby everyone in the population has an equal chance of being selected. This process is like drawing numbers out of a hat; however, numerical techniques can be used to ensure that the sampling is truly randomized. If there were 10,000 people in the population of interest and the desired sample size was 100, then one could systematically count the population, choosing 1 in every 100 people. Alternatively, a computerized data file could randomize the individuals within a population then select the first 100 people on the list. This technique can be time-consuming, and there is the possibility that the sample will have a disproportionate number of participants with certain characteristics (e.g., more males).

Systematic random sampling involves selecting individuals in a numbered population at designated intervals. If, for example, the population of interest is 5,000 hip-hop dancers in New York City, and the desired sample is 200, it would mean that every twenty-fifth dancer is selected starting from a randomly chosen number.

Stratified random sampling is used to prevent over-representation of participants with a certain characteristic in the sample. For example, if the population of interest is a dance audience made up of particular minority groups, simple random sampling may result in no members of the minority groups being chosen. Using stratified random sampling, the sample frame would be divided into relevant groups. Random sampling would then be applied to each group. A specific number of individuals would be selected from each group that is proportionate to the larger population.

Disproportionate sampling provides the opportunity for a group that is under-represented in the larger population to be accurately represented in the sample. If, for example, a researcher were interested in comparing the opinions of a particular minority group with other larger groups, this could be done by over-sampling the smaller categories (Fox et al., 2007).

Cluster sampling allows the researcher to survey a sample of participants in a certain region when it is not possible or practical to access a wider grouping. For example, it may be necessary to survey only people living in a particular town or village or at a certain conservatoire or university. Cluster sampling requires caution when interpreting findings concerning the wider population.

Response Rate

There are various strategies available to enhance the response rate and reduce the number of non-responders, reducing the potential for survey error. First, ensure that the method used to complete the survey (i.e., telephone, internet, mail) is accessible to the whole sample. Second, consider whether respondents understand the questions being asked (i.e., the language is understood by the whole sample). And third, consider whether all those in the sample are available to participate and have the reading or writing skills to respond successfully. For interviews, one way to maximize response rate is to be flexible, offering opportunity for either an in-person or an online interview.

Calculating the response rate is important in evaluating the effectiveness of the data collection procedure. This calculation is simply the number of people who completed the survey divided by the number of possible responders in the sample. For example, if 80 in a sample of 100 responded, the response rate would be 80%.

Non-Response Bias

It is important to report the response rate of a survey when publishing findings so that others may speculate as to how representative the findings are likely to be. A response rate of 90% is more likely to yield better representative values than a response rate of 25%, although it is not always possible to measure the actual effect of non-response. Researchers should attempt to produce a response rate that is deemed reasonable (e.g., 75–80%). This can be done by ensuring that surveys do not take too much time to complete, by making follow-up phone calls at convenient times, or by re-sending internet-based surveys to potential participants, bearing in mind that participation in survey research is usually voluntary.

Measurement

Researchers should carefully consider their questions (sometimes known as items) so that each respondent is asked the same question in the same way, to avoid bias or error. In addition, surveys should be clear in topic or concept, to ensure that responses represent the aims of the research accurately.

When conducting interviews, interviewers need to recognize their role in the survey process and be trained on how to administer an interview to minimize bias. Researchers should create an interview guide with carefully worded questions and sub-questions, and strategies for probing (encouraging a fuller answer) where necessary. Such a guide should also contain scripted introductory and concluding remarks to limit the amount of unstructured discussion that takes place and to ensure consistent data collection across participants. The researchers should pilot test these questions to confirm a consistent understanding of the concepts and terms. The piloting will also help the interviewer to rehearse what will be said and how to determine the likely length of the interview.

Designing Questions

Question development is critical in survey research, and there are key aspects to consider when devising questions. The specific answer to each question is equally important to the researcher in addition to what that answer means in relation to what is being measured. Well-designed questions are *reliable* (consistent) and *valid* (measure what they are supposed to measure).

There are several strategies to ensure reliability and validity of a survey:

- Test-retest reliability—ask participants to answer the survey at different points in time to determine consistency over time.
- Face validity—ask experts whether the answers to questions reflect the variable being measured.
- Conduct a pilot test—ask a group of participants to answer your survey to check if the questions are clear and relevant.

Using the pilot test results to create a database and using statistical techniques such as factor analysis and Cronbach's alpha to assess validity and reliability will identify "weak" questions and adjust the questions based on their "weight" for each factor.

Types of Questions

Generally there are two types of questions: closed-ended and open-ended questions. Closed-ended questions require one-word answers, box ticking, or choosing a value within a range of possible options. Alternatively, open-ended questions do not have pre-defined answers from which to choose and often require a lengthier, more detailed individual response. Likert (1932) proposed the now widely used scaling technique in applied survey research whereby respondents choose a value within a scale that represents, for example, if they most likely agree or disagree with something (see Table 12.1 for examples).

Table 12.1. Examples of Types of Questions

Question	Type of Question
What is your name?	Closed-ended
What do you consider to be the most important aspect of dance training?	Open-ended
How do you feel about classical ballet as a form of training?	Open-ended
Have you taken a hip-hop class before?	Closed-ended
How do you feel about this statement (circle one)? Dancers should take acting classes. Strongly Agree, Agree, Don't Know, Disagree, Strongly Disagree	Scale

Scales are useful to rank the responses of a group of people and may allow comparisons with existing data, but only the values in the scales used are comparable. In Table 12.2 the scales produce a different result from the set of categories used. In the latter scale, the *fair* category is situated closer to the poor category when compared to the former scale when *fair* is situated in between poor and good. The probability of a respondent choosing *fair* would differ according to which scale is used. Therefore researchers should be careful when using scales such as these to make a statement about a certain percentage of the population feeling a certain way about something.

Standardized questionnaires usually contain closed-ended questions and statements so that responses can be analyzed statistically within and between groups of people. Examples of standardized questionnaires used in dance science include the Rosenberg Self-Esteem Scale (Rosenberg, 1965), the Sport Imagery Ability Questionnaire (SIAQ; Williams & Cumming, 2011), and the Dance Imagery Questionnaire developed by Nordin and Cumming (2006a, b).

Developing Questions

If you need to devise a questionnaire for single use for your study, you should carefully consider how you are developing the questions. The type of question, the language used, and the order in which items are presented can

Table 12.2. Scale Responses

Question: How would you rate your current fitness level (circle one)?

Good		Fair		Poor
Excellent	Very Good	Good	Fair	Poor

influence the respondents' answers. Questions should be clear and unambiguous. Questions should use appropriate and simple wording to maximize respondents' understanding and minimize the risk of misinterpretation and error. If two respondents understand the question to mean different things, their answers could be different simply based on their understanding of the questions. A process to minimize this problem is "vetting" by asking experts for feedback using the piloting process explained earlier in the chapter.

Common problems in developing questions include poor formatting and flow, using vague terms, and inconsistent or inappropriate scales. Researchers should avoid incomplete sentences and questions, and questions should not attempt to address more than one topic. Controversial or emotive items should be avoided at the beginning of the survey, and more descriptive questions can be asked at the end. Survey formatting and flow should be carefully considered to ensure that no questions are hanging over several pages or sections; that the instructions are clear; and that the respondents can easily follow each question on their own. The questions should not assume prior knowledge or understanding or lead the respondent to answer in a certain way.

Table 12.3 offers examples of common problems, poorly developed questions, and well-designed alternatives.

Duration of Data Collection

The duration of data collection can vary depending on the aims of the research and the relevance of the timing of the event for the research. For example, a survey on dance students' experiences of the induction activities at the start of the academic year should take place within a short period of time after the event to allow dancers to recall it more accurately. However, in large population survey studies focusing on socioeconomic conditions of workers in the dance industry, data collection can extend for a longer period of time and take place at pre-established time points. Typically, the process often involves a call-out advertisement, an initial mail-out of the survey, and subsequent reminders. A telephone and in-person survey takes less time but usually involves a much smaller sample.

Analyzing Survey Results

In survey research, data analysis often focuses on describing and finding trends in the data. To ensure the accuracy of the findings in survey research, some quality control is needed before data analysis. Whether a dataset is

Table 12.3. Examples of Common Problems, Poorly Designed Questions, and Better Alternatives

Common Problems	Incomplete/Ambiguous	Complete/Unambiguous
Terms are not well-defined/ vagueness	Have you been injured in the last year?	Have you had a musculoskeletal injury in the last 12 months that resulted in not being able to participate in dance?
	Which school did you go to?	Which school did you attend at the age of 10 years old?
	Age?	What was your age on your last birthday?
	When did you start dancing?	When did you start participating in 3 or more dance classes per week on a regular basis?
Inconsistent or inappropriate scales	How many hours of sleep do you have on a typical weekday? 1. 7–9 2. 5–7 3. 0–5	On average, how many hours of sleep do you have on a typical weekday? • More than 9 hours • 7–9 hours • 5–6 hours • Less than five hours
	Are you satisfied with the feedback you were given in technique class? [disagree-agree scale]	Rate your level of satisfaction with the feedback you were given in technique class? [dissatisfied-satisfied scale]
Double-barrelled items	Do you want to perform and choreograph?	Do you want to perform? Do you want to choreograph?
Leading questions	What do you think about these excellent fitness facilities?	How would you describe the overall quality of the fitness facilities? [poor-excellent scale]
Sensitive data	Are you unemployed?	What is your work status? 1. Employed full-time 2. Employed part-time 3. Freelance 4. Unemployed

created manually or extracted from survey platforms, preparing the dataset is a time-consuming but necessary process to ensure the quality of the analysis.

Data Cleaning and Preparation

For quantitative analysis, one of the first steps in data preparation is to replace identifiable information with codes. This deidentification allows for data anonymity and facilitates the statistical analysis using numerical values. Then look for incorrect values (e.g., impossible values due to a typing mistake), irrelevant data (e.g., Internet Protocol [IP address] in online surveys; introductory information on survey interviews), cases that meet exclusion criteria, missing data and outliers, empty or duplicated entries, and different labels for the same variables. These should be removed or replaced as relevant and appropriate. In some cases, scores of scales or subscales of questionnaires need to be calculated for analysis, providing either a mean or total score. Variables may need to be regrouped (e.g., create age groups from age distribution), relabeled to be found more easily, or formatted to adjust to the analysis software. For qualitative analysis, especially when using software to manage and analyze qualitative data, similar procedures can apply.

Data Analysis—Quantitative

The first step in this process is to explore the data, which will allow you to identify and summarize the main characteristics of the sample and give a general description of the results of the survey. This can be done using descriptive statistics, such as frequencies, percentages, calculating means and standard deviations, and using visuals (e.g., charts) to display data characteristics. Exploratory or descriptive analysis also allows one to identify the normality of the distribution of the data, to identify outliers, to explore trends between groups, and to investigate the validity and reliability of the questionnaires. The research question (e.g., to determine relationships or predict outcomes) and the main characteristics of the data inform the inferential analysis to undertake to analyze the data.

Data Analysis—Qualitative

Qualitative data include any descriptive information that is not presented in the form of numbers. These data usually represent subjective experiences, perceptions, attitudes, or behaviors as described by an individual. When using closed-ended and structured interviews in the context of survey research, qualitative data may be codified into quantifiable information, allowing exploratory and inferential statistics. However, data gathered to capture individual experiences are mostly qualitative, and therefore analysis will be focused on describing and making sense of those experiences. There are several ways of analyzing qualitative data, depending on the research methodology and framework (see *Chapter 13*).

For example, Sanna Nordin-Bates and Sofia Kuylser (2020) used questionnaires and interviews to study perfectionism among dance students in a Swedish ballet school. Participants completed a standardized perfectionism questionnaire (Multidimensional Inventory of Perfectionism in Sport, Stoeber et al., 2007) adapted to dance by replacing words related to sports for dance. Following descriptive analysis of the questionnaire results, representatives of each type of perfectionism were identified and invited for an interview. Researchers used both deductive analysis, to identify content related to each type of perfectionism as defined by the theoretical model, and inductive reasoning analyses, to categorize themes that emerged from participants' interview data.

Limitations in Survey Research

Like all methods of gathering data, survey measurement is at risk of error. The procedures used to conduct a survey will impact the extent to which the data can accurately describe the population. One of the common limitations of survey design is social desirability, which happens when respondents provide answers that they believe the researcher wants to hear. This is common when there are certain facts or experiences respondents would rather not speak the truth about in a survey due to associated stigma (e.g., mental illness), socially undesirable behaviors (e.g., drug abuse), or sensitive topics (e.g., legal convictions, religion). These sorts of events and facts tend to be under-reported in survey research (Cannell et al., 1987). Another limitation is dependence upon memory recall. Instead of asking dancers to think back to their past to remember whether they sustained an injury (Laws, 2005), recent studies into injury prevalence in dance have used prospective surveys, which track injuries at the point at which they happen (e.g., Cahalan et al., 2018; Jeffries, 2020). Issues in survey research have also been identified as errors related to *measurement* (non-sampling errors related to what it is that we are measuring and how) and *representation* (sampling errors resulting from sampling design and procedures) (see Figure 12.1) Many of these potential errors and limitations can be prevented or minimized through a well-considered sampling design and questionnaire development process that aligns with a well-formulated research question.

Summary

Overall, survey research is a useful and widely used approach to investigating areas of interest in dance science. Survey research can allow researchers to gather a large amount of data at one time or across multiple time points and offers an alternative approach to interventional research. We have described sampling and the various kinds of sampling and techniques available.

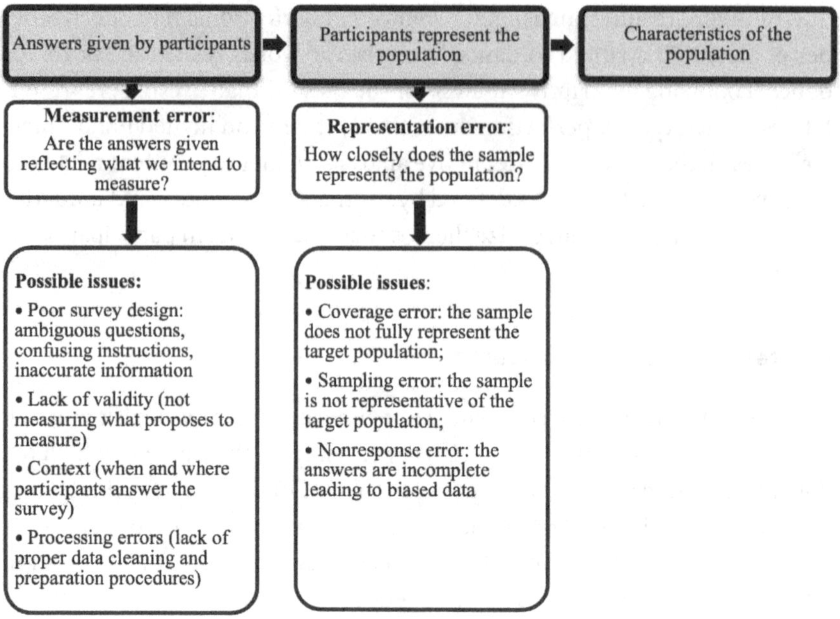

Figure 12.1. Examples of Survey Research Errors. Adapted from Fowler (2014), p. 9.

Questionnaires and interviews are the two main methods of measuring in survey research, and we drew attention to important considerations when devising and interpreting questions. Last, we explained the benefits and limitations of survey research.

Highlights

- Surveys are accessible and time-effective ways of collecting information from samples.
- An appropriate strategy for developing questions is critical for effective survey research.
- Sampling strategies are particularly important for data accuracy and representativeness in survey research.

References

Bakker, F. C. (1991). Development of personality in dancers: A longitudinal study. *Personality and Individual Differences, 12*(7), 671–681.

Bowling, A. (1989). Injuries to dancers: Prevalence, treatment, and perceptions of causes. *British Medical Journal, 298*(6675), 731–734.

Brinson, P., & Dick, F. (1996). *Fit to Dance: The Report of the National Inquiry into Dancers' Health and Injury.* Calouste Gulbenkian Foundation.

Cahalan, R., Kearney, P., Bhriain, O. N., Redding, E., Quin, E., McLaighlin, C., & O'Sullivan, K. (2018). Dance exposure, wellbeing and injury in collegiate Irish and contemporary dancers: A prospective study. *Physical Therapy in Sport, 34,* 77–83.

Cannell, C., Groves, R., Maglev, L., Mathiowetz, N., & Miller, P. V. (1987). An experimental comparison of telephone and personal health interviews. *Vital and Health Statistics, Series 2: Data Evaluation and Methods Research, 106.*

Fowler, F. J. (2014). *Survey Research Methods* (5th ed.). Sage.

Fox N., Hunn, A., & Mathers, N. (2007). *Sampling and Sample Size Calculation.* Trent Research and Development Support Unit (RDSU), NIHR RDS for the East Midlands/ Yorkshire and the Humber.

Gaudreau, P., & Thompson, A. (2010). Testing a 2×2 model of dispositional perfectionism. *Personality and Individual Differences, 48*(5), 532–537.

Fowler, F. J. (2014). *Survey Research Methods* (5th ed.). Sage.

Jeffries, A. C., Wallace, L., Coutts, A. J., Cohen, A. M., McCall, A., & Embellishers, F. M. (2020). Injury, illness, and training load in a professional contemporary dance company: A prospective study. *Journal of Athletic Training, 55*(9), 967–976.

Lakes, K. D., Marvin, S., Rowley, J., Nicolas, M. S., Arastoo, S., Viray, L., Orozco, A., & Jurnak, F. (2016). Dancer perceptions of the cognitive, social, emotional and physical benefits of modern styles of partnered dancing. *Complementary Therapies in Medicine, 26,* 117–122.

Laws, H. (2005). *Report of the Second National Inquiry into Dancers' Health and Injury.* Dance UK.

Likert, R. (1932). A technique for measurement of attitudes. *Archives of Psychology, 22*(140) 55.

Nordin, S., & Cumming, J. (2006a). The development of imagery in dance part I: Qualitative findings from professional dancers. *Journal of Dance Medicine and Science, 10*(1–2), 21–27.

Nordin, S., & Cumming, J. (2006b). The development of imagery in dance part II: Quantitative findings from a mixed sample of dancers. *Journal of Dance Medicine & Science, 10*(1–2), 28–34.

Nordin-Bates, S., & Kuylser, S. (2020). High striving, high costs? A qualitative examination of perfectionism in high-level dance. *Journal of Dance Education,* 1–12.

Rosenberg, M. (1965). *Society and the Adolescent Self-Image.* Princeton University Press.

Tourangeau, R., & Smith, T. W. (1998). Collecting sensitive data with different modes of data collection. In Couper, M.P., Baker, R. P., Bethlehem, J., Clark, C., Martin, J., Nicholls II, W. L., et al. (eds.), *Computer-Assisted Survey Information Collection* (pp. 431–453). John Wiley.

Stoeber, J., Otto, K., Peachick, E., Becker, C., & Oliver, S. (2007). Perfectionism and competitive anxiety in athletes: Differentiating striving for perfection and negative reactions to imperfection. *Personality and Individual Differences, 42*(6), 959–969.

Williams, S. E, & Cumming, J. (2011). Measuring athlete imagery ability: The sport imagery ability questionnaire. *Journal of Sports and Exercise Psychology, 33,* 416–440.

13

Qualitative Methodologies

LYNDA MAINWARING AND IMOGEN AUJLA

Questions to Stimulate Thinking

- How can we explore areas of dance science that have not been examined before?
- How do we study what dancers are thinking or feeling, or aesthetic aspects of dance?
- What are the defining characteristics of qualitative research?
- What are the strengths of qualitative research?
- What are the main types of qualitative research?
- How is the quality of qualitative research assessed?

Qualitative research provides gateways into experiences of dance, dancers, and creative and complex areas of human functioning that might not be captured by traditional research methods. This chapter describes approaches to research that facilitate exploration and description of phenomena such as thoughts, feelings, subjective experience, and artistic expression that are typically qualitative in nature. The various approaches to qualitative research provide rigorous ways to explore areas that have not been explored before or need in-depth examination in context.

Defining Characteristics of Qualitative Research

Qualitative approaches to research emphasize worldviews that value holistic perspectives and the complex intertwined aspects of human functioning. Various philosophical positions including post-positivism, critical realism, interpretivism, constructivism, feminist or critical theory, and postmodernism are identified with psychological and sociological research paradigms that might be categorized as qualitative.

This chapter focuses on constructivism, as it may be most relevant to address the range of research questions that could be devised in qualitative dance science inquiry. Constructivism is based upon the notion that there are multiple realities, each one unique to an individual and that person's particular background and social context (Guba & Lincoln, 1994). The goal of the qualitative researcher is to explore and understand these realities through the research process. Knowledge and meaning are created through the interaction between the researcher and the researched, acknowledging that researchers can never be truly independent from research participants and their context, nor from their own biases and values. Constructivists typically adopt qualitative research methods such as interviews, focus groups, observations, and diaries to situate themselves in the research context and to generate rich data. Researchers who use qualitative methods examine "things in their natural settings, attempting to make sense of, or interpret, phenomena in terms of the meanings people bring to them" (Denzin & Lincoln, 1994, p. 3). The methods of inquiry are particularly well suited to nascent areas of investigation and discovery and may yield new insights into a particular phenomenon. The insights may pave the way for larger hypothesis-driven research but do not need to do so. Regardless, qualitative studies result in rich and thick descriptions of the investigated phenomenon. They provide an avenue for understanding experiences of individuals from their own perspective.

Strengths and Weaknesses of Qualitative Research

Given its exploratory, interpretive nature, qualitative research is ideal for exploring complex or under-researched areas. Rich description, insight, and nuance are valued, and the qualitative research process can provide depth of understanding and meaning. Qualitative research can also be useful for theory testing, particularly in order to uncover the perspectives of "creative and divergent thinkers" (Guba & Lincoln, 1994, p. 106) whose views may not fit a predetermined theoretical perspective.

Qualitative researchers are interested in consensus more than generalization, meaning that collecting and interpreting the data honors diversity of opinion, respect for all participants and collaborators, and shared involvement in the process. It also means that a theme identified in the data may be important even if it is mentioned by only a few participants. When it comes to representing the views of marginalized populations (such as dancers with disabilities), this idea becomes even more important. Qualitative research affords participants opportunity to be heard and offers researchers a chance to explore rich and often complex experiences. Being context-specific, the results of qualitative research are not generalizable, although Guba and Lincoln (1994) argue that data collected in the field as opposed to a laboratory setting may provide greater face validity. Nevertheless, small sample sizes are typical in qualitative research, which also limits generalizability. In addition, qualitative research is somewhat messy, subject to the researcher's biases, beliefs, and values, and given to a lack of precision and dependability (although steps can be taken to reduce these weaknesses; see the section titled Assessing Research Quality near the end of this chapter).

Ethical Considerations in Qualitative Research

In addition to the key ethical requirements for any study outlined in *Chapter 6*, there are specific considerations in relation to qualitative research. It is critical for interview participants to understand that there are no right or wrong answers; that their responses will be treated confidentially; and that they can ask questions throughout; can choose not to answer a particular question if they feel uncomfortable; and can be honest about their experiences. Developing rapport with participants during interviews or focus groups is crucial in order to put them at ease and create feelings of trust, thereby increasing the likelihood of honest and rich responses (Patton, 2015). While the researcher must remain professional and relatively impartial, some degree of empathy and understanding helps to build the relationship. Minimizing possible power imbalances that may be inherent in society—such as academic and non-academic, adult and young person—is fundamental to building rapport, avoiding socially desirable responding, and encouraging in-depth answers to questions. In focus groups, potential power imbalances among participants should be minimized by keeping groups small (4–6 participants are ideal), and ensuring that all present have the opportunity to speak and respond as much or as little as they wish.

Clear and open-ended questions are best adopted to facilitate broad discussion, and the researcher must allow for silences that may represent a participant's focused reflection before answering or continuing to answer a question. Ideally, interviews or focus groups should take place at times that are convenient to participants, and in locations and environments with which they are familiar and comfortable: again, this is likely to enhance feelings of trust and mutual understanding.

Qualitative research may be ideal for researching sensitive topics (i.e., those that could cause distress to the participant, such as eating disorders or injury psychology), because understanding the individual experience is pivotal to the ontological and epistemological roots of qualitative inquiry. Interviews on sensitive topics may result in distress reactions from participants (e.g., crying); this is not necessarily a reason to cease the interview unless the participant requests it. Indeed, such interviews may even be somewhat cathartic for participants, who may otherwise have had little opportunity to discuss their experiences. The reflective process of interviews may enable participants to gain a sense of closure. Nevertheless, researchers investigating sensitive topics must be particularly mindful of factors such as power imbalances, establishing rapport, and giving participants time and space to reflect and respond. In certain instances it is good practice to provide participants with appropriate support service resources if their distress appears to be overwhelming. When conducting research on sensitive topics, there may also be an emotional impact on the researcher, such as feeling uneasy, preoccupation with the topic, or desensitization. Strategies and opportunities for peer debriefing and recording thoughts and feelings in a journal may help to safeguard researchers in these instances.

Finally, the researcher must assure participants of their anonymity during the analysis and write-up of a study. For instance, in an attempt to build rapport, the researcher might be tempted to use a participant's name, which will then feature on the interview transcript (which may have been transcribed by a third party). Where possible, any identifying information should be avoided. There are times when this is difficult: participants may mention particular companies, venues, or people in relation to their experiences, which may easily identify them among informed readers or audiences. However, there may be times when such information is crucial to the overall context of the study. Where this is the case, participants must give consent for identifying information to be published in the research report.

Kinds of Research Questions Answered by Qualitative Methodology

The kinds of research questions typically addressed by qualitative studies include questions that may not have been asked previously and may shed light on human experiences, cognitions, feelings, or behavior from a personal perspective. For example, the researcher may want to understand what it is like for a dancer with disabilities to experience movement that is creative, free-flowing, and expressive rather than functional. Investigating this experience with a qualitative approach to the research acknowledges the importance of personal experience that is complex, relational, multi-faceted, and intertwined with a specific context. Examples of the types of research questions associated with different qualitative methods are presented in Table 13.1.

Types of Research Methods

The term *qualitative methods* is an umbrella term for myriad approaches that examine phenomena and interpret the rich descriptive data collected. There is no precise definition for the research process that involves acquiring an understanding of phenomena that are qualitative in essence. The results are an enhanced understanding of a particular phenomenon and its distinct qualities. The most common approaches to qualitative inquiry include ethnography, grounded theory, phenomenology, narrative inquiry, and case study. Each has specific frameworks for conducting research with characteristic methods of sampling, data collection, analysis, and report writing (see Wertz et al., 2011). Seven common qualitative study approaches are described in this chapter and summarized in Table 13.1. In addition, because the rise in qualitative research and debate has generated interest in combining qualitative approaches to research with more quantitative-based studies, we briefly describe *Mixed Methods* at the end of this section.

Generic Qualitative Research and Thematic Analysis

The growth and appreciation of qualitative inquiry has seen the evolution of various ways of collecting and analyzing qualitative data that differ slightly from more established qualitative approaches described in this section. These generic approaches to qualitative research provide researchers the opportunity to develop a research strategy that might borrow from a particular form of qualitative inquiry (e.g., grounded theory) but not follow its specific structure. For example, in thematic analysis, also known as content analysis, themes and patterns in data are identified and described in detail with quotes from the research participants as exemplars (Braun & Clarke, 2013). Themes reflect patterns of meaning across participants that are described, but not abstracted into a data-driven theory, as in a true grounded theory (Glaser and Strauss, 1967). In this way, analyses may not go beyond the elucidation of themes or patterns in the data but still provide new understanding about a phenomenon.

Table 13.1. Overview of Qualitative Approaches to Research

Qualitative Approach	Illustrative Research Question	Sampling Technique	Sample	Type of Data	Type of Analysis Options	Type of Results	Defining Feature
Generic Qualitative Thematic Analysis	How do aerial dancers prepare psychologically for performance?	Purposive	6 professional aerial dancers	Narratives derived from focus group	Qualitative content or thematic analysis	A description of key themes or patterns arising from the data	Patterns and themes derived from the data are organized into logical hierarchies or relationships
Ethnography	What does dance mean to young break dancers?	Purposive	10 urban break dancers	• Narratives • Observational notes • Artifacts • Photos or videos	Interpretation of the meaning, context, and relationships represented in the data.	A detailed description of the experience from the insider's (e.g., dancer's) perspective	Detailed contextually driven description of personal perspective(s) by participant-observer (s)
Phenomenology	What is the lived experience of a professional ballet dancer?	Purposive	One or two professional dancers	May include narrative accounts or diary records, or photo voice[a]	Depends on the type of approach to phenomenology: empirical, existential, or interpretative	An in-depth personal description of specific dancer's or group's lived experience	A detailed personal account of the lived experience

(continued)

Table 13.1—*Continued*

Qualitative Approach	Illustrative Research Question	Sampling Technique	Sample	Type of Data	Type of Analysis Options	Type of Results	Defining Feature
Grounded Theory There are variations to the original GT by Glaser & Straus, 1967	Why do dancers refrain from visiting medical practitioners when injured?	Snowball	12–20 dancers with lower limb injuries	Narrative accounts of dancers' experience with injury management, taped and transcribed verbatim	Constant Comparative Method of examining data and reexamining it	A substantive theory of why dancers with lower limb injuries do not visit medical practitioners	Data-driven theory formulation
Narrative Inquiry	What are the experiences of dancers who suffer career-ending injuries?	Snowball	8 professional dancers with career-ending injuries	Narrative accounts audio- taped and transcribed verbatim	Identifying patterns in the data Thematic narrative analysis Performative narrative analysis Structural narrative analysis Dialogical narrative analysis	Description of dancers' thoughts, feelings, and experiences associated with injury	Personal stories, their meaning, and contexts
Case Study	What was the impact of a mindfulness intervention at a dance school?	Purposive	One dance school with an established wellness program	A specific instance • documents • policies • interviews	Depends on the approach	A detailed description of the intervention and its outcomes on student wellbeing	In-depth context-bound description that cannot be generalized

Action Research	What is the utility of asking parents of young dancers to refrain from observing training classes?	Purposive	One dance company which is modifying their policies and guidelines.	A specific community company. • policy • interviews	Thematic analysis that is presented to the participants and reanalyzed with the input of participants.	A detailed description of the attitudes, perspectives, and values of parents observing young dancers and recommendations for policy.	Community-based collaboration and research-informed changes to current practices
Mixed Methods	What are the experiences of older dancers taking part in a community dance project?						
1. Survey		1. Stratified Random Sampling	1. N= 50 (older dancers)	Survey of older dancers using valicated questionnaires	Descriptive statistics	Descriptive statistics	Combination of multiple forms of data to provide holistic picture
2. Interviews		2. Purposive	2. N = 8 (older dancers, carers, dance leaders)	Interview data from older dancers, carers, and dance leaders	Thematic Analysis	Narrative descriptions of experiences on the project from multiple perspectives	
					Triangulation		

Notes: This table provides a brief overview of design features for selected qualitative approaches to research. Read each row and imagine how a study using each approach might be conducted. Then scan the columns and notice how approaches differ. Which apprcaches seem appropriate to answer your research question?

a. Photo voice is a qualitative research tool whereby research participants take photographs to reflect their views on a topic.

Ethnography

Ethnography has its roots in cultural anthropology and sociology. It is an approach to research that involves in-depth study of aspects of a particular environment and the participants' culture, goals, motivation, experiences, history, ways of living, and relationships, for example. Typically, the researcher is involved in the community as a *participant-observer* over months or years, so that the nuances of the culture can be experienced, understood, and described fully. The researcher becomes immersed in the ways and experiences of the person or persons and uses multiple methods to document observations, which yield what Geertz (1973) described as a *thick description*. There is always a lens through which experiences are interpreted or integrated, but ethnographers attempt to suspend interpretation from their own perspective to relay an understanding of the phenomenon that is as close to the experience of the participants as possible.

There are various forms of ethnographical research, such as autoethnography, meta-ethnography, performance ethnography, and online ethnography or netnography. Autoethnography explores an individual's particular situation or experience and attempts to analyze the experience systematically. The experience is documented with respect to the traditions of autobiography and ethnography woven together to produce a product that embraces subjectivity, personal perspective, and emotional experience. Where autoethnography documents an experience, meta-ethnography reviews multiple qualitative studies with a set system of specific techniques (see Noblit & Hare, 1988). In performance ethnography, art-based work and narrative text are combined to create an artistic production. For example, Markula (2011) created a dance as a research presentation that portrayed contemporary dancers' injury experiences gleaned from interviews. With technological advancements and the influence of the world wide web and social media, netnography was developed as a form of ethnography to understand contemporary interactions within the digital context (Kozinets, 2010).

Narrative Inquiry

Narrative inquiry involves collecting personal stories through, for example, interviews, diaries, or photographs, and weaving them into a sequential chronicling of evidence that captures an account of the evidence (Dixon-Woods et al., 2005). In its original form, it is not about creating themes and categories of data but rather the story told by the participant(s) as it unfolds.

Various methods, such as one-on-one interviews or focus groups, provide the opportunity to collect personal narratives about experiences related to the target question. When the goal of the research is to understand the participants' perspectives, providing opportunities for them to describe their experiences and thoughts yields important information that may not have been available previously, nor accessible through more traditional research methods.

Grounded Theory

Grounded theory is a well-established approach with specific ways of analyzing data collected typically from interviews or focus groups (Glaser & Strauss, 1967). The emergent rich and thick data from personal interviews are analyzed with a particular structure to the analysis that includes moving the data up levels of abstraction into categories or themes, and then constantly comparing the emerging themes with one another, as well as other possible interpretations. The researcher keeps notes or analytical memos of the research process, and then in a true grounded theory fashion develops an explanatory framework or data-driven "theory"; that is to say, a theory is created from the data.

Phenomenology

Dance and phenomenology have a long-standing history with authors exploring dance, movement, and dancers' experiences (e.g., Albright, 2018). In a phenomenological study the researcher attempts to capture the first-person meaning of a particular phenomenon or "appearance." A phenomenological research approach explores conscious thought, personal meaning, or what is called the *lived experience.* Researchers may collect various forms of data, such as interviews and observations, to examine participants' lived experiences. Phenomenology involves a detailed description about the meaning of a particular experience to the participant(s). A phenomenological study maintains respect for the personal meaning or interpretation of the participant(s) and attempts to examine and report on that personal meaning to gain a deeper appreciation and understanding of the experience.

There are different kinds of phenomenological studies, each incorporating various degrees of interpretation by the researcher. All are conducted with rigor, critical perspective, and in a systematic fashion. One type of phenomenological research that has gained popularity recently, especially in sport science, is *interpretative phenomenological analysis.* This type of phenomenology involves the researcher interpreting the lived experiences of the participants through a lens that is colored by an appropriate (i.e., researcher-chosen with a sound rationale) theoretical framework identified in the relevant disciplinary literature.

Case Study

Case studies are interesting descriptive studies because they can be conducted under either a qualitative or a quantitative research paradigm (see *Chapter 15*). If the case study is conducted with reference to the methodological underpinnings associated with qualitative ontology and epistemology, it will have evolved from participants' perspectives rather than from a preconceived set of questions that shape the findings.

Case studies conducted within a qualitative paradigm in dance describe in-depth the person, group, or phenomenon. According to Creswell (2014), a qualitative case study is structured around the problem, context, and issues of the phenomenon and also includes lessons learned. Multiple data collection methods may be involved in case studies, such as interviews, policy and document analysis, production or evaluation of audio-visual materials, and diary analysis. Case studies provide comprehensive and rich descriptions of a phenomenon that may not be understood well. They are specific to the case that they examine and for that reason are not generalizable.

Action Research

Action research is a form of research aimed at bridging the gap between theory and practice and effecting change based on the needs of a particular community, by employing cycles of planning, action, observation, and reflection throughout the duration of a research project (Baum et al., 2006). It is a form of intervention research whereby the intervention is frequently reviewed, revised, and documented by those involved. Although the methodology should be chosen on the basis of its suitability to address the research question, qualitative methods such as interviews and focus groups may be particularly useful in action research to facilitate discussion among researchers and community members. Action researchers actively involve participants in the research process to ensure that knowledge created can be usefully applied. The exact nature of this involvement will depend on the research question of a particular study but can vary from consultant to co-researcher.

Mixed Methods

Mixed methods research involves the collection of both qualitative and quantitative data, which can offer researchers the opportunity to answer research questions fully, maximizing the strengths and minimizing the weaknesses of each type of research method through triangulation (Johnson &

Onwuegbuzie, 2004). Mixed methods research can be particularly helpful in under-researched subject areas to understand aspects of a phenomenon from different perspectives and to strengthen the confidence with which conclusions are drawn through the combination of breadth and depth of findings. In a mixed methods design, the researcher embraces the strengths of each paradigm of inquiry and uses both quantitative and qualitative methods to optimize the acquisition of new knowledge about a particular phenomenon.

Conducting Qualitative Research (Procedures)

As with any scientific research endeavor, the qualitative research process begins with a well-defined research question. A question informed by reading of the literature and/or careful observation of a situation starts the exciting pursuit of knowledge and understanding. Once the question has been identified, the methodological approach can be determined. The nature of the question, the type of data, and the researcher's philosophical position and worldview guide the determination. With questions that necessitate understanding the holistic, complex experience of an aspect of human functioning, the researcher will likely choose one of the many qualitative approaches to research. Once the approach has been chosen, the researcher decides who the participants will be and the type of sampling strategy to be used. The term *participant* is used in qualitative research rather than the term *subject* to honor the participatory, non-objectified, and key informant role of those being studied.

During the research process, investigators also engage in *reflexivity*—a process of self-awareness and reflection about their role, perspectives, experiences, and biases related to the research endeavor, interpretation of findings, and the co-construction of meaning between the participants and researcher. Engaging in reflexivity and making it explicit acknowledges how relationships, perspectives, and bias might influence the research process.

Choosing a Method

There are numerous methods associated with qualitative research. The more common methods include one-on-one in-depth interviews, focus groups, participant-observation, and researcher notes/memos. There are other creative methods such as using illustrations created by the participants; photovoice, which entails participants taking photographs of their experience in particular situations, or participant journaling or diary completion. The choice of qualitative method is determined by the research question, the researcher's philosophical position, the nature of the data needed to answer the question, and practical considerations (e.g., time, funding, expertise, participant recruitment).

Designing the Research

Embarking on a qualitative study involves engaging in the reflexive process by recognizing the insights and biases one brings to the research process and creating a plan for how the data are to be collected. Will it be a longitudinal study or a cross-sectional design (see *Chapter 10*), and what kind of qualitative data will the researcher collect? Deciding on the timeframe for sampling the data and the kind of data helps in planning the research design. However, a hallmark of qualitative research is the flexibility in the research design, which is usually labeled as *emergent.* That is, the researcher might begin sampling a particular group, but as the research progresses it becomes clear to the researcher that the sample needs to be modified or augmented in a particular direction. The emergent design concept in qualitative research provides opportunities to add, change, or modify the sampling choices to address the research question more fully, or perhaps take the research in a direction that is pertinent to the question but was not obvious until some initial interviews were completed. The researcher typically does not wait until all the data have been collected before beginning to analyze the rich volumes of data. This is because the collected data might inform the research direction, and it is also prudent to begin the time-consuming process of preparing the material collected as data into a format for analysis.

Appropriately designing the questions and prompts to use in gathering qualitative data with participants is critical. The questions can be developed into an interview guide to be used in semi-structured interviews or focus groups. Typically, the questions are developed so that they do not lead participants in a particular direction. Questions should be open-ended, broad, and neutral (e.g., "Can you describe . . ."), and follow-up or probe questions should be more specific to the individual participant in order to glean more detail guided by the researcher's personal interests (e.g., "Could you expand on your last point?")

Sampling

Research questions in qualitative approaches typically target specific situations and people; therefore, the sampling procedures are focused and *purposive.* For example, in a study investigating how modern dance teachers in universities use student-centered learning strategies, the researcher would purposefully seek out teachers who teach modern classes at university-based dance programs. Large numbers of participants are not required. Depending on the approach chosen, the sample size can range from one or two persons to 20 or 30 participants. Each approach and question along with guidance from the literature will inform the choice of sample size. Other approaches to sampling the population to study include *snowball sampling* and *convenience sampling* (see *Chapters 8* & *12*).

Analyzing Data

Large volumes of qualitative data are usually produced from any one study, and thus, preparing and organizing the data for analysis is an important first step. Despite small sample sizes, volumes of data are produced and can be overwhelming to the novitiate investigator.

Rather than a large spreadsheet of variables, data in a qualitative study are words or images found within a particular context. The narrative description of experience may arise from a two-hour audio interview tape, a 50-page portfolio of sketches, or multiple diary entries, for example. As an illustration, preparing an audio recording of an interview typically involves listening to the tape and transcribing it verbatim so that a word processing document is available to read, review, and interpret. A portfolio of sketches may involve scanning them to a digital format and labeling the sketches individually according to a relevant system. Regardless, the data need to be organized for the next step, which usually involves finding meaning and patterns in the collected works. In the case of transcripts, they are often parsed into the smallest meaningful text segment, or unit, which is coded to identify each unit as a meaningful whole identified with a particular participant, time in the interview, and transcript location. Subsequently, the researcher begins the process of examining meaning in the data. The process is iterative, takes time, and uses researcher brain power rather than computing power. However, there are qualitative programs that can help organize the data (e.g., NVivo: http://www.qsrinternational.com/products_nvivo.aspx; The Ethnograph: http://www.qualisresearch.com/).

Each qualitative approach has its own unique way of analyzing data. Table 13.1 depicts the type of analysis typically associated with the selected common approaches to qualitative research described in this chapter.

Assessing Research Quality

Researchers should always endeavor to demonstrate the quality of their research. In qualitative research, various terms such as authenticity, dependability, and validity may be used, but all essentially mean that research must be a credible and accurate representation of participants' realities (Creswell & Miller, 2000). The umbrella term *trustworthiness* is often used, which incorporates:

- **credibility**: confidence in the interpretation of data
- **transferability**: the extent to which findings can be generalized to other contexts
- **dependability**: the stability of findings over time through clarity of research question, methods, and analysis
- **confirmability**: the degree to which findings could be confirmed by other researchers performing the same analysis.

Researchers can use several strategies to enhance the trustworthiness of their work. The particular strategies used will depend on the nature of the research question as well as its ontological and epistemological context (e.g., an ethnographer will rely on extended engagement in the field as one criterion for trustworthiness), but in general researchers should employ multiple strategies for any given project.

Initial steps include assurances of participants' rights, to encourage honest and rich accounts of their lived experiences. Sampling should cease only once *saturation* is reached, whereby further data collection serves only to confirm findings, rather than generate new understanding (Patton, 2015). It can be helpful to keep an *audit trail* of journals, notes, and memos that inform the analysis process, as well as disclosing the researcher's background, values, and biases, as these can influence the outcomes. This in and of itself is not necessarily problematic: a shared contextual understanding between the researcher and participants can aid researchers in empathizing with participants, using appropriate prompts and terminology during interviews, and interpreting the data from an informed perspective.

Following these initial steps, some or all of the following approaches can be adopted to provide further evidence of trustworthiness:

- **Triangulation:** combining multiple data sources (including qualitative and quantitative) to establish commonalities and confirm findings from various perspectives.
- **Inter-rater reliability:** two or more researchers working on the same analysis independently code some or all of the data, then discuss and resolve differences in interpretations.
- **Peer debriefing**: the data and interpretations are reviewed by experts in either the research process, or the topic under inquiry, or

both. The goal is not necessarily to reach a consensus but to reflect on interpretations and develop a more informed understanding.

- **Member checking or member reflection:** member checking can control for researcher bias by returning the data and/or interpretations to the participants for confirmation. An alternative is member reflection, in which researcher and participant engage in a cooperative dialogue about the findings to provide further insight (Braun & Clarke, 2013).
- **Thick, rich description**: when writing up the report, the setting, participants, and findings are described in rich detail to give readers insight into the research context, and to indicate the level of generalizability of the results to other contexts. Participant quotes and field notes should be included in the write-up to illustrate particular themes and demonstrate to readers that data interpretations were appropriate. Further details on writing up a qualitative research report can be found in *Chapter 20*.

There are no strict rules when it comes to writing up qualitative research, so it can be useful to read different qualitative journal articles, paying attention specifically to how the results are reported. Guidelines for qualitative research reports, including mixed methods (e.g., Levitt et al., 2018) may help researchers identify what needs to be included in a qualitative research write-up (see also *Chapter 20*).

Summary

In this chapter we present an overview of the key features of the qualitative research process: procedure, design, choosing a method, sampling, data analysis, evaluating research quality, and writing the research report. We also describe common qualitative approaches and provide examples of research questions. In conclusion, qualitative methodology emphasizes creative and thoughtful ways of acquiring understanding of dance, dancers, and their subjective experiences in context, while maintaining methodological integrity and rigor in the process.

Highlights

- Qualitative methodologies are well suited to exploring participants' experiences and new areas of research.
- Meaning is co-created by the researcher and the participants.

- Qualitative research considers researcher biases, values, and philosophical frameworks.
- There are several approaches to conducting qualitative research and various strategies for quality assurance.

References

Albright, A. (2018). *How to Land: Finding Ground in an Unstable World.* Oxford University Press.

Baum, F., MacDougall, C., & Smith, D. (2006). Participatory action research. *Journal of Epidemiology and Community Health, 60*(10), 854–857.

Braun, V., & Clarke, V. (2013). *Successful Qualitative Research: A Practical Guide for Beginners.* Sage.

Creswell, J. W., & Miller, D. L. (2000). Determining validity in qualitative inquiry. *Theory into Practice, 39*(3), 124–130.

Creswell, J. W. (2014). *Research Design: Qualitative, Quantitative, and Mixed Methods Approaches* (4th ed.). Sage.

Denzin, N. K., & Lincoln, Y. S. (1994). *Handbook of Qualitative Research* (2nd ed). Sage.

Dixon-Woods, M., Agarwal, S., Jones, D., Young, B., & Sutton, A. (2005). Synthesizing qualitative and quantitative evidence: A review of possible methods. *Journal of Health Services Research and Policy, 10*(1), 45–53.

Geertz, C. (1973). *The Interpretation of Cultures.* Basic Books.

Glaser, B. G., & Strauss, A. L. (1967). *The Discovery of Grounded Theory.* Aldine.

Guba, E. G., & Lincoln, Y. S. (1994). Competing paradigms in qualitative research. In N. K. Denzin & Y. S. Lincoln (eds.), *Handbook of Qualitative Research* (pp. 105–117). Sage.

Johnson, R. B., & Onwuegbuzie, A. J. (2004) Mixed methods research: A research paradigm whose time has come. *Educational Researcher, 33*(7), 14–26.

Kozinets, Robert V. (2010). *Netnography: Doing Ethnographic Research Online.* Sage.

Levitt, H. M., Bamberg, M., Creswell, J. W., Frost, D. M., Josselson, R., & Suárez-Orozco, C. (2018). Journal article reporting standards for qualitative primary, qualitative meta-analytic, and mixed methods research in psychology: The APA Publications and Communications Board task force report. *American Psychologist 73*(1), 26–46.

Markula, P. (2011). Dancing the 'data': (Im)Mobile bodies. *International Review of Qualitative Research, 4*(1), 35–50.

Noblit, G., & Hare, R. (1988). *Meta-Ethnography: Synthesizing Qualitative Studies.* Sage.

Patton, M. Q. (2015). *Qualitative Evaluation and Research Methods* (4th ed.). Sage.

Wertz, F. J., Charmaz, K., McMullen, L. M., Josselson, R., Anderson, R., & McSpadden, E. (2011). *Five Ways of Doing Qualitative Analysis* (1st ed.). Guilford Press.

14

Correlation and Regression Analyses

AMIT ABRAHAM AND BOYI DAI

Questions to Stimulate Thinking

- How can you use correlations and regressions in dance science research?
- What are the meanings of, and the differences between, correlation and regression analyses?
- How do regression analyses add to correlational research?
- What does it mean if there is a positive, negative, or no relationship between two variables? How does this translate into dance settings?

This chapter provides an overview of correlation and regression analyses for dance science. Understanding some basic conceptions, tools, and applications of these methods allows readers to interpret dance research and findings better and to appraise them critically. Specifically, this knowledge will assist dance researchers in designing and conducting studies that answer questions. For example, are dancers' height and weight related with dance-related balance measurements? Learning scientific tools to answer such questions will assist dancers, dance educators, and researchers in evidence-based decision-making processes.

What Is Correlation?

Correlation is a term describing the type of association between two phenomena or variables. A correlation analysis illustrates a pattern that links scores on two different parameters, such as between the type of dance training (variable A) and the number of sustained injuries (variable B). While the most common correlation in dance research is between two variables, it is also possible to examine correlations between two sets of variables (with more than one variable in each set).

Types of Correlations

Linear Correlation

A positive correlation (Figure 14.1a) means that the change (either an increase or a decrease) in one parameter is associated with the same type of change (an increase or a decrease, accordingly) in the other parameter. Negative correlation (Figure 14.1b) means that the change in one parameter is associated with the opposite type of change in the other parameter. Positive and negative correlations are both subtypes of linear correlation, in which the data points are scattered around and along a straight line.

Visual representation of the data (scatterplot) is often helpful in highlighting whether correlations between two variables exist, and if so, visually depict the pattern. Further, using scatterplots can help the researcher to detect outliers. Examples are provided in Figure 14a–d. Graph a presents the positive correlation between the dancing experience (in years) and the number of injuries. Graph b presents the negative correlation between the dancers' age (in years) and their parents' involvement (on a scale of 0–10), and graph c presents the lack of correlation (i.e., no clear pattern of association) between the dancers' weekly training routine (in hours) and their mental imagery ability (on a scale of 1–7). In other words, high values of weekly training

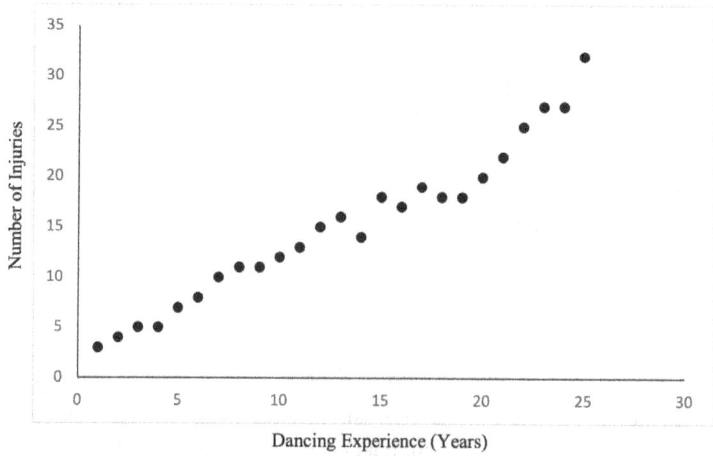

Figure 14.1a. Strong Positive Correlation. Hypothetical data.

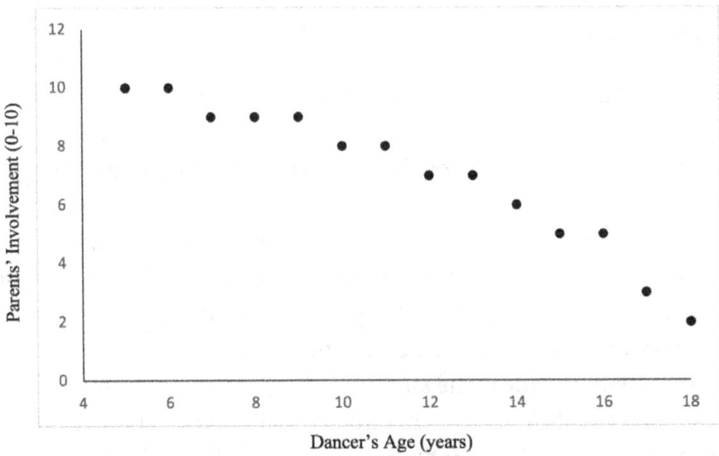

Figure 14.1b. Strong Negative Correlation. Hypothetical data.

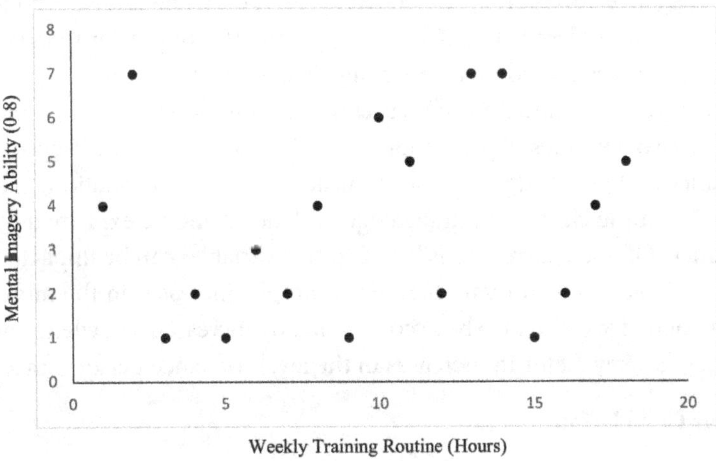

Figure 14.1c. No Correlation. Hypothetical data.

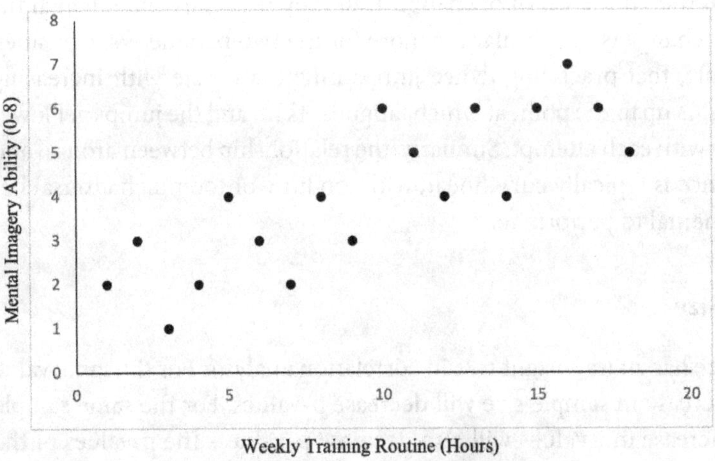

Figure 14.1d. Moderate Positive Correlation. Hypothetical data.

routines are correlated with either high or low values of mental imagery ability.

In a linear correlation, the closer the data points are to a single straight line (Figures 14.1a and 14.1b), the stronger (or higher) is the correlation. Weak (or low) correlation (Figure 14.1d) therefore means that the data points are scattered in a way that does not clearly form a straight line. Moderate (or medium) correlation means that the data points are scattered in such a way that it is not coherent and suggests a correlation that is anywhere between strong and weak (Figure 14.1c). The strength of a correlation can be quantified by Correlation Coefficients (r), a parameter that describes how close the data points are to a perfectly straight line (central tendency line). Correlation takes values between +1 and -1, with the signs "+" or "-" representing positive or negative correlation, respectively. Based on Cohen's suggestions (Cohen, 1988), an r of less than 0.3, between 0.3 and 0.5, or greater than 0.5 are considered as weak, moderate, or strong effect sizes, respectively.

Coefficients of determination (r^2) represent the common variance shared between the two variables. For example, fictional data of r of 0.5 between years of dance and knowledge of dance indicates a positive correlation, and 25% ($r^2 = .25$) of the changes in knowledge of dance could be explained by years of dance. Of note, correlations between two variables can be impacted by a third variable (mediator variable). For example, increases in the number of years of dance could also be accompanied by increased knowledge of dance, which is a key factor in increases in the levels of dance performance.

Curvilinear Correlation

Curvilinear correlation means that the two parameters are following a different pattern than a straight line on the scatterplot. Typically, the variables move with the same pattern of change until they reach a point at which the pattern of change is not similar anymore for the two parameters. Consider, for example, that practicing dance jumps might correlate with increasing jump heights up to the point at which fatigue sets in, and the jumps get lower and lower with each attempt. Similarly, the relationship between arousal and performance is typically curvilinear, with too little or too much arousal being detrimental to performance.

Sample Size

Sample size has an important role in correlation analysis. For the same value of r, an increase in sample size will decrease p-values. For the same sample size, an increase in r values will also decrease p-values. The p-values of the

Type I error rate (see *Chapter 9*) describe whether the correlation is significantly different from 0 (i.e., no correlation between the two parameters). With a large sample size, the strength of the correlation—rather than the p-values—may be more important, and statistically significant but weak correlations may have limited practical value.

Power

Power is the likelihood of detecting a significant difference among conditions/populations when an effect really exists. An a priori (beforehand) power analysis is commonly performed with an estimated *r* to calculate the sample size needed to achieve certain levels of Type I error rates and power.

Intraclass Correlation Coefficients

Correlations are also used to quantify the reliability and validity of new measurement methods or instruments. For example, if two testing sessions (e.g., pre- and post-measurements) or two testing methods (e.g., goniometer versus inclinometer) are highly correlated, indications are that they are more likely to measure the same thing. In the case of more than two assessment sessions or methods, Intraclass Correlation Coefficients (ICC) may be computed. ICC uses unitless values ranging from 0 to 1, with "0" indicating no reliability and "1" indicating perfect reliability. For example, a strong correlation with small errors supports high test-retest reliability. It may also suggest good agreement between novel and existing instruments or between experienced and novice raters.

Canonical Correlation

Researchers sometimes study correlations between two sets of variables. For example, correlations between balance and strength, with both factors being measured with a set of variables (tests). To identify whether strength and balance are correlated, and if so, to what extent, canonical correlation analysis can be used between the two sets of parameters (or tests). Canonical correlations explore the overall relationship between two sets of variables and then identify the most relevant correlation between specific pairs of variables.

Other Considerations

Conducting correlational analyses requires attention to several aspects of the research study, such as the number of participants that the number of correlation tests needed. For Pearson correlation, required sample sizes can be calculated based on estimated correlation coefficients, power, and Type I error rate. The number of correlation tests needs to be pre-planned and controlled. Although a Type I error rate of 0.05 is commonly used for each test, as described in earlier chapters, the Type I error rate will increase with an increase in the number of tests.

Other relevant research considerations for correlational studies include controlling for confounding variables. For example, when studying correlations between dance activities and cardiovascular health in a group of recreational dancers, it is important to include as covariates relevant factors such as other forms of physical activities, age, gender, and history of injury. Because Pearson correlation assumes a linear correlation between two variables, other forms of correlation (e.g., polynomial) should be used when the relationship is not linear. Further, Pearson correlation requires normal distribution and homoscedasticity (i.e., data points are at a similar distance at the correlation line) and is sensitive to outliers. When these assumptions cannot be made, alternative tests should be used, such as the Spearman's rank-order correlation (which analyzes the correlation of the rankings of two variables).

Use of Correlation Analyses in Dance Science Research

One of the main applications of correlation analyses in dance research is to identify relationships among dancers' variables to inform future intervention studies and shape pedagogical and training approaches. For example, Ambegaonkar et al. (2018) examined the correlations between lower extremity horizontal work, vertical power, and balance performance in 61 female collegiate dancers. The researchers found statistically significant strong correlations ($r = .59$, $p < .001$) between horizontal hopping performance and balance performance, and statistically significant moderate correlations ($r = .31$, $p < .02$) between vertical jump performance and balance performance. Similarly, Kotler et al. (2017) investigated relationships between dancers' knowledge of anatomy and multiple variables related to dance experience. The authors reported that factors such as age, completing an anatomy class, years of dance training, and status of being professional dancers correlated with increased knowledge of anatomy.

It is important to mention that weak correlations are also of value. Dai et al. (2019) quantified the correlation among strength and balance performance and asymmetries in 499 collegiate athletes. The researchers found significant ($p < .01$) but weak ($r < 0.3$) correlations between strength and balance performance as well as asymmetries. As the findings do not support a potential cause-and-effect relationship, the researchers concluded that strength and balance were mostly independent.

Summary of Correlation Analysis

Correlation analyses can be applied to data collected cross-sectionally, retrospectively, and prospectively. Correlation studies are typically easier to complete and are commonly performed in the early stages of a specific research area when researchers are interested in gathering preliminary or baseline data. Researchers can use this information to propose additional larger-scale studies to support interventions (e.g., strength training) or other studies (e.g., neuroimaging). Likewise, negative findings—that is, no correlations found among variables—may save researchers from spending unnecessary time and effort on proposing and conducting additional studies among variables that may not be related to each other.

Causality and Interpretations of Causality

It is important to note that just finding a relationship (correlation) between two variables does not tell us anything about the source (or reason) for the correlation. In other words, the correlation analyses do not tell us whether one factor affected the other one. Additional investigations are needed to examine any possible cause-and-effect relationship. Correlation analyses are a first step toward exploring causal (i.e., cause-and-effect) relationships between variables. This causality (or causation) is when one variable impacts another variable and causes a change in the latter. Understanding the differences between correlation and causation is essential when interpreting research findings in correlational and causation research.

For example, Abraham et al. (2018) found statistically significant positive correlations ($r = .68$, $p < .01$) in ankle plantarflexion range-of-motion between static and repeated elevé movements. However, the researchers could not determine whether a change in either task was causing the change in the other. In another (fictional) example, a statistically significant negative correlation between balance ability and severity of current knee injury (both on a scale of 1–10) in dancers would not enable the researcher to conclude

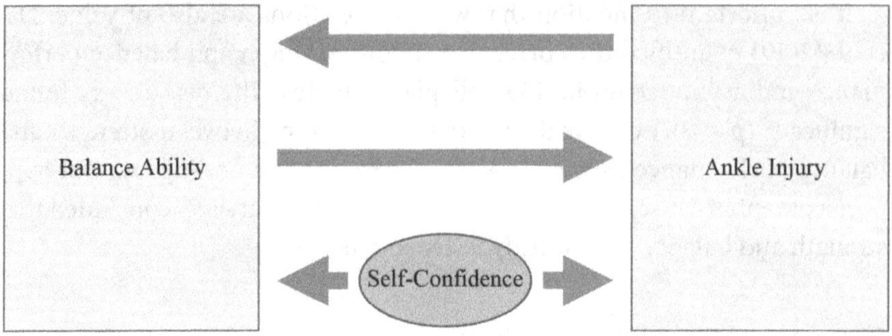

Figure 14.2. Three Possible Directions of Causality. For balance ability and ankle injuries in dancers.

which, if any, of these parameters (either injury severity or balance ability) was causing the change in the other. It might be that the knee injury was causing reduced balance ability, that reduced balance ability was causing the knee injury, or that a third factor, such as low self-confidence, was leading to both knee injuries and low balance abilities. These three potential paths are illustrated in Figure 14.2.

Identifying cause and effect (or *direction of causality*) has scientific and practical importance. In the preceding example, if the researchers find reduced balance causes knee injuries in dancers, they can then design exercise programs to improve dancers' balance to try and reduce dancers' injury risk. In some cases the causal direction may change over time. For example, reduced balance might initially cause knee injuries, but over time sustained knee injuries could themselves, in return, cause further balance deficits. One way to support testable hypotheses about causality is to gather prospective data about the sequence of events and their suggested logic (causality) over time and while using multiple data collection timepoints. However, these research designs are often long-term and resource intensive in terms of time and logistics.

Making Statistical Predictions

If the researchers obtain consistent evidence of causality, then they can start to make future predictions about these variables. How well causality can be predicted can be tested by a statistical method called regression. Regression measures to what extent changes in one variable (bivariate regression) or multiple variables (multiple regression) predict the change in the other variable, the outcome.

The first step in making linear predictions is to decide which variable is being predicted to be the result, also known as the dependent or criterion variable (often labeled "Y") of the other, also known as the explanatory or predictor variable (often labeled "X"). In other words, the researcher needs to use a theory or logic to explain the direction of the causality: which parameter is the cause, and which is the effect. For example, a professional dance company hiring a dancer wants to know which of the many possible factors (i.e., predictor variable) among ones involving the applicant's physiology, psychology, experience, or training, would predict good performance (i.e., criterion variable). Regression analysis should use data that come from the same sample (of dance applicants, in this example).

Linear Prediction Rule

The second step in making predictions is to calculate the linear prediction rule (or linear prediction model), a formula (equation) that describes the relationship between the predictor and criterion variables and thus enables prediction of the criterion variable's values based on those of the predictor one. Using such a linear prediction model allows the researcher to generalize the findings and calculate criterion variable values beyond the studied sample without the need for such data being known. The use of the word *linear* here means that a low score on the predictor variable predicts a low score on the criterion variable, and vice versa. A linear prediction rule is illustrated by the following equation:

$$Y = a + (b)(X)$$

where *a* is a fixed number, termed the *regression constant, b* is termed the *regression coefficient* (or unstandardized regression coefficient; see later discussion), and is the number by which the score on the predictor's variable (X) is multiplied. Y is the sum of the equation and represents the predicted score on the criterion variable. Consider a fictional example in which the linear prediction rule for predicting a dancer's ankle plantarflexion range-of-motion (in degrees) based on dancing experience (in years) is the following:

$$Y = 18.6 + (2.4)(X)$$

The predicted range-of-motion for a dancer with 5 years of dancing experience would therefore be: $18.6 + (2.4)(5) = 18.6 + 12 = 30.6$ degrees

The Regression Line

The linear prediction rule of $Y = a + (b)(X)$ can be displayed graphically as a regression line, with the scores (values) of the predictor along on the

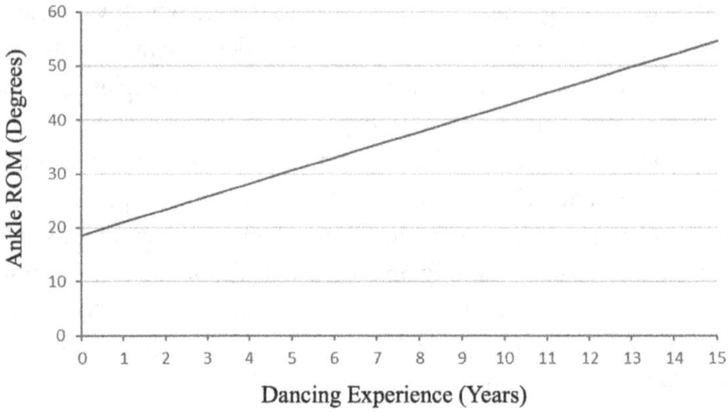

Figure 14.3. Regression Line Example. For dance experience and ankle range of motion (ROM), hypothetical data.

horizontal (X) axis and the predicted scores (or values) of the criterion variable up the vertical (Y) axis. The regression line allows for a visual understanding of the relationship pattern between the two variables. For example, the regression line of Y = 18.6 + (2.4)(*X*) from the preceding example is presented in Figure 14.3.

In the figure the regression line crosses the vertical axis at a point with a Y value of 18.6. This point represents the predicted range-of-motion Y of 18.6 degrees for a dancer with no dancing experience (X = 0). This point is termed the intercept and is equal to the regression constant (*a*). The intercept can be calculated mathematically using the linear prediction rule by using a value of X = 0.

Y = 18.6 + (2.4)(*X*) = 18.6 + (2.4)(0) = 18.6

The steepness (*slope*) of the regression line represents the relationship between the changes in scores of the predictor and criterion variables. In other words, it describes the amount of change (increase or decrease) in the Y value following a one-unit change in the X value. The regression slope is equal to the regression coefficient (or *b*). Therefore, the larger the *b* is, the steeper is the regression line. In our example, the value of *b* is 2.4, which means that for each additional year of dance training, there is an increase of 2.4 degrees in ankle dorsiflexion range-of-motion.

Drawing a regression line based on raw data requires labeling the horizontal and vertical axes (in terms of predictor and criterion variables), drawing a minimum of two points, and drawing a line that connects them. For accuracy, we recommended that researchers choose two points that are distinct from each other (e.g., one point with a low value of the predictor variable and another one with a high value). In a large dataset with many data points scattered widely, a single straight line connecting all points may not be possible, and multiple regression lines can be drawn theoretically. However, the researcher's task is to identify the linear prediction rule that indicates the best route among all data points. This best prediction rule can be thought of as the regression line that comes closest to (and even passes through) the most data points. Such a prediction rule will predict the values of the criterion variable with the greatest possible precision and minimal error from the predicted scores' deviation from the actual (i.e., true) scores.

The difference between the criterion variable score as predicted by the linear regression rule and the actual (i.e., measured) score is termed error and is the difference of Y_{actual}-$Y_{predicted}$. An error can be either positive (when the predicted value is higher than the actual one) or negative (when the predicted value is lower than the actual one). Such a prediction rule has the smallest sum of errors (calculated by adding together all errors for all X values) and thus is most warranted for researchers. To avoid canceling each other out when summing errors of both types (i.e., positive and negative), squared errors are used (by multiplying each value by its own value). By figuring out the sum of all squared errors (sum of squared errors), the level of precision of the prediction rule can be calculated and compared with other linear prediction rules. This process of finding the prediction rule that produces the smallest sum of squared errors, known as the least squares criterion, allows the researcher to decide which prediction rule to choose. An efficient alternative to the process of figuring out the linear prediction rule with the smallest squared errors is with mathematical formulas that calculate the values of the regression constant (*a*) and regression coefficient (*b*) for the linear prediction rule with the smallest possible sum of squared errors (when compared to other possible prediction rules). The regression coefficient (*b*) can be calculated using the following formula:

$$b = \frac{\Sigma[(X-M_x)(Y-M_y)]}{\Sigma[(X-M_x)2]}$$

In this formula X is each individual's score (or value) on the predictor variable, M_x is the mean score of the studied sample on the predictor variable, and X-M_x is the individual's deviation score on the predictor variable. Accordingly, Y is each individual's score (or value) on the criterion variable, M_y is the mean score of the sample on the criterion variable, and Y-M_y is each individual's deviation score on the criterion variable. This formula, therefore, sums the products of the deviation scores of all individuals in the sample of the study.

After calculating b, the regression constant (a) can be calculated using the following formula:

$a = M_y\text{-}(b)(M_x)$

In this formula M_y is the mean of all scores of the studied sample on the criterion variable, and M_x is the mean of all scores of the studied sample on the predictor variable. Of note, b needs to be calculated first and then used for calculating a. Additional formulas for calculating a and b can be found in the statistics literature.

Standardized Regression Coefficients

It is not unusual in dance research to use different outcome measures (or scales) for measuring the same variable, either within the same study or in multiple ones. For example, Abraham and colleagues (2019) used two different questionnaires (with different rating scales) to measure dancers' mental imagery ability. Using multiple measures to assess the same phenomenon makes it challenging to compare results across different linear prediction rules and regression coefficients. One way to address this challenge is by converting regression coefficients into the standardized regression coefficient (β). Like regression coefficients, standardized regression coefficients reflect the predicted change in standard deviations (SD) in the criterion variable with every change of one SD in the predictor variable. When using β, SD units for both the predictor and criterion variables are used instead of the original units or scales used for b. As such, β of different measures (and different studies) allows the researcher to answer questions that compare the different measures through creating a unified scale. For bivariate regression, β has the same value as the correlation coefficient (*r*) discussed earlier. Formula 14.2 shows how to calculate β, the standardized regression coefficient:

$$\beta = (b)\frac{\sqrt{\Sigma[(X-M_x)^2]}}{\sqrt{\Sigma[(X-M_y)^2]}}$$

To date there are no clear recommendations in the dance science literature regarding whether to report unstandardized or standardized regression coefficients. Given the wide heterogeneity in dance-specific operational definitions, outcome measures, and measurement scales, reporting both in research papers is recommended. Whether regression coefficients (both unstandardized and standardized) are statistically significant can be determined using a t-test. A statistically significant regression coefficient means that the value of the predictor variable is useful for predicting the criterion variable.

Multiple Regression

Bivariate regression is insufficient for studying phenomena that are affected by—and thus predicted by—multiple factors. For example, Abraham and colleagues (2018) found that dancers' age and height both predicted ankle plantarflexion maximal range-of-motion of a static elevé task. The process of predicting one variable (i.e., criterion variable) based on multiple variables (i.e., predictor variables), known as multiple regression, can provide a more accurate (and thus realistic) prediction. The multiple regression formula demonstrates the relative effect (i.e., b_1, b_2, b_3, etc.) of each of the predictor variables (i.e., X_1, X_2, X_3, etc.) on the criterion variable. The number of variables included in the equation depends on the number of identified predictor variables and should be a minimum of two. This formula is as follows:

$$Y = a + (b_1)(X_1) + (b_2)(X_2) + (b_3)(X_3) + \ldots$$

Like in bivariate regression, Y represents the predicted value on the criterion variable, a represents the regression constant, and b_1, b_2, b_3 (etc.) represent the regression coefficients for each of the predictor variables. X_1, X_2, X_3 represent the individual's scores (or values) on the predictor variables. The following fictional example of a multiple regression linear prediction rule is for predicting risk for injuries in dance:

Predicted risk = 13.2 + (0.86)(age) + (0.34)(years of dancing) + (0.55)(past injuries)

According to the formula, for a 20-year-old dancer with 13 years of dancing experience and 12 past injuries, the predicted number of injuries throughout their career would be the following:

$$13.2 + 0.86*20 + 0.34*13 + 0.55*12 = 13.2 + 17.2 + 4.42 + 6.6 = 41.42.$$

Standardized coefficients (β) of each of the predictor variables can be calculated as well. The correlation between all predictor variables together and the criterion variable, known as multiple correlation coefficient (R), can also be calculated. The value of R is not equal to—and is usually smaller than—the sum of all *r*s (e.g., $r_1 + r_2 + r_3$) due to some overlap existing between the predictor variables. Once multiplied, the resultant R^2 reflects the amount of variance in the criterion variable that is accounted for by all predictor variables. For example, R^2 of 0.23 means that the predictor variables together explain 23% of the variation in the criterion variable's scores. The statistical significance of multiple regression can also be tested using various procedures.

Limitations of Correlation and Regression Analyses

While correlation and regression analyses are useful, researchers should remember that findings from correlation studies are specific to the studied population and may not be extrapolated to other populations. For regression analysis to represent reality accurately, it requires a few assumptions: (1) the variables of interest are of equal-interval type, (2) the variables are equally distributed, (3) the correlation between variables is linear, (4) the studied sample consists of independent individuals, (5) the studied sample represents the whole available range (e.g., dancing experience, ages, etc.), and (6) no outliers exist. Violation of these assumptions will negatively impact the prediction analysis and potentially make it less accurate. Any failure to meet these assumptions needs to be reported in research reports.

Summary

In this chapter we discuss basic concepts of correlation and regression analyses within the context of dance practice and research. As correlation and regression analyses do not necessarily involve interventions, they can be performed with relatively less time and resources and can serve as a good starting point for researchers to explore novel ideas. Correlation and regression analyses can serve as the first step toward large-scale intervention studies and validating new instruments and tests. When planning correlation and regression studies, researchers need to consider the potential confounding factors of sample sizes and inflated Type I error rates. Additionally, screening for data normality, curvilinearity, and outliers is also warranted.

Highlights

- Correlation can be assessed among individual variables, different sets of variables, and a group of variables.
- Correlation studies serve as a first step to quantify cause and effect relationships and for assessing instrument reliability and validity.
- Correlation does not imply causation.
- Regression describes the predicted way in which one or more predictor variables affect the criterion variable.

References

Abraham, A., Dunsky, A., Hackney, M. E., & Dickstein, R. (2018). Kinematic and kinetic analysis of repeated and static elevé in adolescent female dance students. *Journal of Dance Medicine and Science, 22*(1), 33–43.

Abraham, A., Gose, R., Schindler, R., Nelson, B. H., & Hackney, M. E. (2019). Dynamic neuro-cognitive imagery (dnitm) improves développé performance, kinematics, and mental imagery ability in university-level dance students. *Frontiers in Psychology, 10,* 382.

Ambegaonkar, J. P., Caswell S. V., & Cortes, N. (2018). Lower extremity horizontal work, but not vertical power, predicts balance performance in female collegiate dancers. *Journal of Dance Medicine and Science, 22*(2), 75–80.

Bronner, S., & Rakov, S. (2014). An accelerated step test to assess dancer pre-season aerobic fitness. *Journal of Dance Medicine and Science, 18*(1), 12–21.

Cohen, J. (1988). *Statistical Power Analysis for the Behavioural Sciences* (2nd ed.). Lawrence Erlbaum Associates.

Dai, B., Layer, J., Vertz, C., Hinshaw, T., Cook, R., Li, Y., & Sha, Z. (2019). Baseline assessments of strength and balance performance and bilateral asymmetries in collegiate athletes. *Journal of Strength and Conditioning Research, 33*(11), 3015–3029.

Keener, M. M., Critchley, M. L., Layer, J. S., Johnson, E. C., Barrett, S. F., & Dai, B. (2020). The effect of stirrup length on impact attenuation and its association with muscle strength. *Journal of Strength and Conditioning Research, 35*(11), 3056–3062.

Kotler, D. H., Lynch, M., Cushman, D., Hu, J., & Garner, J. (2017). Dancers' perceived and actual knowledge of anatomy. *Journal of Dance Medicine and Science, 21*(2), 76–81.

Pedhazur, E. J. (1973). *Multiple Regression in Behavioral Research: Explanation and Prediction* (3rd ed.). Harcourt Brace College Publishers.

Walker, I. J., Nordin-Bates, S. M., & Redding, E. (2010). Talent identification and development in dance: A review of the literature. *Research in Dance Education, 11*(3), 167–191.

Additional Resources

Aron, A., Coups, E. J., & Aron, E. N. (2006). *Statistics for Psychology* (6th ed.). Pearson Education.

Cohen, J., Cohen, P., West, S. G., & Aiken, L. S. (2003). *Applied Multiple Regression/Correlation Analysis for the Behavioral Sciences* (3rd ed). Lawrence Erlbaum Associates.

Montgomery, D. C., Peck, E. A., & Vining, G. G. (2012). *Introduction to Linear Regression Analysis* (5th ed). John Wiley & Sons.

Howell, D. C. (2013). *Statistical Methods for Psychology* (8th ed.). Wadsworth Cengage Learning.

15

Case Studies, Case Reports, and Case Series

WENDY C. COATES AND SELINA SHAH

Questions to Stimulate Thinking

- When can a reportable case be useful in dance science?
- How do a single case report, a case series, and a case-based study build on each other?
- What information do I need to record when writing a case-based report?

In this chapter we introduce three related case-based methodologies that may be used to report new discoveries or observations and can serve as a starting point to generate novel research questions across a wide spectrum of methodologies. Case-based reports are particularly appealing for novice researchers, since the subjects studied are familiar and available and the scope is focused and manageable. Writing a case report or a case series can serve as an introduction to the process of conducting research, including: searching the literature; making a research plan; working effectively with subjects, colleagues, and mentors; writing a manuscript; and negotiating the publication process (Florek & Dellavalle, 2016). For more experienced researchers, case-based research can alert the dance science community to new and important observations that warrant further study.

The purpose of case-based reports is to educate the audience regarding a new observation by presenting the findings, to review existing literature, and to place the findings into a broader perspective for the readers based on their expertise. A notable example of a particularly impactful case report was the original description of Acquired Immunodeficiency Syndrome (AIDS) in

Table 15.1. Examples of Case-Based Research

Newly reported entity
New presentation of a previously reported entity
Rare complications of a known entity
New insights into evaluation/treatment of condition[a]
Report on an unusual outcome of standard treatment[a]
Review of relevant literature (for all of the above)

Note: a. Condition and treatment may refer to training or performance standards, curricular practices, medical observations, etc.

1981 (Centers for Disease Control and Prevention). This publication alerted the medical community and the public to a previously unknown disease process and provided a succinct review that enabled readers to recognize the condition, and it set the stage for worldwide research using a variety of research methodologies and policy initiatives to learn more about all facets of AIDS prevention and management (Office of AIDS Research, National Institutes of Health, 2020). Thus astute observers in any discipline can spark a meaningful change by sharing their findings using the case-based approach.

There are three main types of case-based approaches: (1) case report, (2) case series, and (3) case study. A case report generally describes a single observation. A case series describes a particular observation in multiple subjects. A case study focuses on a contextual analysis of several events or conditions and their relationships.

A case study often begins when an interesting case or case series is identified, but instead of simply reporting the case(s) with an accompanying literature review, the author uses the case(s) as a springboard to investigate and report on results of a structured research question, often using qualitative methods (Porcino, 2016; Wilson, 2009). Please see *Chapter 13* for a description of qualitative methodology. The most desirable component of any of the case-based approaches is the novelty of the observation (see Table 15.1). Often, after an observation is noted, a researcher can use other research methods to design an experimental study (e.g., a randomized control trial or a single-case experiment) to examine it further.

The following descriptions and hypothetical examples illustrate essential features of case reports and series.

Newly Reported Entity

For some cases, this means that nobody has reported the phenomenon previously. A hypothetical example could be that university dance majors whose technique class is in the afternoon have a 15% improved aerobic capacity compared to their peers whose technique class is in the morning. While the researcher could speculate on many reasons why this may be true, this is an observation that warrants further study.

Publication as a case series alerts dance departments to pay attention to their own cohorts. There is no clear causality implied, nor is there a means to speculate on associated events. Perhaps the researcher noted that both classes were taught by the same instructor in the same studio with the same warm-up exercises and combinations, using the same musical accompaniment. A next step may be to construct a case study (Alpi & Evans, 2019) using qualitative methods to evaluate characteristics of the students in each of the classes, focusing on their nutritional intake, warm-up regimen, sleep habits, and attitudes. Another approach would be to randomize students into the alternate class time to learn whether the class section explained the phenomenon or if there were other factors influencing the aerobic enhancement. One possibility is to elaborate on the case series by evaluating the effects of a systematically applied intervention with a small number of subjects (see *Chapter 16*). As with any observational or qualitative process, noticing a phenomenon and gathering more information about it sets the stage for the ability to construct a hypothesis-testing study, which may employ a variety of methods.

New Presentation of a Previously Reported Entity

In this case report the researcher has noted that a known phenomenon has presented in a new way. A hypothetical example could be a dancer who sustained a complete rupture of the Achilles tendon during a class several weeks after being treated for an infectious disease with a fluoroquinolone antibiotic. In this example, the dancer had undergone the complete rehabilitation process and had returned to the individual's usual state of fitness and performed routine exercises with proper dance technique after a standard warm-up. This is an example of how the medical community can fully understand the side effects of pharmaceutical drugs outside the clinical trials process and can use collective case reports to generate a meaningful warning for dancers and their providers to make the best medical treatment decisions. This delayed complication is now documented in the side effects profile of this class of drugs as a Black Box warning (Alves et al., 2019).

Rare Complications of a Known Entity

For many scenarios, much is already known. A researcher may notice an instance of an unusual outcome that might herald a new complication that is not yet described in the literature. A hypothetical example could be a dancer who has open blisters on the balls of the feet, from long hours in ill-fitting tap shoes, who is now performing a modern dance piece in an outdoor venue in bare feet and becomes infected with *Clostridium tetani* (tetanus) because infective spores present in the soil have entered his body through the open wounds. This case report could include advice on prevention and treatment of blisters as well as a review of tetanus and vaccination recommendations (Havers et al., 2020).

New Insights into Evaluation/Treatment of a Condition

There may be a standard diagnostic process or treatment or a well-established relationship between two variables. A researcher may have tried a new approach or noticed an unusual outcome that is outside long-standing norms. A case report or case series report would be an ideal way to share this information with the dance science community. One hypothetical example can be drawn from the phenomenon of "tapering" of training of elite athletes as their major competition approaches. Elite swimmers, for example, reduce the amount of training (rehearsal) as their race day nears. Following a strictly prescribed regimen, the coaches keep the swimmers' physiological performance at a peak while enabling body systems to avoid repetitive stress injuries. As the competition begins, their bodies resemble finely tuned instruments that are ready for peak performance.

In contrast, dancers often increase their hours of rehearsal as the performance nears. Frequently this includes unfamiliar class space, which may lack safety conditions, such as sprung floors. Environmental conditions may not be favorable for pre-performance warm-up in a cold theater. In this hypothetical example, perhaps circumstances prevented the usual frenetic preparations leading up to the performance, and the astute artistic director noted a reduction in the typical number of injuries sustained during the performance run. No causal inference can be drawn, but this information can serve as the basis of a deliberate hypothesis-testing study (e.g., alternating treatments design, randomized control trial, or other methodology that allows for experimental analysis) that looks at athletic tapering for dancers leading up to performances.

Report on an Unusual Outcome of Standard Treatment

In this instance a widely accepted standard does not apply to the case in question. A hypothetical example could be an adolescent dancer who meets all standard criteria for pointe readiness as determined by a trained expert, who subsequently suffers an injury that is uncommon in this properly prepared demographic. Perhaps, she has an unrecognized anatomic anomaly or nutritional deficit that contributed to the outcome. Recognition of this risk factor could prevent similar issues in other dancers and could lead to a revision in pointe readiness criteria. However, a formal study would need to be conducted to determine propensity for this underlying condition to lead to the complication.

In summary, case reports and series are helpful to introduce newly observed phenomena to the community of dancers, teachers, and scientists. These observations can stimulate further research to determine if the phenomena are a result of causation or if they correlate with certain risk factors. Determining causation versus correlation is crucial to developing prevention and treatment strategies.

Literature Review

In addition to communicating new ideas to the community, case reports generally provide readers with a comprehensive review of relevant literature. This thorough literature review places the case findings into perspective and allows readers to understand concepts with an associated exemplary and practical case. It can also serve as a comprehensive learning tool for trainees who are interested in the topic. Conducting a literature review identifies what is already known about the phenomenon in question and confirms the novelty of the case.

Several databases are available to search for published and non-published academic resources (as described in *Chapter 4*). If available, the university research librarian is a valuable resource for streamlining the search for relevant materials.

Once a few key references are found, one can search for "related articles" in the databases. This information can shed light on the importance of the case to be reported and can paint a contextual picture for readers. For novice researchers, the literature review phase provides the added benefit of broadening their understanding of the topic to focus their research.

In some circumstances the literature review can reveal that a case is not a new phenomenon and that there are already multiple reports surrounding the topic. In this instance it would be better to consider developing a research project rather than writing another case report that is not likely to be published, given that the findings have already been reported. Thus the literature review can serve to identify potential collaborators to increase the rigor of future study of the case and may direct the researcher to an area ripe for study using qualitative and/or quantitative methods.

Writing the Case Report

After noting the unusual nature of the case (or series of cases) in question and confirming that this is not previously reported in the literature, the author can embark on the actual writing of the case report or series (Biswas & Jones, 2016; Khalil & Mishra, 2016; McCarthy & Reilley, 2000; Vega, 2015; Yin, 1999). Some journals have specific guidelines for the structure of case reports, so authors should identify the target journal and write with these requirements in mind (Akers, 2016).

A sample general format is presented here. Guidance in preparing other types of publications is presented in *Chapter 20.*

Title

The title should convey the essence of the findings clearly. Often, this is constructed as "[Your topic]: A case report and review of the literature."

Abstract

Some journals prefer an abstract for all submissions, while others do not require one for a case report. If an abstract is required, it usually has a specific word limit and contains all the relevant information without detailed explanations (background, summary of observations/possible causes, recommendations) of the sections that follow. It is imperative to convey the most important ideas succinctly in the abstract.

Introduction

This section is designed to introduce the reader to the importance of the case. What is the accepted general knowledge on the topic? What, in general terms, is the significance of the report? Why is this event unique? The detailed analysis will take place in the discussion, so this section should be broad.

Case Report

The exact description of the case will vary based on whether it refers to an individual, a method of training, a medical condition, etc. It should be reported in a way that is informative and easy to follow.

For a case report of a finding in an individual dancer, one may begin by relating the history (e.g., presenting complaint, past medical history, relevant facts about the environment in which the case findings were discovered). Next, the examination findings (e.g., by the physical therapist, physician, exercise physiologist, etc.) should be described, followed by diagnostic testing results (e.g., blood, radiologic imaging, dance wellness screening, performance biometrics), and other outcome measures. A recounting of treatment and outcomes (follow-up) provides closure to the case.

For a case report that describes new outcomes related to an environmental change (e.g., change in rehearsal floor) or a pedagogical change (e.g., increase in weekly training hours), the same sequence can be followed to tell the story. What was the historical standard? What changed? What were the new findings? How did the situation resolve? How did the change affect the dancers/program?

Discussion

This section is the place to put all of the findings into perspective and to share insights with readers about why the case is important and how it could shape the future. Ideally this includes an expansion of the information presented in the introduction. Next, a well-organized and thorough review of the existing literature can focus readers on why this case is very important and what the implications may be. The essential features of the case reported should be placed in context with existing literature. If the case relates to a medical phenomenon, the discussion should contain relevant information about that diagnosis, including the standard presentation, differential diagnosis, and currently accepted treatment regimen. Next, it is important to highlight the unique nature of the case being reported and why it occurred in this new way.

The case report may include theories regarding causation and prevention, if pertinent, but care must be taken to avoid drawing definitive conclusions based on isolated observations. This is an excellent opportunity to pose potential research questions that the reporter or readers may wish to investigate using a qualitative or quantitative research approach. Similarly, speculation on how this case may impact future diagnosis and management is a critical component, as the author's insightful identification of the novelty of the case provides a unique view of the condition that may benefit readers and other researchers.

In many instances trainees seek out case reports in the literature because of their comprehensive review of the topic, and this provides an excellent introduction for them. Many case reports include a section about the lessons that were learned in the process of identifying and reporting the case. The concluding remarks should reaffirm the main message of the report, what can be learned by recognizing such a case, and how this may influence future decisions or treatment regimens.

References

It is useful to list the most important references that were consulted during the literature review. Each journal's style and requirements for citation vary.

Limitations of Case Reports

Case reports and series are generally purely observational in nature, thus there is little standardization of the clinical study environment. There is no external validity assumed about the case(s) in question, and causation versus correlation are difficult to determine. For this reason, many readers may attribute the findings to chance and are not convinced that they should change their practices based on the information provided in the report.

Summary

Case reports, series, and studies provide an important mechanism for the scientific community to become aware of new conditions that fall outside standard practice and previously reported phenomena. An astute researcher who notices a deviation from the norm can use the case report method to alert colleagues of new information. With a thorough review of the literature, readers can be updated at once about the current state of knowledge and can appreciate the novelty of the case that is being described. Scientists can then use the information as a basis to conduct advanced qualitative or quantitative research studies.

Acknowledgments

The authors thank Annabelle Couillandre, PhD for access to information she presented at the 2016 IADMS meeting.

Highlights

- Case reports alert the scientific community to new observations.
- Novice researchers may find that case reports provide a convenient way to learn the basic tenets of academic writing.
- A case report may spur scientists to conduct qualitative or quantitative studies to investigate and explain the novel reported phenomenon.

References

Akers, K. G. (2016). New journals for publishing medical case reports. *Journal of the Medical Library Association, 104*(2), 146–149.

Alpi, K. M., & Evans, J. J. (2019). Distinguishing case study as a research method from case reports as a publication type. *Journal of the Medical Library Association, 107*(1), 1–5.

Alves, C., Mendes, D., & Marques, F. B. (2019). Fluoroquinolones and the risk of tendon injury: A systematic review and meta-analysis. *European Journal of Clinical Pharmacology, 75*(10), 1431–1443.

Biswas, S., & Jones O. (2016). How to write a medical case report. *British Medical Journal (Clinical research edition), 355,* h3731.

Florek, A. G., Dellavalle, R. P. (2016). Case reports in medical education: A platform for training medical students, residents, and fellows in scientific writing and critical thinking. *Journal of Medical Case Reports, 10,* p. 86).

Havers, F. P., Moro, P. L., Hunter, P., Hariri, S., & Bernstein, H. (2020). Use of tetanus toxoid, reduced diphtheria toxoid, and acellular pertussis vaccines: Updated recommendations of the advisory committee on immunization practices—United States, 2019. *MMWR Morbidity and Mortality Weekly Report, 69*(3), 77–83.

Khalil, S., & Mishra, D. (2016). Sharing clinical experience with the scientific community: How to write a case report? *Indian Pediatrics, 53,* 513–516.

McCarthy, L. H., & Reilley, K. E. H. (2000). How to write a case report. *Family Medicine, 32*(3), 190–195.

Office of AIDS Research. (2020, July 7). National Institutes of Health. Retrieved June 19, 2021, from https://www.oar.nih.gov

Pneumocystis pneumonia—Los Angeles. (1981). *Centers for Disease Control MMWR Morbidity and Mortality Weekly Report, 30*(21), 1–3.

Porcino, A. (2016). Not birds of a feather: Case reports, case studies, and single-subject research. *International Journal of Therapeutic Massage and Bodywork, 9*(3), 1–2.

Vega, J. (2015). How to write a case report for publication: A practical guide. *Revista Médica de Chile, 143*(4), 499–505.

Wilson, M. (2009). Dance pedagogy case studies: A grounded theory approach to analyzing qualitative data. *Research in Dance Education, 10*(1), 3–16.

Yin, R. K. (1999). Enhancing the quality of case studies in health services research. *Health Services Research, 34,* 1209–1224.

16

Within-Subject Experiments

TOM WELSH AND STEVEN J. CHATFIELD

Questions to Stimulate Thinking

- What are the fundamental features of within-subject experiments?
- How is experimental control evaluated in within-subject experiments?
- When might it be beneficial to use a within-subject design?
- Are there special requirements for conducting this type of research?

In this chapter we introduce *within-subject experiments* as a research approach that may be used to answer questions of interest to dance science researchers and the constituencies they serve. Such experiments permit researchers to isolate the influence of independent variables in order to identify causal relationships while studying the performance of individual and small groups of dancers. The approach permits researchers to study elite performers and intensive interventions and to observe patterns of response to treatment and training over time and across conditions. Research questions that are well suited for within-subject research include:

- Can targeted training help dancers improve performance of a specific dance skill?
- Is there a typical pattern of acquisition for a newly developing skill, or do different dancers improve at different rates and with different degrees of variability?
- Does a new skill deteriorate when training concludes, and if so, can dancers, teachers, and trainers mitigate the decline?

Within-subject experimental designs have been used for more than 50 years to study phenomena in a diverse collection of fields, such as special education, organizational behavior management, sport psychology, and health promotion. In recent years, within-subject experiments have been used to evaluate interventions designed to improve a number of abilities important to dancers, including back strength, control of turnout, pelvic alignment, and competition dance skills (see Welsh et al., 2020 for a summary).

We describe first the defining features of within-subject experiments and explain how they are applied to designing and conducting a fundamental within-subject study. Then we offer examples of within-subject designs and variations that permit researchers to conduct an experimental analysis of the influence of independent variables (IV) on a dependent variable (DV). The chapter concludes with a discussion of the advantages, requirements, and limitations involved in using within-subject designs.

Features of Within-Subject Experiments

Within-subject experiments (as distinct from pre/post or within-group comparisons) focus on the performance of individuals as the unit of analysis. Other labels used to describe this type of research (*Single-Case Design, Single-Subject Design, Ideographic Study, N=1 Trial—where N = number of participants*) seem to imply that only one participant is studied. While results are analyzed individual-by-individual, it is common for within-subject experiments to involve three to five participants (Hrycaiko & Martin, 1996). The ability to evaluate the causal influence of an IV on a DV using a small number of participants offers some advantages to dance science researchers.

It may be easiest to understand the features of within-subject experiments by contrasting them with the more familiar between-group experiments. In between-group experiments, a sizable number of participants, the research sample, is randomly assigned to one of two or more groups, with each group experiencing a different treatment (IV), one of which may be a no-treatment control condition. Inferential statistics are used to determine the probability that any observed differences in performance (DV) across groups may have occurred by chance. If the chance probability of a difference is less than a predetermined threshold (e.g., less than or equal to 5 chances in 100, $p < .05$), then researchers infer that the treatment is the likely cause of the observed differences.

Within-subject experiments use a different approach to evaluate cause-and-effect relationships. The features described in this chapter allow them to provide a clear demonstration of the influence of an IV on a DV without relying on random assignment and inferential statistics. Some features are particular to the within-subject approach; others involve some augmentation to typical experimental research practices.

First, within-subject researchers measure a performance or capacity (DV) repeatedly (perhaps daily) over several weeks or months. To ensure that the measurements are being made reliably, Bailey & Burch (2018) recommend that a second observer independently assess at least 30% of the observations chosen at random from every experimental condition. If Inter-Observer Agreement (IOA) falls below a predetermined threshold (90% for established operational definitions and 80% for new definitions, see *Chapter 8*), observers are retrained to bring IOA back above the threshold.

Second, within-subject researchers devote extra attention during pilot testing to minimizing within-subject variability in order to establish a visually stable baseline against which any changes in performance that occur during treatment can be compared. In addition, baseline conditions are continued for each participant until performance stabilizes (see later discussion).

Third, each participant's performance under one condition is compared with his or her own performance under the other conditions being evaluated. This feature allows all participants to serve as their own experimental control, thereby removing between-subject variability as a potential confounding variable.

Fourth, performance over time and across conditions is displayed graphically for each participant to reveal any changes in performance. Experimental control is inferred when changes in performance coincide with the application (or removal) of the treatment or training. Replication of the effect within the experiment is required to establish a cause-and-effect relationship. Three replications of the effect, occurring at different points in time, is considered strong evidence of experimental control (Kratochwill et al., 2010). Collecting and graphing frequent measurements of individual participant performance makes it possible to observe within-participant variability directly and to see patterns of change over time.

Finally, within-subject researchers use social validity protocols to allow participants to evaluate the importance of any improvements achieved and the appropriateness of the treatment or training used to achieve them. As explained in *Chapter 10,* social validity may help researchers and practitioners forecast ecological validity.

Simple Comparison: A Quasi-Experimental Design

The *Simple Comparison* (also called an *A-B design;* Miller, 1980) is a quasi-experimental design that offers an intermediate step between an intervention case study and a within-subject experiment. Conducting a simple comparison analysis can allow researchers to develop the background, methodological tools, and many of the skills needed to conduct a convincing experimental analysis. Here is a walkthrough of how a simple comparison study might be conducted.

First, participants are chosen based on their potential to benefit from the intervention under evaluation. For example, to test the effects of a new training protocol designed to improve performance of double pirouettes, a researcher might choose a few dancers who want to improve their performance of pirouettes and who have the potential to improve based, perhaps, on their technique teacher's assessment. A relevant and readily accessible setting for assessing pirouette performance and for administering training (e.g., a small dance studio) might be reserved for two to three weekly sessions. Traveling to a lab that is far from the dancers' training facility may be too much to ask, given the frequency with which measurements are made for within-subject experiments.

Next, as in many approaches to research, an outcome variable is operationally defined (e.g., completing two revolutions and landing without wobbling in the starting position), and a means of assessing the performance is pilot tested and refined to provide a reliable, valid, and responsive assessment. Assessment protocols used for within-subject studies must be especially efficient to administer in order to allow for the frequency of observations needed for analysis. In addition, within-subject researchers must choose a time for making assessments that is convenient for the participants and appropriate to the purpose of the study. For the pirouette study, we might make assessments directly after morning technique classes.

As is true for all within-subject studies, a simple comparison begins with a series of baseline measurements intended to establish the level and variability of performance for each participant prior to applying a training or treatment intervention. If baseline measurements reveal a high degree of variability or the measurements are trending in the direction of the anticipated effect, researchers continue to adjust the environmental conditions until baseline measurements stabilize (Sidman, 1960). Three to five consecutive baseline measurements within a common range, and without trending, are required to establish a stable baseline (Kratochwill et al., 2010).

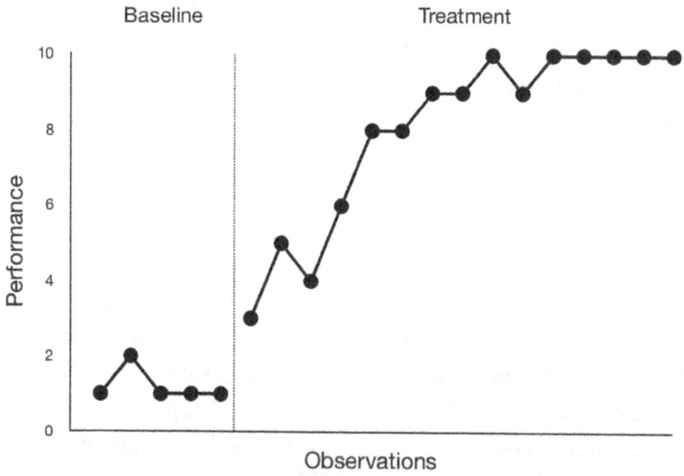

Figure 16.1. Simple Comparison Design. Hypothetical data.

Figure 16.1 displays hypothetical data for a simple comparison study. Applying the pirouette training example introduced here, the *Performance* measurement would be "Number of Successful Double Pirouettes," and the increments on the vertical axis would be labeled 1 through 10 (of 10 trials). *Treatment* is "Pirouette Training," and *Observations* were made twice a week for 10 weeks. The left panel shows the number of successful double pirouettes stabilizing at 1 of 10 tries during the five sessions of baseline assessments (Figure 16.1, left panel).

After a stable baseline is achieved, pirouette training is applied, and measurements continue to be made so that each participant's performance during the training period can be compared to his or her own performance during baseline. Ideally, training continues until the performance stabilizes again, presumably at a new level, as shown in the right panel of Figure 16.1. A discontinuity in performance that coincides with the application of training (either side of the vertical *condition change* line) suggests that training appeared to influence the performance. The absence of baseline trending and the absence of overlap in the level of performance across conditions facilitates seeing a correspondence between the application of training and the change in performance.

It may be tempting to conclude that pirouette training caused the change in pirouette performance, but conclusions must remain tentative because it is possible that some other change may have coincided with the initiation of training (e.g., the dancer may have started training with a new technique teacher). The simple comparison design cannot rule out the possibility that coincidental variables, possibly undetected, may be at least partially responsible for the changes observed. For this reason, the simple comparison is considered a quasi-experimental design (Campbell & Stanley, 1963).

Managing Threats to Internal Validity

This section addresses the misconception that conclusive experiments cannot be conducted with a small number of participants. What distinguishes experimental and quasi-experimental designs is the degree to which the influence of an IV on a DV can been isolated to rule out feasible alternative explanations for an observed effect. In their seminal monograph *Experimental and Quasi-Experimental Designs for Research,* Campbell and Stanley (1963) described eight potential threats to internal validity, which, you will remember, is a fundamental requirement for evaluating experimental control. They pointed out that the *Interrupted Time-Series,* a precursor to within-subject experiments, is "commonly used in the more successful [lab-based] sciences." However, when used in applied and field-based research, the time-series is at risk for two of the eight threats to internal validity—*instrumentation* and *history.* Within-subject researchers have developed strategies to address these threats to internal validity.

Instrumentation can threaten internal validity if a measurement protocol is not certain to produce consistent results when applied repeatedly over time and across conditions. To address this threat, within-subject researchers assess the reliability of their measurement protocols frequently throughout their studies to ensure that observed differences in performance are not simply artifacts of inconsistent measurement.

Within-subject researchers manage the *history* threat by arranging to apply the training or treatment (IV) in a manner that isolates its influence on the performance (DV). The IV is applied more than once and at different points in time to see if changes in the DV coincide with its application. In some within-subject experiments, the IV is applied at different times for a series of participants, performances, or settings. In others the IV is applied and then removed in an alternating pattern. Changes in performance that correspond with the application of the training must be conspicuous for researchers to infer experimental control. Replication of the effect is required to affirm it.

Within-Subject Experimental Designs

The defining features of within-subject experimentation described in previous sections of this chapter achieve what random assignment and inferential statistics are designed to achieve in between-group experiments. Three within-subject experimental designs—*Multiple Baseline, Reversal,* and *Alternating Treatment*—can be configured to produce convincing evidence of the influence of an IV on a DV.

Multiple Baseline Design

Multiple baseline experimental designs are especially useful for studying skill or capacity acquisition. They can be built by combining a series of simple comparisons in a manner that isolates the influence of the intervention in ways that make coincidental influences improbable as explanations for systematic performance changes. It is common for multiple baselines to involve concurrent and repeated measurement for three different performances (e.g., the same skill performed by three different dancers, three different skills performed by one dancer, or the same skill performed by the same dancer in three different settings). Each baseline is called a *leg* of the multiple baseline, and the treatment or training procedure is applied in an offset pattern across legs (see Figure 16.2) to control the threat of history as a potential confounding variable.

As an example, Morgan (2016) developed a field-deployable protocol for measuring the height of six dancers' développé a la seconde (lifting one leg, knee first, to the side). Measurements required approximately 5 minutes of each dancer's time. All six dancers were measured twice a week for the 8-week duration of the study. Training was initiated for three dancers whose performance had stabilized after three baseline observations and for the other three dancers after eight observations. The graphic display of each dancer's développé performance (see Figure 16.2) shows that développé height increased only when each dancer received développé training. Because the change in développé height corresponded with the initiation of training for each dancer, and because training was initiated at different points in time, the simplest explanation for the results (remember parsimony from *Chapter 2)* is that développé training, in all likelihood, caused the change in développé height.

The design just described is called a *Multiple Baseline Across Participants* (dancers in this case). Quinn and colleagues (2015) used a variant called a Multiple Baseline *Across Behaviors,* which showed that the performance of three competition dance skills improved when *an auditory feedback* training

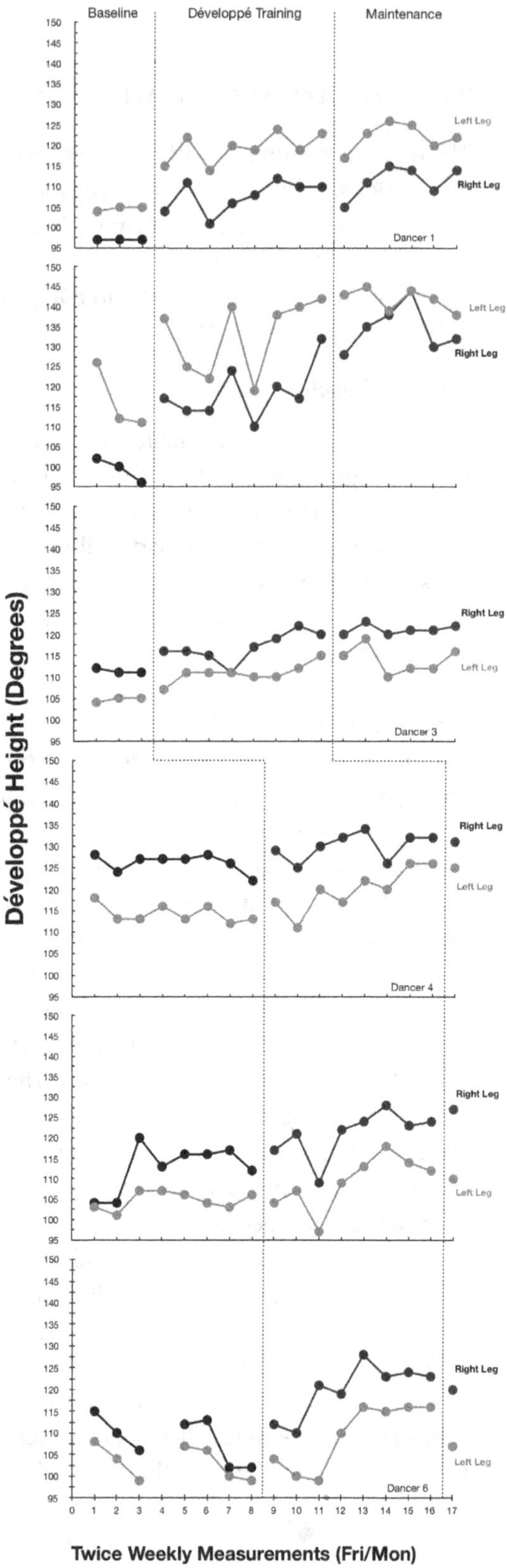

Figure 16.2. Multiple Baseline Across Subjects (Dancers) Design. Display of repeated *développé* measurements for all six dancers. The vertical lines show the initiation and conclusion of training for each dancer. Training for Dancers 1–3 was initiated after three baseline observations. Training for Dancers 4–6 was initiated after eight baseline observations, making this study a two-leg multiple baseline across subjects. "Improving *Développé* Height in University Dancers," Honors thesis, Florida State University. Copyright 2016 by K. Morgan.

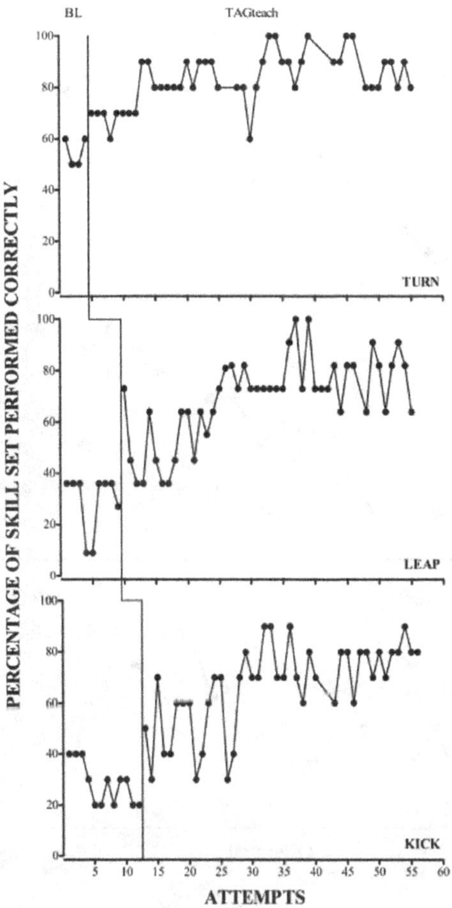

Figure 16.3. Multiple Baseline Across Behaviors (Skills) Design. The percentage of task components performed correctly during baseline and during the TAGteach intervention across three behaviors (turn, leap, and kick) for one of the dancers in the study. "Using TAGteach to Improve the Proficiency of Dance Movements," by M. M. Quinn, R. G. Miltenberger, and V. A. Fogel, 2015. *Journal of Applied Behavior Analysis, 48,* pp. 11–24. Copyright 2015 by John Wiley and Sons.

method (TAGteach) was applied to each skill at different points in time with four young dancers. Figure 16.3 shows results for one of the four dancers who participated in the study. Results for the other three dancers describe a similar pattern.

A third variation is Multiple Baseline *Across Settings.* As a hypothetical example, the teacher of a young dancer who is easily distracted might employ an observer to record the dancer's *percent of time on-task* during (a) tutoring sessions, (b) technique classes, and (c) recital rehearsals. After baseline measurements stabilize in at least one of the settings, the teacher awards points (that can be exchanged for dancewear in the school store) whenever the dancer is paying attention in the tutoring setting. Measurements continue to be made in all three settings. If performance improves during tutoring (see Figure 16.4), the teacher then applies the point system in the second setting,

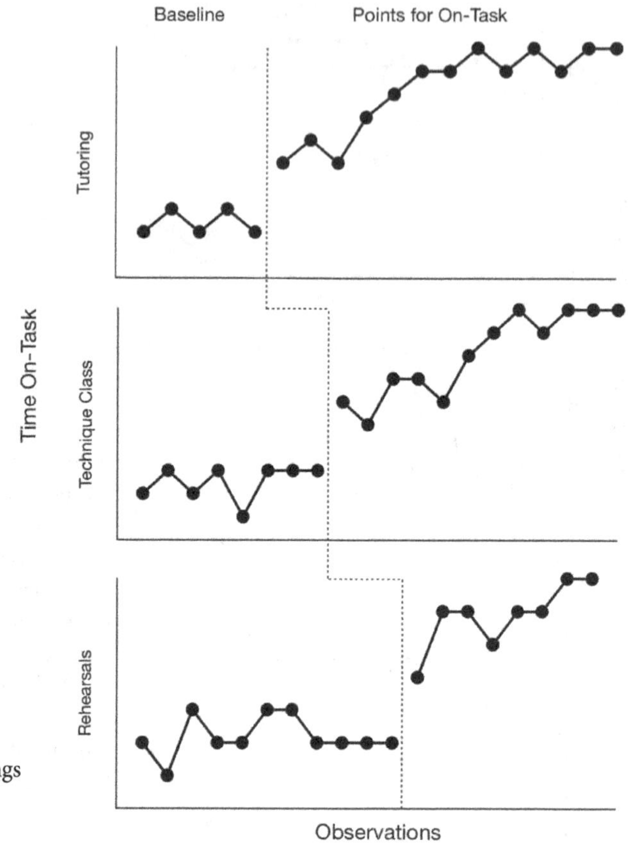

Figure 16.4. Multiple Baseline Across Settings Design. Hypothetical data.

and eventually in the third. If on-task performance improves only when the point system is applied in each setting, it is likely that the point system was responsible for improving on-task performance.

Two special aspects of the multiple baseline design deserve mention. First, the number of baselines *(legs)* used can be increased to allow for more replications of the experimental effect. Three is the number of legs recommended for generating "strong" evidence in a multiple baseline experiment (Kratochwill et al., 2010), but some multiple baseline experiments can provide a convincing demonstration with two legs. Researchers may choose to use more than three legs, but it is rarely necessary if a robust intervention is being evaluated under carefully controlled conditions.

Second, a key requirement of the multiple baseline design is that the baselines (legs) remain independent of one another. Intervening in one leg of the multiple baseline should not change performance in the other legs. To provide a clear demonstration of experimental control, systematic changes

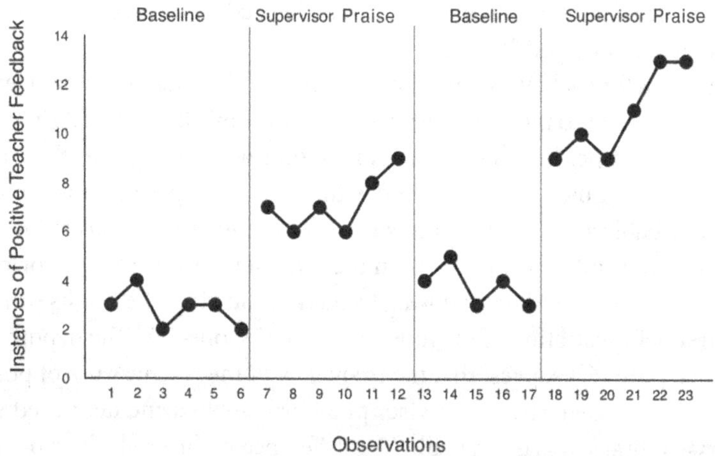

Figure 16.5. Reversal Design. Hypothetical data.

in performance must coincide with systematic changes in treatment or training as it is applied in each leg of the multiple baseline. Other within-subject experimental designs combine the elements in a different pattern to evaluate experimental control.

Reversal Design

The *Reversal* (sometimes called *Withdrawal*) design can be built by combining two simple comparisons end-to-end (Barlow et al., 2009; Portney & Watkins, 2009). This type of experiment can be especially convincing because it amounts to turning the treatment or training on and off repeatedly to see what happens to the performance. Researchers can reverse the conditions as many times as needed to provide a clear demonstration of the influence of the IV on the DV.

A researcher may, for example, count the number of times a teacher provides positive feedback in the first 15 minutes of a beginning ballet class during baseline and later when the intervention is initiated and the teacher receives praise from her supervisor for providing positive feedback to the dancers (see Figure 16.5). Up to this point, the series of conditions constitutes a simple comparison (A-B). To achieve a reversal analysis, both conditions are repeated (i.e., supervisor praise is removed for a number of sessions and then reinstated for a number of sessions). If supervisor praise is responsible for the teacher's increased use of positive feedback, the frequency of positive teacher feedback will decrease to approximate baseline levels when supervisor praise is removed, and then return to a level comparable to that achieved during the first supervisor praise condition when supervisor praise is re-applied. The Reversal design may be labeled an A-B-A-B design, with "A" being baseline and "B" being training.

The rules for evaluating the graphic display of the data in a reversal experiment are similar to those used in evaluating Multiple Baseline data. Changes in the level of performance must correspond with changes in the experimental conditions, and the difference in the level of performance must be distinguishable across conditions with no anticipatory trending during baseline. If there is little or no overlap in the level of the measured performance across conditions, and no opposing trends are showing, the results will offer a clearer demonstration of the influence of the IV on a DV. The hypothetical data in Figure 16.5 suggest that the frequency of the teacher's use of positive feedback increased when supervisor praise was applied and decreased when supervisor praise was unavailable. The difference in the level of performance is clearly distinguishable across conditions in the graphic display, and there is no trend in either baseline that anticipates the change in performance that occurred when the intervention was applied.

For a Reversal design to be informative, the performance must be reversible. Performances that are not likely to reverse when treatment is removed (e.g., acquisition of a basic skill that is likely to be maintained by natural conditions once it is acquired), are not good candidates for a Reversal analysis. Also, performances that should not be reversed due to ethical considerations (e.g., removing a treatment that may be protecting a dancer from re-injury) should not be evaluated using a Reversal design.

These two families of within-subject designs, Multiple Baseline and Reversal, are capable of addressing many of the questions researchers in the dance sciences face. There is a third type of within-subject design that may be useful to dance science researchers.

Alternating Treatments Design

The *Alternating Treatments* design can be viewed as a rapidly changing Reversal. In its ideal form, a targeted performance is assessed until a stable baseline is achieved. Then the treatment is applied in a pattern that alternates session-by-session as determined by a random choice procedure (e.g., a coin flip). The baseline and control condition might be rehearsing new rhythmic hand, foot, and head movement patterns in a beginning Bharatnatyam dance class without vocal self-accompaniment. The comparison condition might be rehearsing the same body movement patterns while reciting previously learned vocal accompaniment that parallels the rhythms in both movements. The level of performance is displayed graphically, and the data points representing performance in one condition (e.g., baseline and control) are

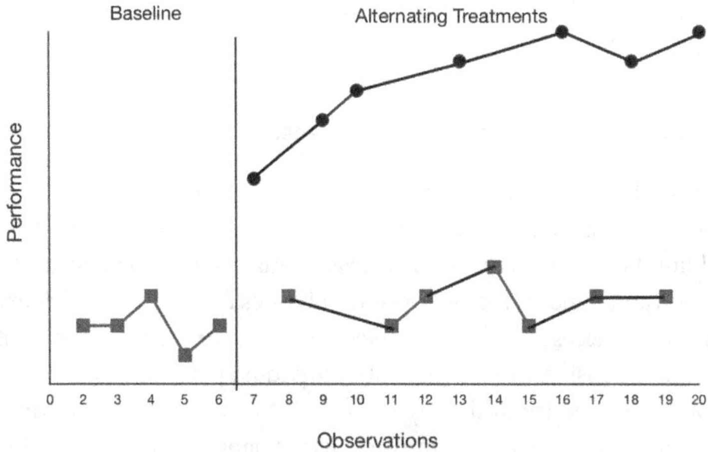

Figure 16.6. Alternating Treatments Design. Hypothetical data.

connected as one dataset, while the data points representing performance under the other condition (e.g., vocal accompaniment) are connected as another dataset (see Figure 16.6). If the connected data sets describe different levels of performance, researchers infer that the differences in treatment are responsible for the differences in performance. As with other within-subject designs, researchers may repeat the changes in conditions as many times as needed to produce a graphically clear demonstration of the influence of the treatment on the targeted performance.

The Alternating Treatments design is especially useful for comparing one procedure (perhaps a new treatment) to another (perhaps a current standard treatment). Variations on this design include *N = 1 Randomized Trials* (Straus et al., 2015) and *Multi-element Baseline Designs* (Bailey & Burch, 2018).

Other Designs

Additional variations on each of the fundamental within-subject designs can be made even during an experiment to isolate the influence of a treatment on a targeted performance without violating the rules for evaluating experimental control. Tailoring a design to match the specific context is both feasible and desirable (Barlow et al., 2009; Kazdin, 2011). The essential requirements are that the performance under one condition must be clearly distinguishable from the performance under the other condition(s), changes in performance must correspond with the condition changes, and threats to internal validity must be managed. Other variations on within-subject experimental designs include the *Changing Criterion* design (see Bailey &

Burch, 2018) and the *Non-Concurrent Multiple Baseline* design (see Portney & Watkins, 2009).

Advantages of Using Within-Subject Designs

A primary advantage of within-subject experimentation is the ability to conduct a demonstrative analysis of the influence that an IV has on a DV with a small number of participants. This feature makes it possible to study elite performers and dancers with exceptional abilities for which a sufficient sample may not be accessible. It also makes it possible to evaluate interventions that require intensive interactions with individual participants.

Second, the practice of displaying frequent assessments of performance makes individual rates and patterns of performance, as well as within-subject variability, conspicuous. This feature can be especially revealing when different dancers may react differently to the same treatment (Chatfield et al., 1996).

Third, interventions used in within-subject experiments are applied in a form that parallels clinical and training practices, which may make it easier for practitioners to see the potential relevance of research findings to their own work (Kazdin, 2011).

Fourth, because the analytic approach requires clear, sizable, and relatively prompt changes in performance to demonstrate experimental control, the use of within-subject experiments may encourage researchers to study robust independent variables that produce clinically and socially significant changes (Baer, 1977).

Finally, it is possible to adapt a within-subject design (even during an experiment) in order to isolate causal variables. The opportunity to do so can expedite efforts to learn how newly researched phenomena operate (Kazdin, 2011; Quinn et al., 2015; Sidman, 1960).

Requirements and Limitations

When conducting within-subject experiments, several requirements must be satisfied. First, you must be able to measure a relevant aspect of the targeted performance repeatedly, efficiently, and reliably. Assessment protocols that require elaborate setup and extensive participant involvement render repeated measurement impractical.

Second, the influence of treatment or training must be robust enough to produce a sizable, consistent, and relatively prompt change in performance.

Weak variables that have small or inconsistent effects are not likely to produce a definitive graphic display of the influence of the IV on the DV.

Third, many interventions that are ripe for within-subject analysis require intensive interactions with each participant for treatment or training and all require frequent contact with observers for making repeated measurements. Researchers should be prepared for both.

Finally, although within-subject experimentation is well established in some fields it is relatively new to the dance sciences. Dance Science researchers who choose to use a within-subject design to answer their research question will have to understand the approach well enough to be able to defend its use.

An important limitation to consider when conducting within-subject experiments is the possible appearance of testing effects (Campbell & Stanley, 1963). Performances that may be influenced by repeatedly being measured may not be good candidates for within-subject experimentation. However, repeated measurements during baseline can reveal such effects when they occur and researchers may be able to minimize the influence by managing the testing schedule. Limitations on specific designs are describe in the sections above.

Summary

Within-Subject experiments offer dance science researchers a method for assessing the causal influence of robust IVs on responsive DVs. You might choose to use a within-subject design if you are interested in studying patterns of individual performance over time and across conditions. To execute a convincing within-subject experiment, you will need access to a practical measurement protocol that can be administered efficiently, repeatedly, and reliably over an extended period of time. You will need to select a research design that is appropriate to the performance you are studying and you will need to tailor it to fit the details of your participants, setting, and contextual constraints.

Because the within-subject approach is new to the dance sciences, it will be particularly important to consider the advice offered in *Chapter 23*. Reading additional references devoted to within-subject research and finding a mentor who has experience conducting within-subject studies will be essential. We suggest starting with a simple comparison study to develop your skills before attempting to use one of the within-subject experimental designs.

Acknowledgments

Keith Miller and Michael Rowley contributed to the conceptualization of this chapter. Jon Bailey and Bruce Thyer helped clarify its presentation. Kaitlin Morgan contributed the hypothetical figures.

Application Exercise

Choose one of the fundamental Within-Subject designs (Multiple-baseline, Reversal, or Alternating Treatments) and outline a basic experiment designed to answer a research question related to your research topic. What special requirements would need to be satisfied to make the study feasible?

References

Baer, D. M. (1977). Perhaps it would be better not to know everything. *Journal of Applied Behavior Analysis, 10,* 167–172.

Bailey, J. S., & Burch, M. R. (2018). *Research Methods in Applied Behavior Analysis* (2nd ed.). Routledge.

Barlow, D. H., Nock, M. K., & Hersey, M. (2009). *Single Case Experimental Designs: Strategies for Studying Behavior Change* (3rd ed.). Pearson Education.

Campbell, D. T., & Stanley, J. C. (1963). *Experimental and Quasi-Experimental Designs for Research.* Rand McNally.

Chatfield, S. J., Barr, S., Sveistrup, H., & Woollacott, M. H. (1996). Electromyographic and kinematic analysis of movement re-patterning in dance. *Impulse, 4,* 220–234.

Hrycaiko, D., & Martin, G. L. (1996). Applied research studies with single-subject designs: Why so few?. *Journal of Applied Sport Psychology, 8,* 183–199.

Kazdin, A. (2011). *Single-Case Research Designs: Methods for Clinical and Applied Settings* (2nd ed.). Oxford University Press.

Kratochwill, T. R., Hitchcock, J., Horner, R. H., Levine, J. R., Odom, S. L., Rindskoph, D. M., & Shadish, W. R. (2010). Single-case designs technical documentation. *What Works Clearing House* (pp. 1–34), https://ies.ed.gov/ncee/wwc/Document/229

Miller, L. K. (1980). *Principles of Everyday Behavior Analysis.* Brooks Cole.

Morgan, K. (2016). *Improving Développé* Height *in University Dancers.* Honors thesis, Florida State University.

Portney, L. G., & Watkins, M. P. (2009). Single-subject designs. In *Foundations of Clinical Research: Applications to Practice* (3rd ed., pp. 235–275). Pearson Education.

Quinn, M. M., Miltenberger, R. G., & Fogel, V. A. (2015). Using TAG-Teach to improve the proficiency of dance movements. *Journal of Applied Behavior Analysis, 48,* 11–24.

Sidman, M. (1960). *Tactics of Scientific Research: Evaluating Experimental Data in Psychology.* Basic Books.

Straus, S. E., Glasziou, P., Richardson, W. S., & Haynes, R. B. (2015). *Evidence Based Medicine: How to Practice and Teach It* (4th ed.). Churchill, Livingstone, Elsevier.

Welsh, T. M., & Chatfield, S. J. (1997). Within-subject research designs for dance medicine and science. *Journal of Dance Medicine and Science, 1,* 16–21.

Welsh, T., Morgan, K., & Williams, G. (2020). Individualizing training for dancers. In Sang-Hie Lee, Merry Lynn Morris, & Santo Nicosia (eds.), *Perspectives in Performing Arts Medicine Practice* (pp. 271–286). Springer.

17

Between-Group Experiments

MATTHEW WYON AND NICO KOLOKYTHAS

Questions to Stimulate Thinking

- When would you need to use a between-group experiment research design in dance science?
- What are the steps you need to follow when designing a between-group comparison study project?
- What types of research questions would be good choices for conducting a between-group comparisons?

Purpose

In the following chapter we introduce between-group comparisons, which are currently the most popular approach to assessing either whether an intervention works (test-retest design) or whether there are differences between two or more independent groups (e.g., differences between males and females). In both these scenarios we are testing differences in a dependent variable (e.g., a performance variable or capacity) made by an independent variable (e.g., a training or treatment protocol). We begin by explaining the key features of the different between-group comparisons before giving examples of research questions and the statistical steps involved in ascertaining whether a statistically significant difference has been achieved. This chapter has two sections: A is focused on "parametric" tests and B on "non-parametric" tests. In *Chapter 9* the differences between parametric and non-parametric data are described, but in instances when there are low participant numbers, potentially causing a non-normal curve, it is better to use a non-parametric test. Power statistics, also described in *Chapter 10,* provide minimum participant numbers according to the proposed methodological

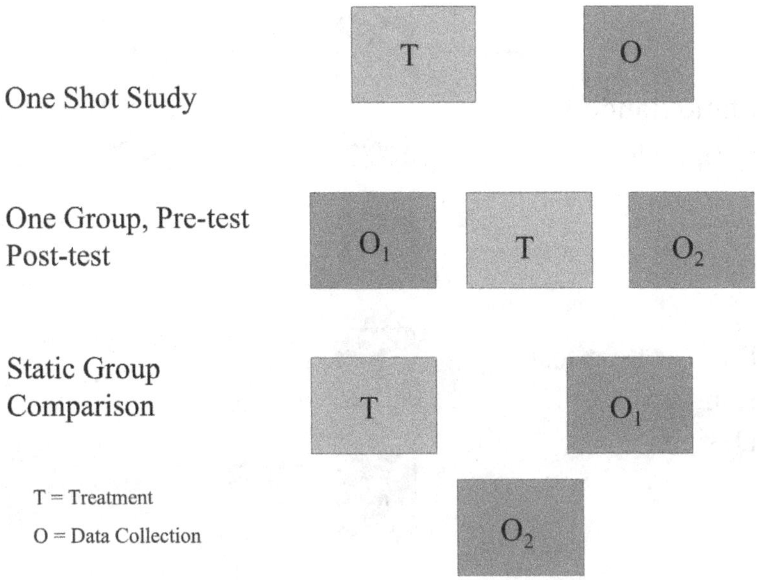

Figure 17.1. Pre-experimental Designs.

design. Non-parametric tests are often considered less powerful than their parametric counterparts; but a test's power is based on the selection of the correct test and based on the characteristics of the data collected. We close the chapter with a discussion of the strengths and limitations of the between-group approach to experimentation and identify the special requirements for conducting these experiments.

Planning your research prior to starting data collection is essential, aimed at eliminating alternative hypotheses being applicable to your research; in other words, the results you obtain are only down to the variables you have manipulated in the specific manner you did. It is often better to take multiple steps to answer a specific question than to take a large leap and end up with multiple reasons for your results. There are a number of ways (designs) to set up your study that range from pre-experimental to experimental to quasi-experimental designs and in all cases examine the extent of difference between the groups. *Pre-experimental designs* (Figure 17.1) have little control of confounding variables (sources of invalidity), and participants are not randomly selected. They include one-shot studies (one group: no pre-test, an intervention followed by a post-test), or one group test-retest designs (one group: pre-test, an intervention followed by a post-test), and static group comparison (comparison between two groups: one group receives an intervention and post-test, second group just receives the post-test).

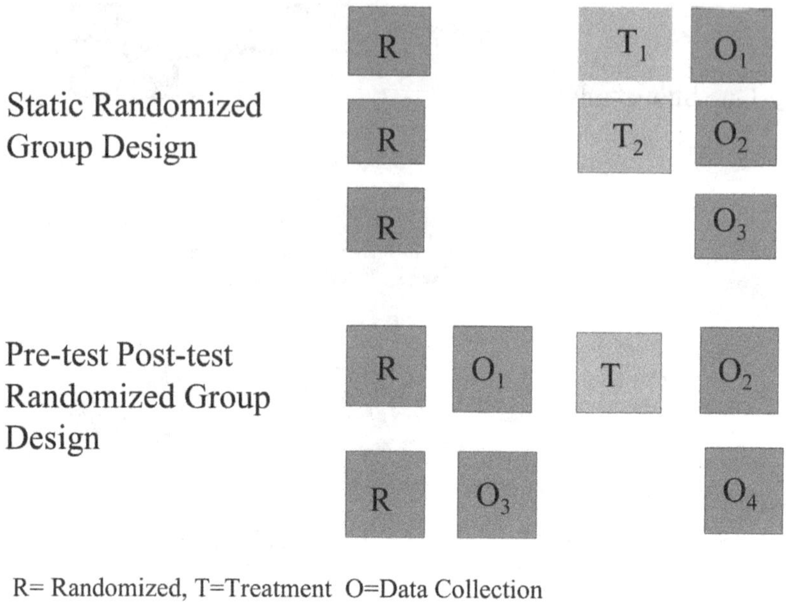

Figure 17.2. Experimental Designs.

Experimental designs require participants to be randomly selected and randomly allocated to different groups. Randomization controls for confounding, including past history, maturation, and non-equivalence between the groups. The experimenter must control the groups to make sure that only the intervention differs between the groups, that testing is accurate, and participant drop-out is not different between the groups. Designs can include static group design if two or more groups are randomly allocated, and test-retest designs, again if the groups are randomly allocated (Figure 17.2). The best true experimental group design is the "Solomon" design that controls for internal and external validity (Allen, 2017).

Quasi-experimental designs cover research that does not meet the criteria of true experimental designs. This often occurs as the researcher tries to make the research more real-world applicable. This increases the external and ecological validity but makes it harder to control for internal validity. Common designs include time-series (one group followed over a long period of time with regular testing, and includes an intervention period; the regular testing before and after the intervention controls for the group's constant rate of change with the tests immediately pre- and post- the intervention, recording the effect of the intervention); non-equivalent control group (this is true experimental test-retest design but when the groups cannot be

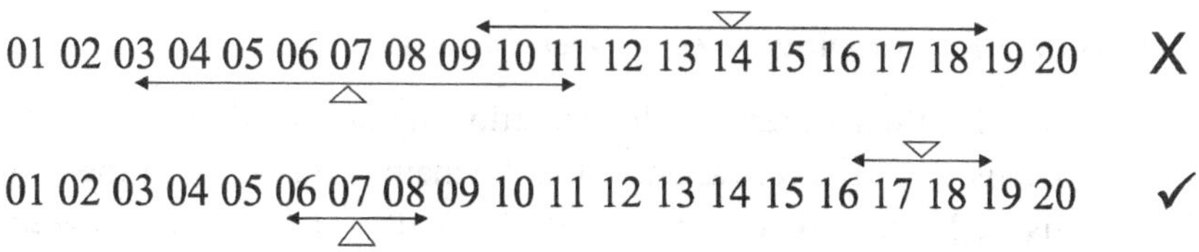

Figure 17.3. Maximizing the Difference Between Central Tendencies. Finding difference that is more than random variation means maximizing the difference between the central tendencies (means [μ]) while minimizing within-group variability.

truly randomly assigned); ex post facto (this is the static group design, but the treatment is not under control of the experimenter; e.g., differences in jump height between males and females). Further quasi-experimental designs include switched replication design and epidemiological designs (retrospective and prospective) (Cook and Campbell, 1979).

To increase the likelihood of uncovering a difference that is more than random variation, you need to maximize the difference between the central tendencies (means [μ]) while minimizing within-group variability (standard deviation) (Figure 17.3). Your methodology and the control of confounding variables (see *Chapters 9* and *10*) will help to increase the likelihood that the data being examined provide a "true" indication of the two groups' characteristics.

Key features of between-group experiments are the following:

- You are trying to find out how an *independent* variable affects a *dependent* variable.
  This could be whether two or more *independent* groups differ in jump height. Jump height is *dependent* (variable) on sex or company rank (*independent variable*), for instance (possible statistical tests: *independent t-test, one-way ANOVA, factorial ANOVA*); or
  How jump height (*dependent variable*) changes because of a strength training intervention (*independent variable*) (possible statistical tests: *dependent t-test, Repeated Measures ANOVA*).
- The participants in a group have been *randomly selected.* If this has not been possible and you have used a convenience sample (dancers from a local school/company), then the results cannot be generalized to a wider population.
- The *dependent data* you have collected are either interval or ratio (*Chapter 8*).
- The data you have collected are *normally distributed* (*Chapter 9*), and the variance (standard deviation) is similar between the groups.

How Do We Test for Differences between Groups?

How we test for these differences has traditionally utilized different t-tests and variations of analysis of variance, and the main aim is to reject the null hypothesis (H_0); that is, there is no *statistically significant* difference between the two methods or groups. Often, however, the data may not reveal a statistically significant difference ($p < 0.05$) between the efficacy of the two methods or between the two groups, therefore the results may support the null hypothesis. It is important to remember that non-statistical significance, however, does not mean no effect or no difference. It is statistically impossible to establish whether the difference or the effect is exactly zero (Lakens, 2017). In the case of the results supporting the null hypothesis, further steps may be needed to evaluate that the difference or effect size is not exactly zero. This can be achieved with equivalence testing, a method of analysis that even though not new, has only just recently come to the fore. This method can prevent the common misinterpretation of non-significant p-values as the absence of an effect, and also probes the researchers to specify the size of the effect they consider worthwhile.

It is good to remember that group difference statistics does not directly use each participant's data but uses the group data to calculate the mean (μ) and standard deviation ($\pm$ 1 SD); therefore, we are actually comparing the estimated central 68% of the groups' populations. There is an increased chance of Type I and II error (*Chapter 9*) if the mean and standard deviation are estimated from low group numbers. The standard error of the mean (SE_M) provides an indication on how much the mean of a randomly drawn group deviates from the larger population mean by chance alone. As explained in *Chapter 9*, the use of power statistics is also a vital component to make sure that there is an increased likelihood the groups' data are normally distributed.

Parametric Statistical Analysis

Two Group Statistics

Dependent and independent t-tests examine the differences between two sample means, one related and the other independent. Underpinning the use of t-tests are some basic assumptions: (a) the population from which the group is drawn is normally distributed; (b) the groups are randomly selected from the larger population; (c) once randomly selected, the variability (standard deviation) within the two groups needs to be approximately equal; and finally (d) the data need to be either interval or ratio (*Chapter 8*). If one or more of these assumptions are not met—for instance, a convenience sample is used, making the design a quasi- rather than a true experimental design—then a more conservative alpha level ($p < 0.01$ rather than $p < 0.05$) or a non-parametric test might be used to reduce the chance of Type I error (*Chapter 9*).

Between-Group Comparison

The assumption for this test is that the two groups are totally independent of each other and are ideally randomly drawn from the same population. Levene's test for Equality of Variance is used to indicate that the standard deviations of the two groups are similar, and if not, some statistics programs provide a different set of results that cater to the difference in variation.

Between-Group Comparison Example A. A researcher wanted to determine if there was a difference in vertical jump height between male and female professional ballet dancers. The null hypothesis was that there is no significant difference in jump height between the male and female ballet dancers. Power statistics (anticipated effect size 0.5; alpha error probability pre-set at 0.05, two-tailed) indicated that 105 participants are needed in each group (check to see how many published studies have achieved these participant numbers). After ethical approval is given, volunteers were recruited from professional ballet companies. Participants signed informed consent prior to carrying out three vertical jumps when standing on a force platform. Jump height was calculated from jump time. The best of three jumps was recorded for each participant.

Table 17.1. Example A Descriptive Data

	Mean (cm)	Standard deviation
Males	51.4	7.47
Females	34.7	5.49

Equation 17.1. Calculating dependent "t":

Standard error of the mean (SE_M) calculation (male)

$$SE_M = \frac{7.47}{\sqrt{105}} = 0.73$$

Standard error of the mean calculation (female)

$$SE_M = \frac{5.49}{\sqrt{105}} = 0.54$$

Calculating the standard error of the difference (SE_D)

$$SE_D = \sqrt{(0.73)^2 + (0.54)^2} = 0.91$$

Calculating t and degrees of freedom (df)

$$t = \frac{51.4 - 34.7}{0.91} = 18.35$$

$$df = (105 - 1) + (105 - 1) = 208$$

The *t* value (18.35) is then cross-referenced with the degrees of freedom (*df*) (208), and in this instance the difference is significant to $p < 0.01$. This would be reported in the text as $t = 18.35$, df = 208; $p < 0.01$.

Within-Group Comparison

Within-group comparison is when the same group is tested twice, such as a pre-post comparison, and they are not independent of each other. Underpinning this concept is that there is a relationship between the pre- and post-test scores in that a person's post-test scores are partially dependent on their pre-test scores. The correlation is an important aspect of the test as it will indicate how the participants reacted to the intervention; a strong correlation (above 0.8, though this is discipline dependent) indicates that the participants all reacted in a similar fashion. Calculating the standard error of the difference (SE_D) (Equation 17.2) needs to be adjusted to take into account the correlation between the pre- and post-tests. A positive *r-value reduces* the SE_D and increases *t*, thereby increasing the chance of significance; a negative *r* is unlikely for within-group comparisons.

Equation 17.2

$$SE_D = \sqrt{(SE_{MPre})^2 + (SE_{MPost})^2 - 2\,r\,(SE_{MPre})(SE_{MPost})}$$

Between-Group Comparison Example B. A researcher implemented a 4-week whole body vibration intervention on a group of female professional modern dancers to see if it would improve vertical jump height. The null hypothesis was that there is no significant difference in jump height pre- and post- a whole body vibration intervention. Power statistics (effect size 0.5; alpha error probability 0.05, two-tailed) indicates that 54 participants are needed (again check to see how many published studies have achieved these participant numbers). After ethical approval was given, 58 volunteers were recruited from professional modern companies. Participants signed an informed consent form prior to carrying out three vertical jumps using a jump mat before and after the intervention.

Table 17.2. Example B Descriptive Data

	Mean (cm)	Standard deviation
Pre	41.0	4.66
Post	45.3	4.95

Equation 17.3. Calculating independent "t":

$$SE_D = \sqrt{(SE_{MPre})^2 + (SE_{MPost})^2 - 2r(SE_{MPre})(SE_{MPost})}$$

Standard error of the mean calculation (pre-test)

$$SE_M = \frac{4.66}{\sqrt{58}} = 0.61$$

Standard error of the mean calculation (post-test)

$$SE_M = \frac{4.95}{\sqrt{58}} = 0.64$$

Calculating the standard error of the difference

$$SE_D = \sqrt{(0.61)^2 + (0.64)^2 - 2\,(0.94)\,(0.61)(0.64)} = 0.048$$

$$(r = 0.94 \; p<0.001)$$

Calculating t and degrees of freedom

$$t = \frac{41-45.3}{0.048} = 89.58$$

$$df = (58 - 1) = 57$$

Once again, the *t* value (89.58) is then cross-referenced with the degrees of freedom (57) within a table, and in this instance the difference is significant to $p < 0.001$. This would be reported in the text as $t = 89.58$, $df = 57$; $p < 0.001$.

Familywise Error Rate: Are t-Tests Enough to Analyzed All Comparisons?

At the beginning of the chapter, we explained that t-tests compared their findings to the probability of chance; 19 to 1 for a confidence level of 95% ($p < 0.05$). When multiple t-tests are conducted on randomly drawn samples from the same population, the chance of getting an incorrect conclusion increases. This phenomenon is exacerbated when multiple t-tests are carried out on a single randomly drawn sample, in that the sample becomes less and less representative of the larger population, and generalization of the results to the larger population is harder to justify.

The use of 95% confidence levels implies that if 20 t-tests were carried out on a sample, then one test would be significant just by chance alone, and it would not be possible to know if it was a true difference or if a Type I error has occurred (*Chapter 9*). This predicament is called the familywise error rate. The equation to compare a single test error rate ($\alpha = 0.05$) with a number of comparisons (C) to produce the familywise error rate (FWα) is:

Equation 17.4. Familywise error rate equation:

$$FW_{\alpha} = 1 - (1 - a)^{c}$$

If there were three groups, then three separate t-tests would need to be done to make all possible comparisons; if four groups, then six t-tests would be required. This would raise the chance of a Type I error occurring from $p < 0.05$ for one test (1 to 19 odds) to $p < 0.15$ for three tests (1 to 7 odds) or $p < 0.26$ for 6 tests (1 to 4 odds). To avoid this increase in Type I error, either a Bonferroni adjustment of the single test alpha level is required ($p < 0.05$/ number of tests) or the use of appropriate analysis of variance tests.

Between-Group Analysis for 3+ Groups

Analysis of Variance (ANOVA)

Analysis of variance (*ANOVA*) allows the comparison between three or more independent groups by comparing the between-group variability (groups' mean) with the within-group variability (groups' standard deviation). Underpinning this is if the between-group variability is greater than the within-group variability by more than chance alone, it means that one group differs significantly from another. It does not indicate where the difference lies, though. *Post hoc* tests, such as *Bonferroni* or *Scheffe's confidence interval,* can then be utilized to ascertain where the difference lies between the groups. These are like t-tests but contain an alpha error correction within them. The *ANOVA* assumes that the participants in each group are randomly drawn from a larger normally distributed population, though a valid *F* ratio will still be produced with samples drawn from a non-normal population. In an ideal world the variance (standard deviation) within the groups would be the same, but again the robustness of the *ANOVA* will allow one group's variance to be approximately half the size of another's.

In a contrived situation, three groups are randomly selected from the same population and their means and standard deviations should be relatively similar. In this situation the between-group and within-group variance between the groups should be the same.

Calculating ANOVA *F* Ratio

Rather than going through the entire formulae for *ANOVA*, which although mathematically simple, are very long, we offer this explanation in text. There are two main equations to determine *F*, the *definition method* and the *raw score method*. The definition method uses deviation of each group mean from the grand mean via a "sum of squares" method, while the raw score method used the raw data. In summary, the *F* ratio of the average of the between-group variance divided by the average within-group variance should be 1.00. If the ratio exceeds this, by more than chance alone, then there is a statistically significant difference between the group means.

ANOVA Example. A researcher wanted to see if there was a difference in aerobic fitness between female recreational, university, and professional modern dancers. The null hypothesis is that there is no significant difference in aerobic fitness between the three groups. Power statistics (effect size = 0.25; alpha error probability 0.05, number of groups 3) indicates that 252 participants are needed in total. After ethical approval was given, 266 volunteers were recruited across the groups (recreational, university, and professional). Participants signed informed consent prior to carrying out a maximal aerobic test.

Reporting the statistics results in the text follows the following formula F (Between-group *df*, Within-group *df*) = *F* ratio; p-value; or in this instance $F(2,263) = 66.64$; $p < 0.001$. In this example, both post hoc tests indicate there are statistically significant differences between the groups; recreational

Table 17.3. Analysis of Variance (ANOVA) Descriptive Group Data

Aerobic Capacity	Recreational (n = 86)	University (n = 93)	Professional (n = 87)
VO_2 (ml.kg^{-1}.min^{-1})	30.4 ±3.80	36.8 ±5.87	42.4 ±10.12

Table 17.4. Analysis of Variance (ANOVA) Outputs Including Sum of Squares, Mean Squares, and F-Ratio

Source of Variance	Sum of Squares	df	Mean Square	F	Significance
Between groups	6565.63	2	3280.82	66.64	p < 0.001
Within groups (error)	12947.21	263	49.23		
Totals	19508.84	265			

Table 17.5. Post-Hoc Tests Outputs

(I) group	(J) group	Mean Difference (I-J)	Std. Error	Sig.
Tukey Honestly Significant Difference (HSD)				
recreational	university	-6.19692	1.04965	p<0.001
	professional	-12.31743	1.06690	p<0.001
university	recreational	6.19692	1.04965	p<0.001
	professional	-6.12050	1.04651	p<0.001
professional	recreational	12.31743	1.06690	p<0.001
	university	6.12050	1.04651	p<0.001
Scheffe				
recreational	university	-6.19692	1.04965	p<0.001
	professional	-12.31743	1.06690	p<0.001
university	recreational	6.19692	1.04965	p<0.001
	professional	-6.12050	1.04651	p<0.001
professional	recreational	12.31743	1.06690	p<0.001
	university	6.12050	1.04651	p<0.001

vs. university ($p < 0.001$), university vs. professional ($p < 0.001$) and professional vs. recreational ($p < 0.001$) (Table 17.5). The mean difference column indicates the direction of the difference.

Analysis of Variance with Repeated Measures (RM-ANOVA)

The *RM-ANOVA* is a development of the dependent group t-test when the same group is tested three or more times. The total variability between the data groups is expected to be less than that between independent groups because participant inter-individual variability has been removed. This means that the participants act as their own control, and any observed changes in the means must be due to (a) the intervention, (b) variations within the participants, as not all the participants will be identical, or (c) error. This does not mean that a control group should not be used as this would emphasize (a) and help to control (b) and (c). Fundamental to *RM-ANOVA* is the assumption of sphericity; this requires that the repeated tests demonstrate homogeneity (similarity) of variance and covariance. The former refers to the standard deviation, in that it should stay a similar size over the multiple tests; the covariance refers to the correlation between the tests (as performed during dependent t-tests) being equal. Between the two of them it will indicate if all the participants reacted in a similar fashion to the intervention.

Calculating the RM-ANOVA *F* Ratio

The variability due to the intervention is calculated and is examined against the variability among the participants. As this is a repeated measures scenario, degrees of freedom are calculated similarly to the dependent t-test. There are two main degrees of freedom to calculate; *test* (column) *degrees of freedom* being number of tests minus one, *and error degrees of freedom,* which is test *df* multiplied by participant *df.*

Correcting for Violations of Sphericity Assumptions. The RM-ANOVA equations require that the variance (sphericity) does not change between the tests and that the correlations between the trials are also equal. If the assumption of sphericity is not met, then there is an increased chance of Type I error. Two equations have been developed to adjust for this violation, Greenhouse-Geisser and Huynh-Feldt. The Greenhouse-Geisser adjustment assumes maximum violation of the assumption of sphericity, therefore when the violation is minimal, there is an increased chance of Type II error (*Chapter 9*). The Huynh-Feldt adjustment attempts to correct for the amount of violation that has occurred. These adjustments might mean that *F* becomes statistically significant, but the confidence in the statistically significant result is less and it might be better to reduce the alpha value to $p < 0.01$ rather than $p < 0.05$.

Comparing Between the Repeated Measures. Contrast analysis allows comparison between the measures (tests) to see if there are different changes between the measures. There are several different equations available, and choosing the right one depends on the initial research question. Polynomial, difference, Helmert, and repeated contrasts compare measure 1 with 2, 2 with 3, 3 with 4, etc., while simple and deviation contrast allows a reference category to be defined (first or last measure) and the other measures compared directly with the mean.

RM-ANOVA Example. A researcher wanted to see if a 6-week whole body vibration (WBV) intervention increased hip flexibility in preprofessional male jazz dancers. The null hypothesis was that there was no significant difference in hip flexibility over the intervention period. Power statistics (effect size = 0.25; alpha error probability = 0.05, number of measures = 3) indicated that 43 participants were needed in total. After ethical approval was given, 46 volunteers were recruited and signed informed consent. Participants were tested at the start, after 3 weeks and at the end of the intervention. The intervention consisted of 10-minute WBV training 4 times a week.

Mauchly's Test of Sphericity is similar to Levene's test for Equality of Variance and indicates a significant difference in the assumption of sphericity (*p*

Table 17.6. Repeated-Measures Analysis of Variance (RM-ANOVA) Descriptive Data for the Three Measures

	Hip flexibility (degrees)		
	Pre	Mid	Post
Dancers (n = 46)	103.9 ±16.37	105.5 ±17.38	111.9 ±16.69

< 0.01). Greenhouse-Geisser and Huynh-Feldt both reported no statistically significant differences ($p = 0.786$ and $p = 0.81$, respectively); therefore, in this instance Huynh-Feldt equations were used (in bold italics in Table 17.7). The same formula is used to report ANOVA results in the text; F (1.619, 72.866) = 20.99; $p < 0.001$.

Deviation contrasts (see Comparing the Repeated Measures section) were used to compare pre- with mid- and pre- with post-tests. In this example statistically significant differences were reported between pre- and mid-test ($p < 0.01$) and pre- with post-test ($p < 0.001$).

Factorial Analysis of Variance

Multiple Factorial Analysis

The previous section looked at analysis of variance with just one factor: either one factor tested across multiple groups (*ANOVA*) or one factor tested multiple times within one group (*RM-ANOVA*). Research questions are often not that simple, and more complex analyses are required that account for more than one factor. For instance, the difference in jump height between sexes (one factor) and ranks (second factor with 4 levels) within a ballet

Table 17.7. Repeated-Measures Analysis of Variance (RM-ANOVA) Outputs Including Sum of Squares, Mean Squares, and F-Ratio

		Sum of Squares	df	Mean Square	F	Sig.
Factor	Sphericity Assumed	1606.043	2	803.022	20.993	.000
	Greenhouse-Geisser	1606.043	1.572	1021.542	20.993	.000
	Huynh-Feldt	***1606.043***	***1.619***	***991.850***	***20.993***	***.000***
	Lower-bound	1606.043	1.000	1606.043	20.993	.000
Error (factor)	Sphericity Assumed	3442.623	90	38.251		
	Greenhouse-Geisser	3442.623	70.748	48.660		
	Huynh-Feldt	***3442.623***	***72.866***	***47.246***		
	Lower-bound	3442.623	45.000	76.503		

Table 17.8. Deviation Contrast Outputs

		Sum of Squares	df	Mean Square	F	Sig.
Factor	Mid- vs. Pre-	115.848	1	115.848	8.245	.006
	Post- vs. Pre-	1033.130	1	1033.130	40.735	.000
Error (factor)	Mid- vs. Pre-	632.263	45	14.050		
	Post- vs. Pre-	1141.314	45	25.363		

company could be answered by 2 *ANOVAs* (between the ranks separated by sex) and then a series of independent t-tests comparing sex differences between the ranks, or factorial analysis could be used. In this example it is a 2 × 4 factorial analysis (sex x rank), a *between-between* design; it could be expanded to include another ballet company (possibly with different characteristics or from a different country); this would become a 2 × 2 × 4 factorial analysis.

Generally, the number of factors examined is limited to three. Multiple factorial analysis (*MANOVA*) is the global umbrella term for these, but within methodologies it is better to reference the number of factors being compared. A study that tests two groups (intervention and control) repeatedly is a development on the *RM-ANOVA* and is classified as *between-within* design, between groups and within tests. The statistical test produces an *F* ratio for each factor and an *F* ratio for the interaction of the factors. The interaction is the combined effect of the factors on the dependent variable. The underlying equations are complex and are not be presented here, but the inputs (group descriptive data) and outputs are reviewed and discussed.

Between-Between Example. A researcher wanted to see if there was a difference in maximum aerobic capacity with classical ballet dancers comparing sex (male and female) and rank (principals, soloists, and artists). The null hypothesis was that there is no statistically significant difference in maximum aerobic capacity between rank and sex. Power statistics (effect size = 0.25; alpha error probability = 0.05, number of measures 2 × 3) indicated that 42 participants were needed in total. After ethical approval was given, 74 volunteers were recruited and signed informed consent. Participants underwent a maximal aerobic test on a treadmill until voluntary cessation or specific physiology criteria were met.

In this example both sex and rank indicate individual statistical significance; these are reported in the same manner as ANOVA tests; sex $F(1,68) = 35.39$; $p < 0.001$ and rank $F(2,68) = 8.778$; $p < 0.001$. The interaction between the two factors is not statistically significant ($p = 0.47$). As a statistically

Table 17.9. Between-Between Factorial Analysis Descriptive Data for Groups

	Maximal Aerobic Capacity (ml.kg^{-1}.min^{-1})		
	Principal (n = 19)	Soloist (n = 29)	Artist (n = 26)
Male (n = 24)	50.5 ±1.61	46.1 ±2.86	50.1 ±5.22
Female (n = 0)	45.9 ±2.73	38.5 ±4.64	42.9 ±3.91

Table 17.10. Between-Between Factorial Output

Source	Sum of Squares	df	Mean Square	F	Sig.
Sex	581.295	1	581.295	35.399	p<0.001
Rank	288.282	2	144.141	8.778	p<0.001
Sex by Rank	25.021	2	12.510	.762	.471
Error	1116.639	68	16.421		
Total	145891.323	74			

Table 17.11. Post-Hoc Tests for Rank

(I) Rank	(J) Rank	Mean Difference (I-J)	Std. Error	Sig.
Tukey HSD				
artist	soloist	6.438090	1.0944545	**p<0.001**
	principal	-2.111093	1.2230517	.203
soloist	artist	-6.438090	1.0944545	**p<0.001**
	principal	-8.549183	1.1960432	**p<0.001**
principal	artist	2.111093	1.2230517	.203
	soloist	8.549183	1.1960432	**p<0.001**
Scheffe CI				
artist	soloist	6.438090	1.0944545	**p<0.001**
	principal	-2.111093	1.2230517	.233
soloist	artist	-6.438090	1.0944545	**p<0.001**
	principal	-8.549183	1.1960432	**p<0.001**
principal	artist	2.111093	1.2230517	.233
	soloist	8.549183	1.1960432	**p<0.001**

significant difference was reported between the ranks, post hoc tests (Tukey HSD and Scheffe CI) should be carried out to compare the difference between the ranks.

Both post hoc tests indicate statistically significant differences between principal and soloists ($p < 0.001$), and between artists with soloists ($p < 0.001$), but not between principal and artists ($p = 0.23$).

Between-Within Factorial Analysis (RM-ANOVA with Multiple Groups)

With *RM-ANOVA,* changes in the dependent variable are compared over time in one group. With the addition of another group to be able to ascertain the effect of the independent variable, it is important that groups are similar at the beginning of the study. One way of achieving this is by matched pairs; this method manipulates the groups at the start of the study so that the two groups (intervention and control) are comparable in terms of characteristics that may influence the variables under study, and the means, and standard deviation of the dependent variables are the same.

There are several ways of doing this, but the simplest is to rank all the participants from highest to lowest and then either alternate putting the participants into the different groups (e.g., odds in one group, evens in the other, or the ABBA method, where highest goes in first group, then next 2 in second group, then next 2 in first group, etc.). Both these methods will result in a high correlation between the groups pre-test, but often result in a smaller post-test difference; therefore, the observed difference is more likely to be due to the intervention (independent variable). This is not always possible, especially when a double-blind method is used (this is when the testers and statistician are blind to the study and the intervention).

RM-ANOVA (with multiple groups) Example. A researcher wanted to see the effect of strength training on jump height on male dancers over a 6-week period. The null hypothesis is that a 6-week strength training intervention will have no statistically significant difference on jump height between the intervention and control groups. Power statistics (effect size = 0.8; alpha error probability 0.05, number of groups 2, number of measurements 3) indicates that 28 participants are needed in total. After ethical approval was given, 36 volunteers were recruited (it is always better to recruit more participants than required for long-term intervention studies to account for participant drop-out) and signed informed consent. Participants were randomly assigned (1:1 allotment) to either the intervention or the control group by the sealed envelope method. All participants had their vertical jump height measured at the start, mid-point, and end of the 6-week period. The intervention consisted of two weight training sessions a week for the intervention group, while the control group carried out additional dance lessons that equated to a similar time period.

The first statistical output is the *main effect* for time and group (Table 17.13); the statistically significant F for Time indicates that for all three measures combined, the intervention group scored statistically significantly better than the control group. The statistically significant F for the interaction

Table 17.12. Between-Within Factorial Analysis Descriptive Data for Groups

	Jump height (cm)		
Group	Pre	Mid	Post
Intervention	42.4 ±3.79*	43.9 ±4.35	46.6 ±4.11
Control	40.6 ±2.67*	40.8±3.91	39.8 ±3.41

Note: No significant difference between the groups at start of study.

Table 17.13. Main Effect for Time and Group

Effect		Value	F	Hypothesis df	Error df	Sig.
Time	Pillai's Trace	.438	13.236[b]	2.000	34.000	p<0.001
	Wilks' Lambda	.562	13.236[b]	2.000	34.000	p<0.001
	Hotelling's Trace	.779	13.236[b]	2.000	34.000	p<0.001
	Roy's Largest Root	.779	13.236[b]	2.000	34.000	p<0.001
Time * Group[a]	Pillai's Trace	.674	35.199[b]	2.000	34.000	p<0.001
	Wilks' Lambda	.326	35.199[b]	2.000	34.000	p<0.001
	Hotelling's Trace	2.071	35.199[b]	2.000	34.000	p<0.001
	Roy's Largest Root	2.071	35.199[b]	2.000	34.000	p<0.001

Notes: a. The asterisk in "Time * Group" means interaction.
b. Significant difference at the 0.05 level.

between Time by Group indicates that the intervention group reacted differently than the control group.

Sphericity can be assumed as Mauchly's Test of Sphericity indicated no statistically significant difference ($p = 0.187$). A statistically significant difference was noted over time (across the three measures) for the combined groups, $F(2, 70) = 17.251$; $p < 0.001$.

The interaction between time and groups indicated a statistically significant effect $F(2,70) = 40.27$; $p < 0.001$, which is reinforced by Figure 17.4.

Equivalence Testing

The term equivalence is not used in a literal sense; rather, it means that the effect or the difference between two groups is close enough that superiority or inferiority cannot be established. A simple way to approach equivalence

Table 17.14. Within-Subjects Effects Output

Factor		Sum of Squares	df	Mean Square	F	Sig.
Time	**Sphericity Assumed**	**52.719**	**2**	**26.359**	**17.251**	**p<0.001**
	Greenhouse-Geisser	52.719	1.828	28.838	17.251	p<0.001
	Huynh-Feldt	52.719	1.979	26.642	17.251	p<0.001
	Lower-bound	52.719	1.000	52.719	17.251	p<0.001
Time* Group	**Sphericity Assumed**	**123.061**	**2**	**61.530**	**40.270**	**p<0.001**
	Greenhouse-Geisser	123.061	1.828	67.317	40.270	p<0.001
	Huynh-Feldt	123.061	1.979	62.191	40.270	p<0.001
	Lower-bound	123.061	1.000	123.061	40.270	p<0.001
Error (Time)	**Sphericity Assumed**	**106.957**	**70**	**1.528**		
	Greenhouse-Geisser	106.957	63.983	1.672		
	Huynh-Feldt	106.957	69.257	1.544		
	Lower-bound	106.957	35.000	3.056		

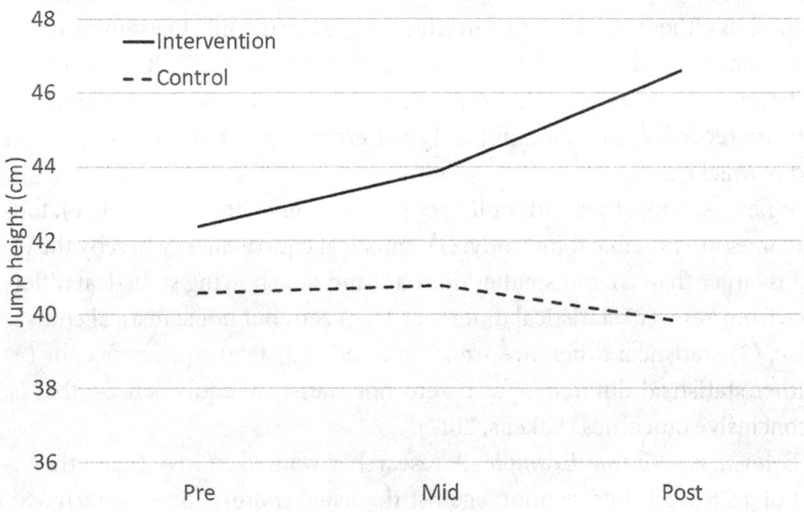

Figure 17.4. Between-Within Factorial Analysis Plot.

testing is the two one-sided test (TOST) procedure (Schuirmann, 1987). Equivalence (Δ) is established at the α significance level if (1–2α) x 100% confidence interval for the difference in the effects (new-usual) is inside a pre-specified equivalence (Δ) margin (e.g., Δ*Lower* = -0.3 to Δ*Upper* = 0.3). The reason why the confidence interval is (1–2α) x 100% instead of the usual (1–α) x 100% is because this method is equal to performing two one-sided tests. This means that the use of 90% confidence interval yields a 0.05 significance level for testing equivalence.

The Δ margin defines the range of values whereby the effect or difference is "close enough" to be equivalent (Walker & Nowacki, 2011). This margin not only determines the result of the test but also gives credibility to the study; if the value is too small, it is harder to establish equivalence, and therefore, it is a very important step. The difference Δ can either be defined by standardized differences (e.g., Cohen's *d*) or raw differences (e.g., 0.3 scale point on a 5-point scale) (Lakens, 2017). The equivalence of a new intervention is established when there is enough evidence that its efficacy is within Δ units from the usual training (Walker & Nowacki, 2011), which means that the between-group difference is smaller than the smallest effect size of interest (Lakens et al., 2018). Whereas in null hypothesis testing the burden of proof rests on the existence of a statistically significant difference in the efficacies of two interventions (i.e., the alternative hypothesis), in equivalence testing the burden of proof rests on demonstrating equivalence. The null hypothesis, therefore, is the presence of a true effect of Δ*L* or Δ*U*, and the alternative hypothesis is either the absence of an effect that is worthwhile to examine or the effect that is within the equivalence bounds (Lakens, 2017). It is important to determine the values of the equivalence upper and lower limits before the data are recorded, to maintain the Type I error at the desired level (Walker and Nowacki, 2011).

When used together with null hypothesis significance tests (NHST), four outcomes are possible for a study: (1) statistical equivalence, whereby the effect is larger than Δ*L* but smaller than Δ*U* and not showing statistical difference from zero, (2) statistical difference from zero but not statistical equivalence, (3) statistical difference from zero and statistical equivalence, or (4) neither statistical difference from zero nor statistical equivalence—that is, inconclusive outcomes (Lakens, 2017).

Equivalence Testing Example. A researcher wanted to investigate the effect of an 8-week intervention against the usual training on countermovement jumps (CMJ), in elite adolescent ballet dancers. After ethical clearance was obtained, power statistics indicated that with a power of 80%, an alpha of 5%, and using the group post- means and standard deviations, sample

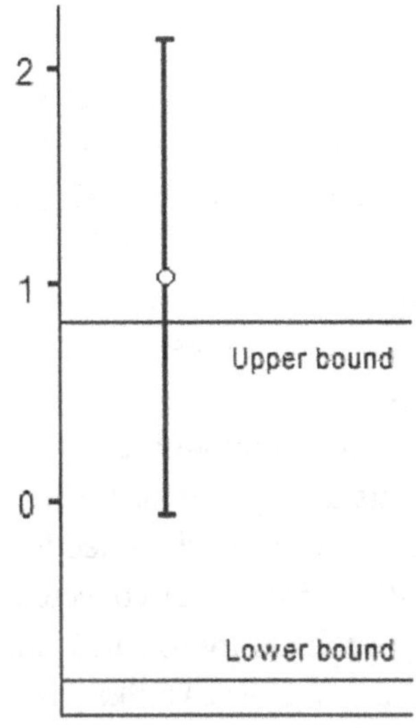

Figure 17.5. Delta of Counter Movement Jump.

calculation revealed that 55 participants in each group (n = 110) would be required for the study. Seventy-five elite ballet dancers were recruited for this study, and they were randomly assigned to either the intervention or the usual training group.

An *RM-ANOVA* revealed no statistically significant effect of the intervention on CMJ compared with the usual training F (1, 68) = 2.45, p = 0.12, ή = 0.04; the data, therefore, support the null hypothesis. However, with the lower equivalence bounds set at -0.3, the upper equivalence bound set at 0.3, and the alpha level at 0.05, the results indicate that equivalence cannot be established; therefore, considering the low power due to the small sample size, further investigation is needed.

Table 17.15. Two One-Sided Test Results

		t	df		
Delta_CMJ	t-test	1.57	66.9	Note. Welch's t-test	
	TOST Upper	0.310	66.9		
	TOST Lower	2.82	66.9		
				90% Confidence interval	
Equivalence Bounds		Low	High	Lower	Upper
Delta_CMJ	Cohen's d	-0.300	0.300		
	Raw	-0.829	0.829	-0.0679	2.14

Non-Parametric Statistical Analysis

There are a series of alternative statistical tests when data do not meet the fundamental requirements of the preceding (parametric) tests. When the data are nominal or ordinal, and normal distribution of ratio or interval data is not achieved due to too few data points or the data has skewness or kurtosis, then non-parametric statistical tests should be used. These tests can also compare ranked or category data such as mood rating on a Likert or visual analog scale or data from questionnaire categories.

Non-parametric tests are often considered less powerful than their parametric counterparts; but a test's power refers to its ability to reject a null hypothesis that is false. Therefore, a test's power is based on the selection of the correct test and based on the characteristics of the data collected. Some statistical computer programs may not have the ability to carry out the non-parametric equivalent of multivariate parametric tests. Unlike parametric tests that focus on the differences between the central tendency and its within-group variabilities, non-parametric tests use all the data points within their equations.

Psychological and educational researchers have often used parametric tests on the premise that because there are a lot of data points, normal distribution can be assumed even when the basic data are not considered interval or ratio, or that other published studies have used parametric tests. The main non-parametric difference tests are Chi-square, Mann-Whitney *U* test and Kruskal-Wallis ANOVA, examples of which follow.

Chi-Square Test

This test examines the discrepancy between the observed and the expected results. The expected frequency is subtracted from the observed frequency, then squared, thereby removing any negative data, before being divided by the expected frequency before being summed.

Equation 17.5. Chi-square equation:

$$\chi^2 = \Sigma \frac{(O-E)^2}{E}$$

Chi-Square Example. A clinical director at a ballet company wanted to see if the dancers were getting more injuries at certain touring venues. The total number of injuries observed (O) was recorded at each venue; the expected (E) value is calculated by the total number of injuries divided by the number of venues; there should be an equal probability of getting injured at each venue. The null hypothesis stated that there is no statistically significant

Table 17.16. Chi-Square Example Data

Venue	1	2	3	4	Total
Observed injuries (O)	15	10	12	23	60
Expected injuries (E)	15	15	15	15	60

Table 17.17. Chi-Square Calculations

Venue	1	2	3	4	Total
Observed injuries (O)	15	10	12	23	60
Expected injuries (E)	15	15	15	15	60
(O-E)	0	-5	-3	+8	
$(O\text{-}E)^2$	0	25	9	64	
$(O\text{-}E)^2/E$	0	1.66	0.6	0.27	
χ^2					2.53

difference in injury incidence between the venues. Alpha error of probability was set at 0.05.

Degrees of freedom for this analysis is number of counts (n-1 = 3), and it is reported in text as $\chi2\ (3,60) = 2.53$, $p < 0.05$, and therefore the null hypothesis is accepted.

Mann-Whitney U Test

This test can be used with a small and large number of groups and requires two independent groups and one dependent variable. There are several methods of calculating *U*, but a ranking method is often used.

Equation 17.6. Mann-Whitney U test equation:

$$u = n_1 n_2 + \frac{n_1(n_1+1)}{2} - \Sigma R_1$$

$$z = \frac{u - \frac{u-(n_1)(n_2)}{2}}{\sqrt{\frac{n_1 n_2 (n_1+n_2+1)}{12}}}$$

Mann-Whitney U Test Example. A tester wanted to examine whether a dancer's happiness varied between a Skinner Release technique class and Martha Graham technique class. Participants were given a mood questionnaire after each class. The null hypothesis was there is no statistically significant difference in mood. Thirty dancers were recruited and signed informed consent; 15 completed a questionnaire immediately after a release class, and 15 completed the same questionnaire after a Graham class.

This indicates a statistically significant difference between the two groups $z = 4.09$; $p < 0.001$.

Table 17.18. Raw Data for Mann-Whitney U Test

Release	Graham
2.00	8.00
4.00	9.00
8.00	4.00
3.00	7.00
6.00	8.00
2.00	6.00
4.00	8.00
5.00	9.00
6.00	8.00
2.00	9.00
4.00	9.00
5.00	8.00
1.00	7.00
6.00	8.00
3.00	9.00

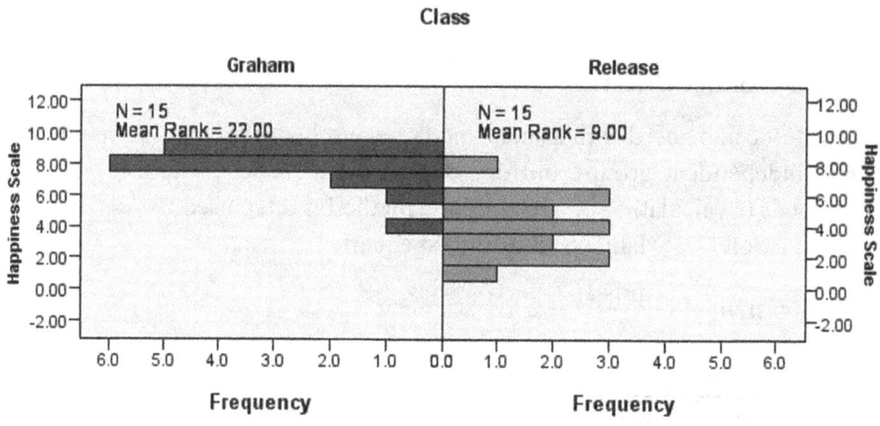

Total (N)	30
Mann-Whitney	210.00
Standard Error	23.83
Standardized Test Statistic	4.091
Asymptotic significance (2-sided test)	<0.001
Exact significance (2-sided)	<0.001

Figure 17.6. Example of Mann-Whitney Test Output.

Kruskal-Wallis ANOVA (by Ranks)

This is the non-parametric version of a one-way ANOVA in that it is used for 3+ independent groups. Like other non-parametric tests, it has ranking of the data at its core.

Equation 17.7. Kruskal-Wallis test equation:

$$H = \left[\frac{12}{N(N+1)}\right] \times \Sigma\left[\frac{(\Sigma R_i)^2}{n_i}\right] - \alpha(N+1)$$

N = the total number of scores; ΣR^2 the sum of the squared sum of the ranks for the three groups; n = the number of scores in a group; α = the number of groups.

Kruskal-Wallis Example. A researcher wanted to see whether Pilates was better than the Gyrotonic method at improving développé height. Nineteen participants were recruited, 7 in each of the Pilates and Gyrotonic groups and 5 in the control group. All signed consent forms prior to the 5-week intervention. Développé height was measured by 2-d video analysis.

Strengths, Limitations, Requirements

Between-group experiments, whether true or quasi-experimental designs, are designed to test the statistical difference between variables. The statistical result is solely a mathematical computation of the differences between variables and needs to be reviewed in relation to its biological/psychological/

Table 17.19. Kruskal-Wallis Data Ranking and Calculations

Pilates	(rank)	Gyrotonic Method	(rank)	Control	(rank)
112	2	133	14	116	5
134	15	145	19	113	3
132	12.5	132	12.5	101	1
128	9.5	128	9.5	121	7
119	6	126	8	114	4
136	17	131	11		
141	19	135	16		
Sum of rank	81		90		20

Calculate $\frac{(\Sigma R_i)^2}{n_i}$ for each group $\frac{(81)^2}{7} + \frac{(90)^2}{7} + \frac{(20)^2}{5} = 2174.43$

[12/19(19+1)] × 2174.43 = 0.032 × 2174.43 = 68.67
68.67–3(N+1) = 68.67–3(19+1) = 8.67
Asymptotic significance (2-sided test) at $p<0.05$

biomechanical reality. In fact, the application of statistical tests cannot make up for poor methodological design or protocols. The hypothesis determines the research design, which in turn determines the statistical test, so in fact prior to collecting data you should know the statistical test you are going to use!

Real-world human research is fraught with difficulties, but as a researcher engaged in rigorous quantitative science you need to try and control as many confounding variables as possible, from the homogeneity of the test participants (*Chapter 8*) to the accuracy of the equipment used to collect data, and the reliability of the tester to use it (*Chapter 8*), to the time of day the test is carried out. Getting enough participants, keeping them, making sure they do turn up and do what you want or don't do what you don't want--these are always among the most difficult tasks. Power statistics will provide the minimum number of participants for parametric tests, but remember that data reported in research articles have often been collected over a number of years, and they do not have the time constraints of a degree program. If you do not manage to recruit enough participants, do not worry; just adjust your chosen statistical test appropriately (parametric to non-parametric) or do not do statistics at all and report the observed trends.

The main focus of group research is to examine how the group responds to an intervention, or is different from another group, and how these results can be generalized to the wider population from which the participants were randomly selected. It is not about individuals and how they react to an intervention or differ from other individuals; this is the realm of case studies or single-case designs (*Chapters 15* and *16*), which rely on replication to determine the generality of their findings.

Summary

Between-group experiments are reductionist in nature in that we often focus only on one variable at a time and try to control other variables so the changes we observe are more than likely due to how we have manipulated the independent variable to show its influence on the dependent variable. As with most quantitative dance medicine and science research, planning is fundamental, and the development of a research proposal is essential. It will allow you and your supervisor to follow a step-by-step protocol of what you intend to do and the order in which you intend to do it. The importance of pilot testing should not be underestimated, as it will help iron out any glitches and highlight any areas of concern. Ethical approval through an IRB or Ethics Committee (*Chapter 6*) is central to human research and it will also allow more established researchers to examine what you intend to do and whether it is possible or relevant (all part of a good ethical review). Finding statistical significance is not vital to research but understanding your results in light of previous research and being able to explain what you have found is germane.

The axioms "less is more, simple is better" (Cohen, 1990) and "contact between people testers and people subjects in people experiments should be minimized, standardized, and randomized" (Martens, 1973) are fundamental to successful research. To become a researcher is a lifetime of striving and reflection, of perfecting techniques, just like becoming a dancer.

Highlights

- Your methodology and participant numbers will determine the correct statistical test.
- Parametric tests are no "better" than non-parametric tests; which is the best test is determined by the data and the research question.
- Highlight your independent and dependent variables; it helps you and the reader.
- Plot your data as it helps you understand the data and the statistical output.

References

Allen, M. (2017). *Solomon Four-Group Design.* Sage.

Cohen, J. (1990). Things I have learned (so far). *American Psychologist, 45*(12), 1304–1312.

Cook, T., & Campbell, D. (1979). *Quasi-Experimentation: Design and Analysis Issues for Field Settings.* Houghton Mifflin.

Lakens D. (2017). Equivalence tests: A practical primer for *t* tests, correlations, and meta-analyses. *Social Psychological and Personality Science, 8*(4), 355–362.

Lakens, D., Scheel, A. M., & Isager, P. M. (2018). Equivalence testing for psychological research: A tutorial. *Advances in Methods and Practices in Psychological Science, 1*(2), 259–269.

Martens, R. (1973). People errors in people experiments. *Quest, 20*(1), 16–20.

Schuirmann, D. J. (1987). A comparison of the Two One-Sided Tests Procedure and the Power Approach for assessing the equivalence of average bioavailability. *Journal of Pharmacokinetics and Biopharmaceutics, 15,* 657–680.

Walker, E., & Nowacki, A. S. (2011). Understanding equivalence and noninferiority testing. *Journal of General Internal Medicine, 26*(2), 192–196.

Zhu, W. (2012). Sadly, the earth is still round ($p < 0.05$). *Journal of Sport and Health, 1,* 9–11.

Additional Resources

Atkinson, G., & Nevill, A. M. (1998). Statistical methods for assessing measurement error (reliability) in variables relevant to sports medicine. *Sports Medicine, 26*(4), 217–238.

Nevill, A. M., Holder, R. L., & Cooper, S. M. (2007). Statistics, truth, and error reduction in sport and exercise sciences. *European Journal of Sport Science, 7*(1), 9–14.

Vincent, W. J. (1999). *Statistics in Kinesiology* (2nd ed.). Human Kinetics.

18

Systematic Literature Reviews

MANUELA ANGIOI AND CLAIRE E. HILLER

Questions to Stimulate Thinking

- What are the essential features of systematic reviews?
- How does a systematic review of the literature help researchers assess available evidence in an objective, unbiased way?
- What are the advantages of using the systematic review approach to reviewing the literature in dance medicine and science?

The purpose of this chapter is to introduce systematic literature review as a methodology that you may use to resolve debates or answer a specific research question. As described in earlier chapters, before conducting any research, you should conduct a review of the literature to understand what is known about your topic of interest, so that you can plan your project appropriately. In the past, researchers learned about prior work in their area of interest by reading traditional narrative literature reviews. These narrative reviews were often written by subject matter or content area experts. While narrative reviews can stimulate good debates and provide good background information, they may be at risk for selection and researcher bias. A systematic review can address these limitations.

A systematic review is defined as "a review of the evidence on a clearly formulated question that uses systematic methods to identify, select and appraise relevant primary research, and to extract and analyze data from the studies that are included" (Adams et al., 2016). A systematic literature review describes, compares, contrasts, and evaluates relevant original evidence (or knowledge) with the goal of providing an objective comparative summary using a systematic approach. The main functions of this review are to summarize and evaluate the literature using objective tools for analysis to reduce the risk of researcher bias. There is a trade-off here, as this reduction may sometimes increase the risk of selection bias by missing the opportunity to include relevant studies that do not meet sometimes misguided filtering criteria (e.g., sample size or age minimum or maximum cutoffs, random selection, or random assignment). Still, a systematic review provides an objective evaluation of available evidence in relation to a specific question or topic in the dance sciences. Thus, a systematic review is not an annotated bibliography, a summary of sources, or a descriptive summary of chosen papers. The three main activities that characterize systematic reviews are examining, evaluating, and establishing the relevance of research findings.

Systematic reviews of qualitative research are known by various names, such as Meta-synthesis, Meta-ethnography, Qualitative Evidence Synthesis, Qualitative Meta-Synthesis, or Meta Narrative Review. They aggregate qualitative data to generate an evidence-informed new interpretation or summary, which helps in building theory and deep understanding about a specific phenomenon. There is no hierarchy nor are there rules of evidence for qualitative studies. However, in 2013 the Cochrane Qualitative and Implementation Methods Group (CQIMG) created a template for qualitative research synthesis (http://epoc.cochrane.org/epoc-specific-resources-review-authors) as part of the GRADE CERQual (Lewin et al., 2015) approach to assessing confidence in qualitative research findings.

Before we describe the essential features of the systematic review, it is important to describe what constitutes research literature. Often the inclusion criteria in a systematic review list original research but do not adequately explain how the studies selected qualify for inclusion in the review, which in turn can lead to the inclusion of journal articles only. Rather, it is important to recognize that research literature includes academic journal articles, internet sources, conference proceedings, government and corporate reports, theses and dissertations, and books.

In simpler words, a systematic review is a methodological tool that can be used by researchers, students, and clinicians to evaluate the evidence objectively on a given topic. A systematic review is pre-planned with the steps fully described in advance. All actions are transparent and reproducible. The clear advantages of the systematic review are that it reduces the risk of researcher bias, is easier to replicate, is designed to resolve controversy (rather than stimulate it), and offers a reliable basis for decision-making. However, as with all research methodologies, it has limitations. For example, a systematic review can provide a specific answer only to a specific, well-defined question. In addition, some topics may have too few studies to form a convincing conclusion, and some studies do not present sufficient data or information to answer the research question.

Reasons to Perform a Systematic Review

A systematic review can provide clarity for dance science. To do so, an objective approach that selects and evaluates the relevant evidence is vital to synthesize a large body of research, particularly when conflicting results are presented in the literature (see Table 18.1). A specific example in dance could be reviewing the evidence on whether supplementary fitness training has positive effects on performance and/or injury reduction.

Research Questions that Systematic Reviews Can Answer

Researchers can use systematic reviews to answer several different types of research questions (Munn et al., 2018). See Table 18.2 for specifics of the common systematic review types. In dance science, the most common is to examine the effectiveness of interventions on outcomes, usually using original research reports (e.g., randomized controlled trials). Systematic reviews can answer questions relating to: prevalence, risk, prognosis, diagnosis, psychometric properties, experiences, expert opinion, costs, and methodology (Munn et al., 2018). Reviews can also synthesize expert opinion or policy, costs or cost effectiveness of a condition or intervention, or current research methods. Sometimes a review aims to synthesize information on a topic,

Table 18.1. Summary of Essential Methodological Features of Systematic Reviews

Feature	Requirement
Define a question	Essential
Writing a protocol	Essential
Methodology	Rigorous
Searching	Exhaustive search strategy
Definition of inclusion/exclusion criteria	Essential
Screening	Systematic/cross checked
Quality assessment	Essential
Data extraction	Essential
Analysis and synthesis	Quality assessment and meta-analysis
Replication	Explicit—replicable
Timescale	To be planned

Table 18.2. Examples of Types of Systematic Reviews

Type of Review	What It Examines	Example
Traditional	Effectiveness of an intervention	The Effectiveness of Dance Interventions on Physical Health Outcomes Compared to Other Forms of Physical Activity (Fong Yan et al., 2018).
Prevalence/Incidence	Occurrence or rate of something, such as a health condition or injury	Injuries in pre-professional ballet dancers: incidence, characteristics, and consequences (Ekegren et al., 2014).
Risk	Factors associated with the development of something, often a health condition or injury	Risk factors for musculoskeletal injury in pre-professional dancers (Kenny et al., 2016)
Prognostic	How an injury, disease, or condition usually changes over time	Return to Sport-Specific Performance After Primary Anterior Cruciate Ligament Reconstruction (Mohtadi & Chan, 2018).
Diagnostic	Which type of clinical or other test is the most accurate to diagnose a particular condition, or using reliable diagnostic tests.	An updated Systematic Review of Turnout Position Assessment Protocols used in Dance Medicine and Science Research (Angioi et al., 2021)
Experience, or qualitative, reviews	People's experiences of something or the meaning to them of having a condition or receiving an intervention	Creative Arts-Based Therapies for Stroke Survivors: A Qualitative Systematic Review (Lo et al., 2018)

or the review question does not fit neatly into a particular "type" of review. However, the robust methodology of a systematic review and a strong review question are still needed to address the topic. All these types of systematic review can answer a well-developed and focused question by synthesizing evidence from several studies. The way the findings from the studies are extracted and synthesized may vary slightly based on the questions.

Stages of a Systematic Review

The accompanying flow diagram offers a summary of stages that you might want to follow when conducting a systematic literature review.

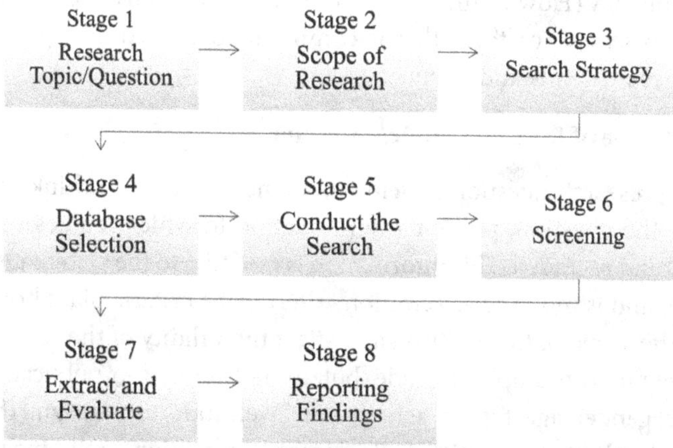

Figure 18.1. Stage Sequence Template of Systematic Review Process.

Stage 1: Research Topic/Question

Similar to other research types, a systematic review should aim to provide a valid answer to a specific question. In clinical research, defining a clinical question in terms of the specific patient problem aids the researcher in finding relevant evidence in the literature. It is important to spend time formulating the question, as clarity of thinking at this stage can save time later in the process. One useful tool is to formulate the question using the PICO model (or one of its variations, depending on the type of systematic review you are doing):

- Population (e.g., Ballet dancers)
- Intervention (e.g., Training Type)
- Comparison HIIT (High Intensity Interval Training) versus Circuit Training
- Outcome (Muscular power)

All these lead to the research question: "Is HIIT more effective than circuit training at improving muscular power in ballet dancers?"

It is also important to think of and define the research question to be reviewed in terms of whether the overall aim is to investigate:

- Etiology (What causes this condition?)
- Diagnosis (Is this method/strategy valid to detect an X condition?)
- Prognosis (What is the probability of developing an X outcome?)
- Intervention (What are the effects of treatment X on outcome Y?)

- Frequency (How common is condition X) of a given condition (clinical research) or topic that is commonly observed or reported in dance medicine and science?

Stage 2: Scope of Research Topic/Question

Once the research question is defined, the next stage is to think about the scope of the question, particularly in relation to which studies should be included and excluded. This information is essential to the systematic review approach and is reported as selection criteria. The criteria place boundaries around the scope of the question and affect the validity of the results. Some commonly used examples include, but are not limited to publication date, language, gender, age, type of activity and level, and study design. This stage refers to the ability to identify if all relevant information would include academic journal articles only or other sources such as reports and guidelines (including reports from national health bodies or government agencies). It is essential to remember that type of information includes the type of study designs to be included. For example, if the research question is to explore the effects of supplementary fitness training on muscular power in ballet dancers, only intervention and selected types of observational studies should be included in the search, as cross-sectional studies cannot detect causality, to reduce selection bias.

Example

Publication dates from inception to XXX OR from January 2000 to current.

No restrictions on language OR language restricted to articles written in English. Peer-reviewed original articles.

Inclusion criteria: male and female; ballet training full-time; professional; circuit training; HIIT training; intervention, case-control, cohort studies.

Exclusion criteria: other forms of dance; ballet dancers not in full-time training; ballet dancers not employed by a professional company; other forms of exercise training; cross-sectional; case studies; case series studies.

Stage 3: Search Strategy

Key words are essential to ensure that the search will be sensitive and specific to retrieve all relevant evidence. In this case, sensitivity refers to the ability of the search to identify correctly the manuscripts that are relevant to the research question, and specificity refers to the ability to identify correctly the evidence that is specific to the inclusion criteria. Search techniques

Table 18.3. MeSH (Medical Subject Headings) Terms Search Example

"muscles" [MeSH Terms] OR "muscles" [All Fields] OR "muscle" [All Fields] AND "power" [MeSH Terms]

OR

"power" [All Fields] AND endurance [All Fields] AND strength [All Fields] AND fitness [All Fields] AND High Interval Intensity Training [All Fields] AND "circuit-based exercise"[MeSH Terms]

OR

("circuit-based" [All Fields] AND "exercise" [All Fields])

OR

"circuit-based exercise" [All Fields]

OR

("circuit" [All Fields] AND "training" [All Fields])

OR

"circuit training" [All Fields] AND ("dancing" [MeSH Terms])

OR

"dancing" [All Fields]

OR

"ballet" [All Fields] AND dancers [All Fields] AND ("exercise" [MeSH Terms])

OR

"exercise" [All Fields]

include considering the use of synonyms and suffixes. Medical Subject Heading Terms (MeSH) are a key feature of Medline (a bibliographic database of life sciences and biomedical information that includes bibliographic information for articles from academic journals covering medicine, nursing, pharmacy, dentistry, veterinary medicine, and health care), which allows a searcher to identify relevant articles. Baumann (2016) describes MeSH terms as "labels assigned to each article in Medline to describe what the article is about. When labeling an article, indexers select terms only from the official MeSH list—never other spellings or variations." This guideline allows for efficient searching. Once the relevant MeSH term is identified, this will include variants such as spelling and lay terms.

Example

"Is HIIT more effective than circuit training at improving muscular power in ballet dancers?"

Search terms: Ballet or dancers or classical dancers or ballet dancers AND muscular power or strength or endurance or physical fitness, or see Table 18.3 for MeSH terms:

Stage 4: Database Selection

Then you should identify the most relevant databases that are likely to contain the research reports that will answer your question. Also check which databases overlap, so you do not do more work than needed. This stage allows you to access relevant information that has been identified in Stages 2 and 3. Most searches can be done using online libraries or databases, such as CINAHL (Cumulative Index to Nursing and Allied Health Literature), PubMed, Ovid, SportDiscus, and others. You might want to consider whether you can access the full-text versions of the literature, and what to do if you do not have access. One good strategy is to contact the corresponding author(s) but also to access academic portals that allow for article sharing, such as Research Gate, Academia.edu, or LinkedIn. Review one database first before completing all searches. This will ensure that you see how many reports are found (and whether your search yield is too large or small) and ensure that all important studies are identified.

You should also consider registering your Systematic Review in an open register. PROSPERO register (Centre for Reviews and Dissemination, University of York, www.crd.york.ac.uk/prospero/) allows for researchers to update the progress of a given registered title or topic. This process is now common as many journals are starting to publish only a protocol that can be checked, so that there is lower risk for duplication in systematic reviews.

Stage 5: Conduct the Search

During this stage it is advisable to consider using reference management software (e.g., EndNote, Mendeley, Zotero, RefWorks) as described in earlier chapters to save relevant literature and to cite and format references as needed based on the outlet where you are planning to publish this work. This is an important stage, the first moment of pause and reflection since you have embarked on a systematic review journey. Have you found too many or too few references? Considering that dance medicine and science are relatively new areas for research, we suggest that if your search generates fewer than five research reports, review the selection criteria to increase sensitivity, and if it generates more than 100 reports, increase the specificity of your search. Specificity and sensitivity are determined in Stages 2 and 3, when key words are determined in relation to the lists (selection criteria) you have decided on to answer a given research question. Consider the topic and the study design(s) that are most suited to provide an objective answer to your research question.

Stage 6: Screening

At the start of this stage, you should have all the research literature that matches your inclusion criteria uploaded to your reference management software. Screening of titles and abstracts should be by multiple reviewers (usually two per report). Make sure you capture all relevant reports, and include all chosen reports from each screener in the next screening. Retrieve and store full-text copies of all studies included. Then screen the full text and note reasons why studies are not included. You may not be able to access all full texts, often due to the age or location of the reports. Note this limitation in your reasons for excluding reports in your final analyses.

According to the Preferred Reporting Items for Systematic Reviews and Meta-Analyses (PRISMA) guidelines (www.prisma-statement.org), to ensure sensitivity and specificity of results, and avoid selection bias, multiple independent reviewers need to compare search results from the identified number of articles found in multiple databases and the reference list of each identified article. If the reference list check reveals a high number of missed articles (defined as articles that are relevant to the research topic but that have not been depicted by the search terms used in the databases), you should consider going back to Stages 2 and 3 and reviewing the selection criteria, and identify a key study to compare the databases and search terms used. A third researcher is usually employed to solve discrepancies between the main reviewers (usually two or more reviewers). These steps help strengthen the objectivity and replicability of the systematic reviews and so should be carefully followed when conducting a systematic review.

Stage 7: Extract Data and Evaluate the Evidence

Then check the information you need to report to answer your research question. Make up a spreadsheet or other document with the data you wish to extract from each paper or source, along with clear guidelines of what should be extracted for each point. For example, you may want to extract characteristics of the population being studied—does this include gender, age (mean and standard deviation or standard error, or range), weight, and height? Try your data extraction template on several studies. This will help avoid the need to return to all the included studies again to extract forgotten data.

To provide an objective evaluation of the studies included in the review, Quality Assessment Tools (QAT) and risk of bias need to be considered thoroughly at this stage. Similar to Stage 6, PRISMA guidelines include the use of relevant QAT to be used by multiple reviewers. Quality Assessment (QA) is necessary because understanding strengths and weaknesses of a given study design can influence your interpretation of the results. QA can determine how much you can trust the results of a particular study and can assist with data analyses and guide future research. The main components of QA include appropriateness of study design, risk of bias, choice of outcome measure(s), statistical choices, and quality of reporting and quality of intervention (if relevant to research topic/question). To improve reporting, several groups of experts have developed guidelines and statements to standardize reporting quality, such as Consort for RCTs; TREND for non-randomized trials, and STROBE for observational studies. These guidelines can be accessed via the Enhancing the QUAlity and Transparency Of health Research (EQUATOR) website (www.equator-network.org). When conducting a systematic review, PRISMA guidelines suggest the use of QAT to assess objectively the overall quality of the studies included.

As many scales exist, it is good to keep in mind that: (1) multiple scales should be used within one systematic review if multiple study designs were included; (2) the scales allow for a numerical overall score that should guide in understanding the level of evidence. For example, the PEDro scale covers internal validity for included studies very well, but it is only suitable for RCTs and randomized crossover trials and might not be suitable for observational studies.

The overall aim of QA is to describe objectively the methodological features of the retained studies, and therefore authors should consider types of studies that are intended to be used in the review and the quality assessment strategy. Some assessment tool examples are included in Table 18.4. Please note this is not an exhaustive list, and researchers should also be aware of

Table 18.4. Selected Quality Assessment Tools (QAT) According to Study Design

Quality Assessment Tool	Study design
PEDro scale (Physiotherapy Evidence Database, www.pedro.org.au/english/downloads/pedro-scale)	Trials
Delphi list (Verhagen et al., 1998)	
Modified Downs and Black scale (Morton et al., 2014)	Non-randomized interventional studies; cross sectional studies
Downs and Black scale (Downs & Black, 1998)	
Newcastle-Ottawa Scale (Wells et al., n.d.)	Observational studies (cohort study and case-control study)
CASP checklists (Critical Appraisal Skills Programme, casp-uk.net/casp-tools-checklists)	

re-validation of existing QA tools such as the Downs and Black tool, originally developed in 1998 and later modified and revalidated. QA assists the researcher by providing an indication of the overall quality of the findings reported in a systematic review. These assessments may also help reviewers make decisions about which studies to include and exclude in quantitative analyses such as meta-analyses.

Stage 8: Reporting Findings

The PRISMA flow diagram should always be included in the systematic review. The diagram allows for an accurate and concise representation of the search from beginning to included studies for qualitative and quantitative (if applicable) evaluation. The diagram includes four stages: identification, screening, eligibility, and inclusion (www.equator-network.org/reporting-guidelines/prisma).

It is possible to undertake a meta-analysis of data extracted from your included studies. A meta-analysis is a statistical approach that combines the results of similar studies (or outcomes or populations) to achieve a larger sample size and thereby, it is hoped, increase statistical power, improve effect size estimations, and resolve uncertainty. How to conduct a meta-analysis is outside the scope of this chapter, and excellent guides are available elsewhere (Cleophas & Zwinderman, 2017). However, we offer an example of a meta-analysis conducted with dancers near the end of this chapter. It is important to consider objectively which studies are appropriate to combine in your analysis. For example: Are the studies comparing the same populations or intervention types? Are the outcomes the same? Once you have included all the data in your statistical software package, be sure you check (and report!) that the studies are also statistically similar. This is usually done by reporting the Chi-square statistic (for whether the studies are homogeneous), or how large the variation is (heterogeneity, I^2 statistic) if they are not homogeneous.

Systematic Review Example

One example of a dance systematic review is "The effectiveness of dance interventions on physical health outcomes compared to other forms of physical activity: A systematic review and meta-analysis" by Fong Yan et al. (2018). This review followed the PRISMA guidelines and the authors registered the protocol before commencement with PROSPERO. The aim of the systematic review was to investigate the effectiveness of structured dance interventions, in comparison to structured exercise programs, on physical health outcome measures.

There were two hypotheses: (1) that dance would be as effective as structured exercise programs on measures of physical health, and (2) that there would be increased adherence with dance interventions compared to structured exercise programs. The authors named key databases and described their search strategy. The authors provide very clear definitions for the key terms of "dance," "dance class," and structured exercise, along with inclusion and exclusion criteria. The authors did not have language restrictions to minimize potential reporting bias, but they chose not to search the gray literature, defined as "a range of documents not controlled by commercial publishing organizations" (Adams et al., 2016). The authors used a validated tool, the MINORS (Methodological Index for Non-Randomized Studies) to examine reporting bias. Two researchers independently screened and extracted data. Overall, in this systematic review, the authors were able to synthesize the included studies descriptively and performed a meta-analysis for some variables where more than four studies reported the same outcome.

Advantages of Systematic Reviews

Overall, there are several advantages of systematic reviews:

- The methods used to find and select studies reduce bias and are more likely to produce reliable and accurate conclusions.
- Systematic reviews summarize findings from multiple studies. This makes the information easier for the end user to read and understand.

Systematic reviews can also:

- Give an idea of how well findings might apply to everyday practice
- Identify knowledge gaps or reveal discrepancies that call for more research
- Offer new insights when drawing conclusions, as it takes in a range of views and findings

Systematic reviews are also less costly to carry out than a new set of experiments and take less time.

Other Considerations

As will be evident by now, you will need to assemble a team of people to help you answer your question with a systematic review. The team can include fellow students, supervisors, friends, etcetera. You might also consider having a librarian on your team, to assist with searching different databases and extracting the studies chosen to review, and consult a content expert if this aspect is not represented on the research team, to ensure that inclusion and exclusion criteria will not inadvertently result in a selection bias that may compromise the validity of the review. Ways of searching often change in databases, databases may change affiliations, and the requirements of the software you use to collate the studies may also change as the programs are updated. A research librarian is often knowledgeable across all of these features, and you can save a great deal of time and stress by consulting one. Managing your record keeping, databases, and analysis software is very important, so ensure that you have a clear view of the steps and have appropriate software to help you. It is possible to do a systematic review without specialized software, but it is easier to manage if you have access to specialized software that will keep records of your steps as you go. If you decide to do a meta-analysis, then this will require the use of other specialized software, such as RevMan by the Cochrane collaboration. It is available free (training.cochrane.org/online-learning/core-software-cochrane-reviews/revman) and can be used even if you are not doing a Cochrane review, provided you reference its use.

Limitations of Systematic Reviews

A systematic review can be limited by several aspects. The authors of the included studies and sources may have provided limited information in the manuscript. The authors may not have included their study designs. In these cases you may not have evidence, or you cannot combine the results because the methodologies, outcomes, or populations are too different. Only published research reports are usually included, but there may be a selection bias because studies with negative findings may not be selected for publication. The question you want to answer may not suit the necessary narrowness of a systematic review question. Additionally, a systematic review takes time and resources, and you may need a quicker answer to your question.

Summary

Overall, systematic review is a robust research methodology that offers dance science researchers an opportunity to review the evidence objectively around a given research question or topic. Using systematic reviews to analyze the literature reduces the risk of bias. Clear descriptions of the processes make it easier for other researchers to replicate the review if needed. Finally, a well-organized review can resolve controversy about a topic and offer practitioners a reliable basis for decision-making. Researchers may benefit from using a systematic review approach when interested in evaluating the evidence around a given topic open to debate or containing contrasting results. To perform a systematic review, multiple researchers will follow the PRISMA guidelines to allow for reduction of bias in selection, screening, and evaluation of the identified relevant evidence. The results will undergo a quality assessment, which should be specific to study designs that have been included for the purposes of the review. Researchers who succeed in conducting a systematic review are likely to make a significant contribution to the field by informing an overall objective evaluation of current knowledge and identifying gaps that provide opportunities for future research.

Highlights

- A systematic literature review is a systematic approach that describes, compares, contrasts, and evaluates relevant original evidence (or knowledge) with the goal of providing an objective answer to a specific question.
- As with any other original research, the most time and energy in a systematic review should be invested in setting up the question and the methods.
- Sometimes a review aims to synthesize information on a topic, or it does not fit neatly into a "type." However, the robust methodology of a systematic review and the need for a strong research question still apply.
- The main functions of a systematic review are to summarize and evaluate the literature by employing objectivity tools for analysis and reducing the risk of bias.

References

Adams, J., Hillier-Brown, F. C., Moore, H. J., Lake, A. A., Araujo-Soares, V., White, M., & Summerbell, C. (2016). Searching and synthesizing 'grey literature' and 'grey information' in public health: Critical reflections on three case studies. *Systematic Reviews, 5,* 164.

Amorim, T., Wyon, M., Maia, J., Machado, J. C., Marques, F., Metsios, G. S., Flouris, A. D., & Koutedakis, Y. (2015). Prevalence of low bone mineral density in female dancers. *Sports Medicine, 45*(2), 257–268.

Angioi, M., Hodgson, K., & Kryger, K. (2021). An updated systematic review of turnout position assessment protocols used in dance medicine and science research. *Journal of Dance Medicine & Science, 25*(1), 55–71.

Baumann, N. (2016). How to use the medical subject headings (MeSH). *International Journal of Clinical Practice, 70,* 171–174.

Cleophas, T. J., & Zwinderman, A. H. (2017). *Modern Meta-Analysis Review and Update of Methodologies.* Springer International Publishing.

Downs, S. H., & Black, N. (1998). The feasibility of creating a checklist for the assessment of the methodological quality both of randomized and non-randomized studies of healthcare interventions. *Journal of Epidemiology and Community Health, 52,* 377–384.

Ekegren, C. L., Quested, R., & Brodrick, A. (2014). Injuries in pre-professional ballet dancers: Incidence, characteristics, and consequences. *Journal of Science and Medicine in Sport, 17*(3), 271–275.

Fong Yan, A., Cobley, S., Chan, C., Pappas, E., Nicholson, L. L., Ward, R. E., Murdoch, R. E., Gu, Y., Trevor, B. L., Vassallo, A. J., Wewege, M. A., & Hiller, C. E. (2018). The effectiveness of dance interventions on physical health outcomes compared to other forms of physical activity: A systematic review and meta-analysis. *Sports Medicine, 48*(4), 933–951.

Glenton, C., Bohren, M. A., Downe, S., Paulsen, E. J., & Lewin, S., on behalf of Effective Practice, and Glenton, C., Bohren, M. A., Downe, S., Paulsen, E.J., & Lewin, S., on behalf of Effective Practice and Organisation of Care (EPOC). (2021). EPOC Qualitative Evidence Syntheses: Protocol and review template v1.2.

Kenny, S. J., Whittaker, J. L., & Emery, C. A. (2016). Risk factors for musculoskeletal injury in preprofessional dancers: A systematic review. *British Journal of Sports Medicine, 50*(16), 997–1003.

Lewin, S., Glenton, C., Munthe-Kaas, H., et al. (2015). Using qualitative evidence in decision making for health and social interventions: An approach to assess confidence in findings from qualitative evidence syntheses (GRADE-CERQual). *Public Library of Science Medicine, 12*(10), e1001895.

Lo, T., Lee, J., & Ho, R. (2018). Creative arts-based therapies for stroke survivors: A qualitative systematic review. *Frontiers in Psychology, 9,* 1646.

Mohtadi, N. G., & Chan, D. S. (2018). Return to sport-specific performance after primary anterior cruciate ligament reconstruction: A systematic review. *American Journal of Sports Medicine, 46*(13), 3307–3316.

Morton, S., Barton C. J., Rice, S., & Morrissey, D. (2014). Risk factors and successful interventions for cricket-related low back pain: A systematic review. *British Journal of Sports Medicine, 48,* 685–91.

Munn, Z., Stern, C., Aromataris, E., Lockwood, C., & Jordan, Z. (2018). What kind of systematic review should I conduct? A proposed typology and guidance for systematic reviewers in the medical and health sciences. *BioMed Central Medical Research Methodology, 18*(5).

Verhagen, A. P., de Vet, H. C., de Bie, R. A., Kessels, A. G., Boers, M., Bouter, L. M., & Knipschild, P. G. (1998). The Delphi list: A criteria list for quality assessment of randomized clinical trials for conducting systematic reviews developed by Delphi consensus. *Journal of Clinical Epidemiology, 51*(12), 1235–1241.

Wells, G. A., Shea, B., O'Connell, D., Peterson, J., Welch, V., Losos, M., & Tugwell, P. (n.d.). *The Newcastle-Ottawa Scale (NOS) for Assessing the Quality of Nonrandomised Studies in Meta-Analyses.* Ottawa Hospital Research Institute. http://www.ohri.ca/programs/clinical_epidemiology/oxford.asp

Additional Resources

Bastian, Hilda (2017, July 3). 5 tips for understanding data in meta-analyses. *Public Library of Science.* https://blogs.plos.org/absolutely-maybe/2017/07/03/5-tips-forunderstanding-data-in-meta-analyses/

Campbell, M., McKenzie, J. E., Sowden, A., Katikireddi, S. V., Brennan, S. E., Ellis, S., Hartmann-Boyce, J., Ryan, R., Shepperd, S., Thomas, J., Welch, V., & Thomson, H. (2020). Synthesis without meta-analysis (SWiM) in systematic reviews: Reporting guideline. *British Medical Journal (Clinical Research Edition), 368,* l6890.

Centre for Reviews and Dissemination, University of York. (2009, January). *Systematic Reviews CRD's Guidance for Undertaking Reviews in Health Care.* http://www.york.ac.uk/inst/crd/SysRev/!SSL!/WebHelp/SysRev3.htm

19

Epidemiology Research in Dance

ROGIER M. VAN RIJN, SARAH J. KENNY,
AND JANINE H. STUBBE

Questions to Stimulate Thinking

- How should dancer health problems (e.g., injury and illness) be defined?
- How much of a problem are injuries and illness for dancers?
- What are the best ways to measure injuries and illness among dancers?

This chapter on research in dance epidemiology uses multiple approaches (e.g., observational, analytical, qualitative), and we hope this will encourage you to integrate concepts covered in previous chapters. While reading, consider how different research methodologies can be brought together, or used in series, to address various aspects of health-related issues that are relevant to dancers.

What Is Dance Epidemiology?

Dance epidemiology is the dance science that studies the patterns, causes, tracking, and prevention of health-related events among dance populations. The objective of dance epidemiology research is to minimize the risk of injury and illness experienced by dancers, thus protecting their overall health and well-being. Table 19.1 describes some important terms that are relevant to understanding dance epidemiology.

Table 19.1. Key Terms Used in Dance Epidemiology

Term	Definition
Epidemiology	The study of the distribution (patterns), determinants (causes), surveillance (tracking), and deterrents (ways of preventing or reducing) of health-related events in specific dance populations
Burden	The overall impact of a health problem on a given dance population
Prevalence	At any given point in time, the number of dancers who currently have a health problem divided by the total number of dancers who are at risk of the same health problem
Incidence	The number of new health problems that occur during a specified period of time
Severity	The degree of a health problem
Prevention	Taking action to reduce the risk for a health problem
Mechanism	Reasons or causes for why a health problem occurred
Risk factor	Something that increases a dancer's chance of developing a health problem
Intrinsic factor	A factor acting from within a dancer; related to the internal characteristics of the individual dancer
Extrinsic factor	A factor that has influence from outside the dancer; related to the external environment of the individual dancer
Interaction	The action or influence of multiple risk factors on each other
Exposure	The time during which dancers are at risk of a health problem
Outcome	Health problem of interest (i.e., injury or illness)

Sport Injury Prevention Models

Several models in sport have been used to provide a framework to study dance epidemiology. One of the most common models often referred to in the literature is called the *Sequence of Prevention of Sports Injuries Model* proposed by van Mechelen and colleagues (1992). This model describes a four-step process to understand prevention of sports injuries, though it can be applied to all health problems in a variety of contexts. First, the size (or magnitude) of the health problem must be identified and described in terms such as prevalence, incidence, and severity. Second, the factors (i.e., intrinsic and extrinsic) and possible causes (or mechanisms) that play a part in the occurrence of the health problem must be identified. Third, strategies are introduced to reduce the risk of the health problem. These strategies should be based on information about the possible causes identified in step two. Finally, the impact (or effect) of the risk reduction strategies must be evaluated by circling back to the start of the model and repeating step one.

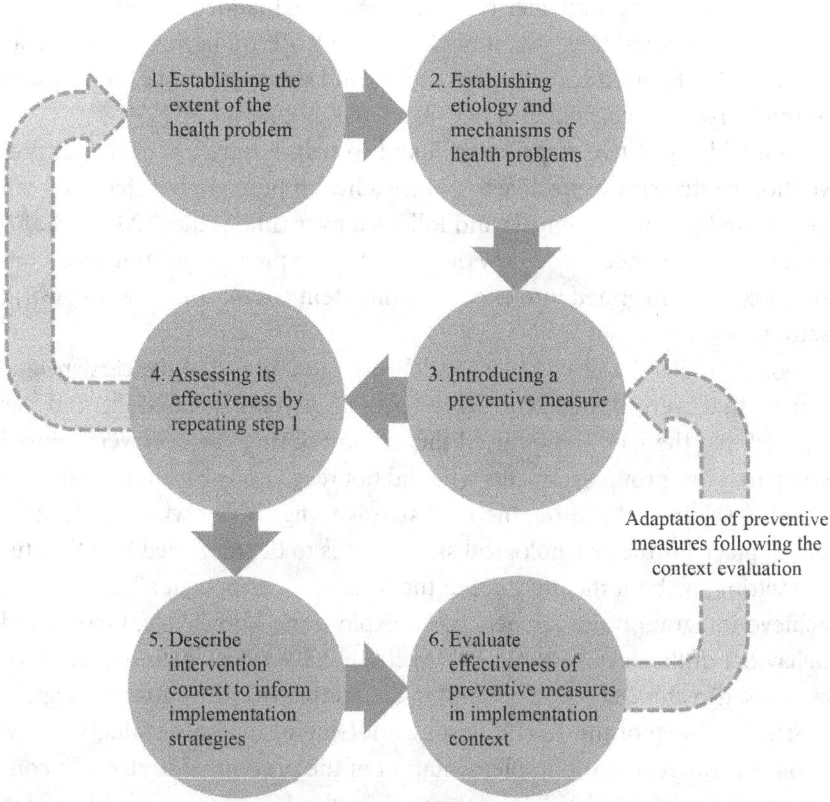

Figure 19.1. An Adapted Model for Translating Research into Injury Prevention Practice (TRIPP) Framework. Adapted from Finch (2006).

A critical step of this model is identifying all possible causes of health problems. This includes gathering information on why a particular dancer may be at risk in a given situation (e.g., what are the risk factors?), and how health problems happen (e.g., what are the mechanisms?). With this comprehensive understanding of health problem causation, researchers can then figure out how best to reduce the burden experienced by dance populations. In current dance epidemiology research, studies are largely focused on step one of this model. While progress is being made to investigate the application of this model (e.g., van Seters et al., 2020; van Winden et al., 2020; van Rijn et al., 2021), much remains unknown (Kenny et al., 2016).

A further development to the van Mechelen model was introduced by Finch in 2006. The *Translating Research into Injury Prevention Practice* (TRIPP) framework (see Figure 19.1 for dance-specific adaptation) emphasizes the need for research that focuses on the introduction (or implementation) of effective preventive strategies (Finch, 2006). Finch added two steps to van Mechelen's original four-step model, so the TRIPP framework includes six steps needed to build good understanding and strong evidence for how best to minimize the occurrence of health-related problems.

The additional two steps (steps 5 and 6) help ensure not only that prevention strategies are effective to reduce a health problem but that they can be adopted (or implemented) and followed over time (Finch, 2006). Step 5 ensures that we understand how the results of the prevention strategies from step 4 can be integrated into effective, consistent approaches in a real-world setting.

For example, if a researcher has delivered to a group of dancers specific classes that were designed to develop their psychological skills and has learned that these classes reduced the amount of time dancers were injured compared to a group of dancers who did not receive any psychological skills training (Noh et al., 2007), the next step is to figure out what needs to be put in place for the psychological skills classes to be integrated into the studio setting, without the presence of the researcher, to benefit all dancers. To achieve integration, researchers often explore the knowledge, beliefs, and behaviors of those who would be directly involved (e.g., dancers, teachers, coaches, directors, choreographers) in facilitating the preventive strategy.

The final step of the TRIPP framework (step 6) involves evaluating how good was the real-world implementation of the preventive strategy. In contrast to the original TRIPP framework, a feedback loop was added in which adjustments can be made to the preventive measures based on the evaluation in step 6. Using the current example related to psychological skills training, in step 6 we might ask questions like:

- Did the teachers deliver the psychological skills training as they were designed?
- Did the dancers engage with the training as they were instructed?
- Was there an overall positive impact on reducing the burden of injury (the extent of the problem) among most of the dancers?

Additional key sport models to be aware of when studying dance epidemiology relate specifically to the possible factors that put dancers at increased risk for health problems (i.e., step 2 of the van Mechelen model). In 2007 Meeuwisse and colleagues compiled a model called the *Dynamic, Recursive Model of Etiology in Sport Injury* that highlights the importance of examining both intrinsic factors and extrinsic factors, as well as any possible interaction between intrinsic and extrinsic factors that may make a dancer susceptible to a health problem before it happens. The model also recognizes that the occurrence of a health problem (e.g., getting injured or experiencing illness) is not always an endpoint or final outcome, since rehabilitation and recovery are part of a continuing process (Meeuwisse et al., 2007). A dancer's risk for injury or illness is described as being dynamic, something that can change throughout the time of participating in dance. In other words, by simply dancing, some intrinsic risk factors (e.g., those that are modifiable, like a dancer's muscular flexibility, level of depression, etc.) can change, which in turn changes the dancer's susceptibility to injury or illness. The dancer can also be exposed to the same or different extrinsic risk factors and, again, can have a different susceptibility (Meeuwisse et al., 2007). This model (see Figure 19.2 for adapted dance-specific version) illustrates these susceptibilities with circular arrows, incorporating the consequences of repeated participation in dance, both with and without injury or illness, into the risk profile (or predisposition) of dancers (Meeuwisse et al., 2007).

The dynamic, recursive model also illustrates that sometimes a cause of a health problem (the mechanism or inciting event) may not always lead to a health problem. For example, if a musculoskeletal injury does occur, dancers may no longer participate in dance if there is no recovery (i.e., they are removed from dancing). However, they may also have partial or full recovery from the injury, which will facilitate their coming back to dance classes (i.e., repeated participation), leading to a change in their status as a "predisposed dancer." In most circumstances, participation does not result in a health problem but may produce an adaptation (or maladaptation) to their muscle tissues, for instance, which would modify their intrinsic risk factors on the basis of the interaction (influence) of the characteristics of the event and pre-existing risks.

Understanding the occurrence of health problems in a dance setting is complex and involves multiple factors. The models mentioned help researchers understand these multiple factors and how they interact to protect the health and well-being of the dancer. These models also highlight the strong need to design dance epidemiology studies with rigor so that both study designs and data analyses acknowledge and account for the changing nature of risk within dancer populations.

Methodological Considerations

A general understanding of epidemiological study designs is necessary to answer our health-related research questions efficiently and appropriately. Please refer to previous chapters for descriptions of these specific study designs. Of particular importance to epidemiological research is the study

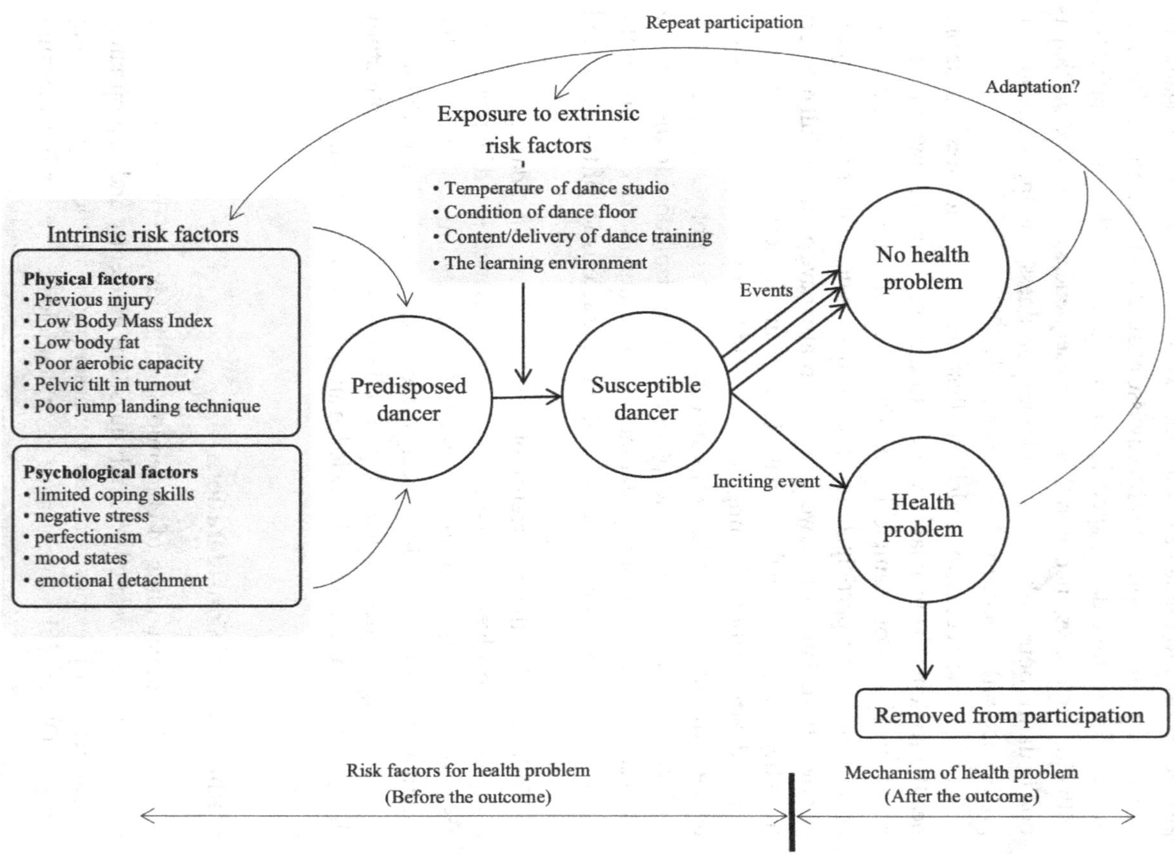

Figure 19.2. An Adapted Dynamic, Recursive Model of Etiology in Dance. Adapted from Meeuwisse et al. (2007).

design that is traditionally prospective in nature—an investigator identifies a study group and follows them forward in time while collecting data on possible risk factors and health problems as they occur. The prospective component of a study design is critical because it enables a clear sequential relationship between an exposure (e.g., potential factor), which occurs first, and an outcome (e.g., specific health problem), which follows.

Reporting Dance Epidemiology Data

Important Definitions

Following consensus statements established in sport epidemiology, a health problem is an inclusive term to describe any condition that reduces a dancer's normal state of full health, where health is understood to be "a state of complete physical, mental and social wellbeing" (World Health Organization, WHO, 1946). It is important to recognize that health problems include both injuries and illnesses and depict conditions regardless of their impact on a dancer's ability to train, rehearse, or perform or need to seek medical care (Bahr et al., 2020). Burden describes the overall impact of a health problem (i.e., injury or illness) on a specific dance population.

Put simply, a dance-related injury is any physical damage to body tissue(s) that leads to an impairment of physical function because of dancing. Illnesses are complaints or disorders experienced by a dancer that can relate to physical health (e.g., getting the flu), mental health (e.g., bouts of anxiety), or social well-being (e.g., feelings of loneliness).

There are further operational definitions of injuries and illnesses that are commonly used in sport epidemiology, and now in dance, since the Standard Measures Consensus paper was published in the *Journal of Dance Medicine and Science* in 2012 (Liederbach et al., 2012). Figure 19.3 illustrates the

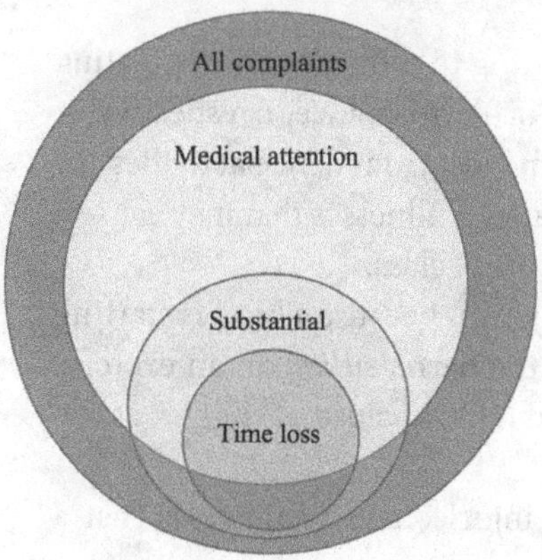

Figure 19.3. Distribution of Various Dance-Related Health Problems Definitions. Circle size represents the relative number of incidents likely to be registered. Adapted from Clarsen et al. (2014).

relationship between definitions, ranging from broad to narrow, that are based on the number of health problems that each is likely to capture (Bahr et al., 2020; Clarsen & Bahr, 2014).

Applying these health problem definitions to a dance setting has resulted in the following operationalization (Kenny et al., 2018a; Kenny et al., 2018b; Clarsen et al., 2013):

1. **Time-loss**—an anatomical tissue-level impairment that results in a dancer not able to complete a class, rehearsal, or performance or a subsequent class, rehearsal, or performance one or more days beyond the day of onset.
   - The dancer must have taken time off from dancing because of the injury.
   - For example, sitting out of an entire class for one or more days due to a severe ankle sprain.
2. **Substantial**—health problems leading to moderate or severe reductions in how much a dancer is dancing (dance training volume), or moderate or severe reductions in dance performance, or complete inability to participate in dance.
   - For example, a dancer who has not stopped dancing may partake only in certain classes, rehearsals, or performances, or certain sections of class, rehearsal, or performance, so as to manage recovery from a substantial health problem.
3. **Medical attention**—an anatomical tissue-level impairment that results in a dancer seeking care from a medical practitioner.
   - Depending on what country you are from, this health care professional may be a physiotherapist/physical therapist, an osteopath, a physician, acupuncturist, etc.
   - The dancer *specifically* takes the initiative to make an appointment to get help for the injury.
4. **All complaints**—any complaint leading to difficulties participating in a normal dance class, rehearsal, or performance, irrespective of the need for medical attention or time lost from dance activities.
   - Complaints describe the injuries and/or illnesses that may cause a dancer to modify particular sections of class.
   - For example, not participating in jumps because a knee is starting to twinge; instead of standing at the barre, *sitting* on an exercise ball to practice upper body movement because of niggles at the ankle joint.
   - All complaints can also include injuries and/or illnesses that a

dancer just copes with, or perseveres through, *not* modifying anything at all.

Across both dance and sport populations, it is now recognized that these different definitions impact our understanding of how much of a problem the injuries and/or illnesses are to the dancer. Each definition has its strengths and limitations (see Table 19.2). Knowing the advantages and disadvantages of each, can guide:

- How researchers design their next research question appropriately.
- How artistic staff or medical providers decide what they want to know about the health status of their dancers.

Surveillance of Health Problems

Another important piece of the puzzle is the surveillance system—or the way in which injuries and/or illnesses are acknowledged, counted, or captured, regardless of how they are defined.

1. Traditional, standard surveillance systems register health problems by way of a *third person*
    - In a dance context, this may be a licensed health care professional or a dance teacher.
2. Health problems can also be captured through *self-report*, where the dancers themselves are prompted to tell us about their injuries and/or illnesses.
    - This may be done through a specific tool or survey, an online questionnaire, or a smart phone app (e.g., Karreman et al., 2019).

Underlying the choice of health problem definition and surveillance strategy are several practical and theoretical issues, including the duration and setting of surveillance, the available resources, the type of injuries and illnesses of interest, how data are to be collected, and what they are to be used for (Clarsen & Bahr, 2014).

Severity of Health Problems

The most widely used measure of severity in dance epidemiology research is the duration of **time-loss**, which can be measured as the number of days that the dancer is not able to dance in class, rehearsal, or performances, from the date the health problem started until the dancer is fully able to take part in class, rehearsal, or performances. The number of time-loss days should be counted from the day after the onset (start) that the dancer is unable to

Table 19.2. Advantages and Disadvantages of Three Operational Definitions of Dance-Related Health Problems

Definition	Advantages	Disadvantages
Time-loss: An anatomical tissue-level impairment that results in a dancer not able to complete a class, rehearsal, or performance or a subsequent class, rehearsal, or performance 1 or more days beyond the day of onset.	Relatively easy to collect	Likely to miss less severe or overuse injuries
	Can be recorded by anybody (e.g., coaches, parents)	Dancers often continue to train and compete despite an injury
	Relatively reliable	Self-management strategies of injury may not lead to time loss
	Most severe injuries likely captured	Difficult to apply to a setting where dancers may simply choose to modify their training in response to injury
	Captures injuries that influence the ability to participate in dance	Threshold for time loss varies according to the importance of the injured dancer and the time of the season the injury occurs (e.g., soloist vs. corps)
	Permits comparison of data across disciplines	Can lead to underestimation of true health status
	Most feasible in long-term, large-scale studies	
Medical attention: An anatomical tissue-level impairment that results in a dancer seeking care from a medical practitioner.	Likely to capture greater number of conditions compared to time loss	May be underreported due to dancers not wanting to disclose injury for fear of being told to stop dancing

	Provides a more complete picture of the status of injury	Places large demands on team medical staff
	Important when the aim of surveillance is to assist in the allocation of medical resources	Susceptible to systematic bias due to non-uniform and inconsistent access to medical support
	Used most when there is relatively uniform access to medical care	There may be differences in who is qualified to provide medical attention
		There may be differences in the interpretation of what constitutes medical attention
		Completeness and accuracy of data may be compromised
All complaints: Any complaint leading to difficulties participating in normal dance class, rehearsal, or performance, irrespective of the need for medical attention or time lost from dance activities.	Data may provide good representation of the total burden of injury (full health status of dancer)	Dependent on dancer responses to regular surveys (e.g., susceptible to recall bias)
	Captures subclinical problems (e.g., overuse, chronic, recurrent, less severe) before they become more substantial/severe	Susceptible to systematic bias and false positives due to interpretation of what constitutes a recordable complaint (e.g., delayed-onset muscle soreness)
	When self-reported, mitigates dancers' fears of being told to stop dancing	
	Results can assist dance schools and companies to allocate health care provision	

Note: From Kenny et al. (2018b).

participate (day 1), through to the day before the dancer is fully able to dance in class, rehearsal or performances. Fuller and colleagues (2006) recommended categorizing health problem severity by using the following groupings: minimal (1–3 days), mild (4–7 days), moderate (8–28 days), severe (>28 days), and career ending.

Subsequent Health Problems

Subsequent health problems can be categorized by using the flowchart from Hamilton and colleagues (2011) (see adapted version in Bahr et al., 2020). The occurrence of a first health problem during the study period is called the index. Any injury or illness that occurs after the index is a subsequent health problem. A subsequent injury or illness that happens at the same location or to the same system as the index (e.g., same type or diagnosis) is called a recurrent health problem. A recurrent injury or illness is further classified into an exacerbation (i.e., if the index injury or illness is not fully recovered at the time of the recurrence) or a re-injury or repeated illness (i.e., if the index injury/illness is fully recovered at the time of the recurrence).

Basic Measures of Health Problem Occurrence

There are different quantitative measures (estimates) used to describe the extent of health problems occurring in dance populations. Definitions, formulas, and examples of basic estimates typically reported in dance epidemiology are detailed in Table 19.3. The example used to illustrate each estimate follows the story of a professional contemporary dance company made up of 82 dancers, who experience a total of 103 injuries among 65 of the 82 dancers during a full, contracted season. Of the 82 dancers, 55 dancers had a previous injury in the last one year, of whom 48 were injured, reporting a total of 93 injuries; 27 dancers had no previous injury in the last one year, of whom 17 were injured, reporting a total of 30 injuries. The total exposure of the contemporary dancers is 92,000 dance hours (i.e., approximately 28 hours/week for 40 weeks of their professional contract). For dancers with a previous injury in the last one year, the total exposure was 61,707 dance hours: [(92,000 dance hours / 82 dancers) * 55 dancers with previous injury]. For the 27 dancers without a previous injury, the total exposure was 30,293 dance hours.

Summary

Overall, dance epidemiology research is is important. When conducted conducted with sound and comprehensive methods, it allows us to study the

Table 19.3. Measures of Occurrence Commonly Used in Dance Epidemiology

Estimate	Definition	Formula	Example
Prevalence proportion	The percentage (or proportion) of dancers in a defined population who have an injury/illness at any given point in time; a report on all existing health problems at any point in time	Number of dancers who experience a health problem (new or subsequent) during a season divided by the total number of dancers at risk	
		Point prevalence—specific point in time	Point prevalence—At the start of the contracted season, 28% of dancers were currently seeking medical attention for an injury.
		Period prevalence—specific duration of time	Period Prevalence—In the previous one-year, 38% of dancers reported experiencing performance anxiety.
		Lifetime prevalence—any time in life up to the point of measurement	Lifetime prevalence—During their respective lifetimes, 48% of dancers reported sustaining at least one time-loss injury.
Incidence proportion	The proportion of a dance population at risk who become injured/ill during a given period of time	Number of dancers who experience a new injury/illness during a season divided by the total number of dancers at risk	65 injured dancers / 82 dancers at risk = 0.793 (79.3%)
Incidence rate	The occurrence of new injuries/illnesses among a dance population per unit of exposure (time dancing)	Number of new injuries/illnesses during a season divided by the total exposure of dancing during that season	103 injuries / 92,000 dance hours = 0.00112 or 1.12 injuries per 1,000 dance hours

(*continued*)

Table 19.3—*Continued*

Estimate	Definition	Formula	Example
Clinical incidence	The average number of injuries/illnesses per dancer during a given period of time	Number of total injuries/illnesses during a season divided by the number of dancers at risk at the start of the season	103 injuries during the season / 82 dancers at risk at the start of the season = 1.26 injuries per dancer
Odds ratio	The probability of an injury/illness occurring compared to the probability of it not occurring in a specific dance population	The odds of exposure to a suspected risk factor among those with an injury compared to the odds of exposure among those without an injury	
Risk ratio	The number of injuries/illnesses among dancers exposed to a suspected risk factor compared to dancers who are not exposed to the same factor	The incidence proportion of dancers with a previous injury in the last one year (suspected risk factor) divided by the incidence proportion of dancers without a previous injury in the last one year. If RR = 1.0, no increased risk of injury due to exposure; no association. If RR < 1.0: decreased risk of injury due to exposure; exposure may be protective. If RR >1.0: increased risk of injury due to exposure; exposure may be hazardous	(48/55) / (17/27) = a risk ratio of 1.38. This means that dancers with an injury in the previous one year are 1.38 times more likely to sustain a new injury compared to dancers without an injury in the previous one year.
Rate ratio	The incidence rate of injury/illness among dancers who are exposed to a suspected risk factor compared to the incidence rate of injury/illness among dancers who are not exposed to the same factor	The injury incidence rate of dancers with a previous injury in the last one year (suspected risk factor) divided by the injury incidence rate of dancers without a previous injury in the last one year	1.51 injuries per 1,000 dance hours / 0.99 injuries per 1,000 dance hours = a rate ratio of 1.53. This means that dancers with an injury in the previous one year are 1.53 times more likely to sustain a new injury compared to dancers without an injury in the previous one year.

overall impact of health problems (i.e., injury and/or illness) that are specific to dancer populations. Research in this field focuses on describing patterns of health problems, understanding why these health problems may be happening, and then ultimately, informing the development and implementation of effective ways that will improve and better protect the overall health of the dancer. Finally, when dancers are healthy, their potential to perform at their best, and to enjoy a long and enriching career, is enhanced.

Highlights

- Dance epidemiology research aims to understand the impact of health problems (i.e., injury and/or illness) on dancer populations to inform us on how best to protect dancer health.
- A dancer may be affected by a spectrum of injury- and illness-related health problems, so how these health problems are defined in research matters.
- Methods used to capture, analyze, and report dancer health problem data also matter, not only to facilitate comparison across different dancer populations but to aid in the development of prevention programs.

References

Bahr, R., Clarsen, B., Derman, W., et al. (2020). International Olympic Committee consensus statement: Methods for recording and reporting of epidemiological data on injury and illness in sport 2020 (including STROBE Extension for Sport Injury and Illness Surveillance (STROBE-SIIS)). *British Journal of Sports Medicine, 54*, 372–389.

Clarsen, B., Myklebust, G., & Bahr, R. (2013). Development and validation of a new method for the registration of overuse injuries in sports injury epidemiology: The Oslo Sports Trauma Research Centre (OSTRC) overuse injury questionnaire. *British Journal of Sports Medicine, 47*(8), 495–502.

Clarsen, B., & Bahr, R. (2014). Matching the choice of injury/illness definition to study setting, purpose and design: One size does not fit all! *British Journal of Sports Medicine, 48*(7), 510–512.

Finch, C. (2006). A new framework for research leading to sports injury prevention. *Journal of Science and Medicine in Sport, 9*(1–2), 3–10.

Fuller, C. W., Ekstrand, J., Junge, A., Andersen, T. E., Bahr, R., Dvorak, J., Hägglund, M., McCrory, P., & Meeuwisse, W. H. (2006). Consensus statement on injury definitions and data collection procedures in studies of football (soccer) injuries. *Clinical Journal of Sport Medicine, 16*(2), 97–106.

Hamilton, G. M., Meeuwisse, W. H., Emery, C. A., & Shrier, I. (2011). Subsequent injury definition, classification, and consequence. *Clinical Journal of Sport Medicine, 21*(6), 508–514.

Hill, A. B. (1965). The environment and disease: Association or causation? *Proceedings of the Royal Society of Medicine 58*(5), 295–300.

Karreman, D. E., Keizer-Hulsebosch, S. C., & Stubbe, J. H. (2019). Performing artist and athlete health monitor: User experience, content and conditions for use of an online dance-health surveillance system in a professional ballet company. *British Medical Journal Sports and Exercise Medicine, 5*(1), e000566.

Kenny, S. J., Palacios-Derflingher, L., Whittaker, J. L., & Emery, C. A. (2018a). The influence of injury definition on injury burden in preprofessional ballet and contemporary dancers. *Journal of Orthopaedic and Sports Physical Therapy, 48*(3), 185–193.

Kenny, S. J., Palacios-Derflingher, L., Whittaker, J. L., & Emery, C. A. (2018b). The use of a broad or narrow definition of injury in dance surveillance [Author response to letter to the editor]. *Journal of Orthopaedic and Sports Physical Therapy, 48*(5), 423–424.

Kenny, S. J., Whittaker, J. L., & Emery, C. A. (2016). Risk factors for musculoskeletal injury in preprofessional dancers: A systematic review. *British Journal of Sports Medicine, 50*(16), 997–1003.

Liederbach, M., Hagins, M., Gamboa, J. M., & Welsh, T. M. (2012). Assessing and reporting dancer capacities, risk factors, and injuries: Recommendations from the IADMS standard measures consensus initiative. *Journal of Dance Medicine and Science, 16*(4), 139–153.

Meeuwisse, W. H., Tyreman, H., Hagel, B., & Emery, C. (2007). A dynamic model of etiology in sport injury: The recursive nature of risk and causation. *Clinical Journal of Sport Medicine, 17*(3), 215–219.

Noh, Y. E., Morris, T., & Andersen, M. B. (2007). Psychological intervention programs for reduction of injury in ballet dancers. *Research in Sports Medicine, 15*(1), 13–32.

van Mechelen, W., Hlobil, H., & Kemper, H. C. (1992). Incidence, severity, aetiology and prevention of sports injuries: A review of concepts. *Sports Medicine (Auckland, N.Z.), 14*(2), 82–99.

van Rijn, R. M., Stubbe, J. H., (2021). Generalized joint hypermobility and injuries: A prospective cohort study of 185 pre-professional contemporary dancers. *Journal of Clinical Medicine, 10*(5), 1007.

van Seters, C., van Rijn, R. M., van Middelkoop, M., & Stubbe, J. H. (2020). Risk factors for lower extremity injuries among contemporary dance students. *Clinical Journal of Sport Medicine: Official Journal of the Canadian Academy of Sport Medicine, 30*(1), 60–66.

van Winden, D., van Rijn, R. M., Savelsbergh, G. J. P., Oudejans, R. R. D., Stubbe, J. H. (2020). Limited coping skills, young age, and high BMI are risk factors for injuries in contemporary dance: A 1-year prospective study. *Frontiers in Psychology, 11,* 1452.

World Health Organization (WHO). (1946). Preamble to the constitution of the WHO as adopted by the International Health Conference, New York. Retrieved March 16, 2022, from https://www.who.int/governance/eb/who_constitution_en.pdf

V

Influencing the Field

20

Proposing and Reporting Research

JEFFREY A. RUSSELL AND JATIN P. AMBEGAONKAR

Questions to Stimulate Thinking

- To what audiences do I want to propose or report my research?
- What pieces of information are most important for me to communicate to others about my research so that they are able to understand my work?

Communicating your research—proposing research you want to conduct or reporting research you have done—is an essential step in the research process. However, this does not simply mean putting words on a page. There are several important components that go into proposing and reporting research that we will explore in this chapter to provide a framework for presenting your efforts in written form. Remember, if a reader does not understand well, the author has not communicated well!

When proposing research and soliciting support for your research, you will attempt to present the need for the study (the Why?) and describe its value (the So What?). When reporting research, your focus will be to interest readers in your work and facilitate their understanding of it, so that they can use it to help themselves or others. There are several groups to whom you will communicate (see Table 20.1).

When writing a proposal or report, quantitative and qualitative approaches have some similarities and differences. We try to highlight places in this chapter where different techniques are needed, but please consult other methodology chapters in this textbook as they offer tailored guidance for reporting additional types of research. While we are writing from our experiences primarily as quantitative researchers, the intentions of all

Table 20.1. Stakeholders in a Research Project, Items to Consider When Proposing and Reporting Research

Stakeholders	Items to Consider when Proposing and Reporting
Dancers	Most important
	Often serve as participants
	Must see that the research is meaningful and any burden placed on them during participation is worthwhile for them, peers, and dance as an art form
Educators and Choreographers	More interested in whether the research project will improve their ability to teach dancers and students and improve dancers' abilities
	Less interested in whether you personally can complete the study
Clinicians and Practitioners (e.g., medical doctors, athletic trainers, physical therapists, psychologists, fitness trainers, dietitians, and others)	Interested in how they can use findings to improve the healthcare they provide
Academics (e.g., Research Mentors and Committee Members)	In an academic environment you must convince your mentors that you can successfully undertake the project
	Need to see that research will make a worthwhile contribution to the field and be feasible to complete.
	Do not like surprises, and they may be too busy to fix problems you create for yourself because you did not ask for their advice ahead of time or did not heed it!
	Write and submit whatever documents these individuals require and do so clearly and concisely for a mutually beneficial effort for all involved parties
Organizations and Funders (e.g., Professional Organizations, Journals, and Funders, Institutional Review Board [IRB] or Research Ethics Committee [REC])	Require you to justify your project's value before they consider it for a crowded conference program, journal, or grant competition
	Review committees receive many more applications than they can support, so your application must stand out from the crowd to be funded
	Obtain proposal guidelines and follow them precisely
	Write so that the proposal fits into the organization's criteria
	Do not waste their time or yours by applying outside their criteria or by submitting a poorly written application.
	IRB or REC approve research with human participants. More information in *Chapter 6.*

research reports are the same—accurately reporting what you did, what you observed, and the meaning of your work.

General Format of a Research Document

In research documents there is a typical way of communicating portions of a project that helps maintain consistency. The acronym for the basic arrangement is "AIMRaD," which stands for Abstract, Introduction, Methods (or Materials and Methods), Results, and Discussion. Of course, when you write a proposal (what you intend to do) there will be no "RaD" because you will not yet have results to report or discuss. When writing a proposal, you must take "AIM" at your project, so that others will be willing to invest the time, effort, and material resources necessary for success. We suggest you write in the future tense for all proposals, as the research has yet to be conducted, and in the past tense for reports after the research has been conducted.

Cuschieri and colleagues (2019) provide an excellent guide for writing a scientific paper using this AIMRaD arrangement. If you follow a few fundamental steps, your writing will fit the requirements for most proposals and journals. Abstracts prepared for conferences often incorporate a similar structure. In a journal article an introduction often includes the purpose statement or hypotheses, and the methods section will contain research variables and participant inclusion and exclusion criteria. However, a proposal—particularly a formal one for a thesis—may need separate sections for these items. Additionally, note that an introductory section for some journals may not be labeled "Introduction." However, it will always be the first portion of the main body of a journal article.

Abstract

The abstract is extremely important because it provides the summary of your research. This is what people read first to make a judgment about whether they want to read more of the paper. Thus it is essential that the abstract accurately and succinctly portrays what you intend to do (in a proposal) or have done (in a report). While organizations' or journals' requirements may vary, abstracts usually range from 250 to 400 words.

Start by writing a paragraph that summarizes your intended or completed work. Using your word processor's word count function will give you an idea of its length. Some journals may require you to write what is known as a "structured abstract." This means that you will use section headings, such as: Background, Methods, Results, and Conclusions. Other required headers may include Setting, Research Design, Analysis, and Discussion.

Once you have your initial attempt written, edit it for clarity and length. Word economy is of prime importance; wordiness is your enemy in writing an abstract. Place your purpose statement at the end of the Background. Include all the important parts of your project and avoid trying to conserve space by using statements like, "Results are presented in the main article." Only the primary findings will fit, so write as concisely as possible. This section may also include key words that are not in the title but are related terms that increase the indexing reach of the manuscript for online searches in databases.

Introduction Section

Literature Review

The initial portion of a proposal is a literature review or background section. *Chapter 7* explains how to prepare this important foundation. Do not shortchange this portion of your writing because just as a house stands more solidly when built on a firm foundation instead of on sand, so will your document stand more solidly when the literature foundation is solid.

Purpose

Also in *Chapter 7,* the importance of a purpose statement is promoted. Without a concrete purpose, how can you conduct research or write about it? The purpose statement, then, is the overall theme on which you should focus in preparing your work; it states what you want to accomplish. Figure 20.1 gives an example of a purpose statement. For a qualitative study, the purpose may read: "Our purpose is to assess the extent to which injured university dancers feel emotionally and socially distant from their uninjured dance classmates." In either case, these statements clearly indicate what the intention is for a research project.

You may find it helpful to begin your purpose statement with a phrase like this: "Therefore, in light of the gap in knowledge about this topic, the purpose of this study is (or was) . . ." You can also have more than one purpose in a study. This often happens in studies with two or more related portions. In such a situation, use plural statements to identify the purposes by writing your statement this way: "Therefore, our purposes were to ______ and to______."

Placement of your purpose statement is also important. The very end of the Introduction section is an ideal place for your purpose statement because

Purpose: To determine the extent to which hip-hop dancers are affected by upper extremity injuries differently than are modern dancers.

Specific Aim 1: To establish an injury recording system and database of injuries in a group of hip-hop dancers.

Specific Aim 2: To establish an injury recording system and database of injuries in a group of modern dancers.

Specific Aim 3: To use valid and reliable instruments to assess patient-oriented outcomes of upper extremity injuries in hip-hop and modern dancers.

Figure 20.1. Example of Purpose Statement and Specific Aims.

it launches from your review of what has been studied before and carries the reader directly into the way that you will conduct your study. In addition, write the exact same statement in your Abstract (between the background information and the methods) and at the beginning of your Discussion to help keep you and your reader focused on the purpose for your study.

Specific Aims

After stating the purpose, most research proposals are well served by presenting specific aims. While seemingly similar in many ways to a purpose, specific aims usually are more precise. They are target pieces of a study that all feed into achieving the overall purpose. You might consider them as steps or segments of your investigation. In Figure 20.1 we present specific aims to accompany the purpose statement. While this is a quantitative study example, you can apply a similar structure to qualitative studies.

Hypotheses

Your purpose(s) and specific aim(s) provide structure for developing the hypothesis or hypotheses to predict the results you expect from your quantitative study. For example, based on the study proposed in Figure 20.1, you may hypothesize, "Hip-hop dancers experience a higher incidence of upper extremity injuries than do modern dancers." This is a testable hypothesis; that is, you can record the number of injuries sustained by the two groups of dancers, and then compare these between the groups. Note that in a qualitative study you provide a rationale for your research but not a hypothesis.

Methods or (Materials and Methods) Section

This section is where you tell readers what you are going to do (for a proposal) or what you did (for a report). It may be helpful organizationally to place subtitles over the various sections, such as: Research Design; Participants, Setting, and Equipment; Procedures; Data Analyses. You must give enough detail about your study for others to replicate it. Remember that while the steps in your study may seem second nature to you, they may not seem that way at all to others, so extreme clarity is important when writing.

Participants, Setting, and Equipment

In this section describe details about your participants, the setting where you conducted your study, and any equipment that you used to complete the study.

Inclusion and Exclusion Criteria

In an experiment with human participants, it is important to describe what participants you wish to study. Identifying participants' characteristics helps the study remain focused, limits unwanted variability, and assists recruitment (because participants can be either retained or excused based on the inclusion and exclusion criteria). Thus your report should contain a list of these criteria and state very matter-of-factly how participants were included or excluded.

Independent and Dependent Variables

Variables in quantitative research are exactly what you would imagine them to be: facets of a research study that vary in one way or another. However, you must be precise about the variables that are being considered. Section III of this textbook provides more detail about variables than our writing focus allows here. Still, in whatever way you conduct your study, identify your independent and dependent variables. This is important so that you can indicate two main qualities to the reader of your proposal or report. First, you must show that you understand your project's design. Second—especially for a proposal—this portion will help substantiate (or not) that the research is feasible and has a reasonable chance of succeeding.

Equipment, Software, and Survey Instruments

Equipment and software descriptions are the most common methods-related items that should follow a specific convention in their presentation. For each piece of equipment or software, the name of the company that produces that item and the company's office location in parentheses. To illustrate, consider some sample sentences we might find in a Methods section:

- Participants entered their demographics data into a survey created in Qualtrics (Qualtrics, Provo, Utah).
- Biomechanical data were collected with a Noraxon MR3.16 system (Noraxon, Inc., Scottsdale, Arizona).
- Data analysis was conducted with MAXQDA software (VERBI Software GmbH, Berlin, Germany).

If you incorporate survey instruments (questionnaires, standardized data recording sheets, etc.) in your study, cite their source; for example, a journal article where the instrument was first published or the professional association that published it. In all cases, unless you develop an original instrument, you must cite the sources for all of your research tools.

Procedures

Overall, start constructing this portion by putting yourself in the place of participants and try to map the path you would want them to follow as you perform your study. How will they interface with your study information? Where will they go? What will they do? How will equipment be applied to them? Will they complete a questionnaire or interview, or will you derive data by measuring them when they dance while being filmed or while wearing sensors? Or will they stay in a certain position while you measure them with a device, rating scale, or clinical test?

Whatever the process, write it down from the participant's perspective and then edit it from your point of view as a researcher who will guide them through the steps you ask them to follow. You may also find another step advantageous to finalizing your description: give your methods section to someone unfamiliar with the study (but still generally aware of the dance science field) to see if the steps are adequately understandable.

Data Analyses

In this section, describe how you extracted the raw data (for example, biomechanical data from the software, or transcripts from interviews). Then explain how you analyzed the data. For example, in quantitative research, describe how you ensured that data were normally distributed (e.g., checked for statistical normality), and what statistical tests (e.g., t-tests, ANOVA) you used to answer the research questions as described in your purpose. Likewise for qualitative studies, explain the coding techniques and approach (e.g., grounded theory) that you used to arrive at your major themes. In quantitative research, you may also detail information such as your power analysis that informs the number of participants you enroll and the values you chose to determine statistical significance.

Results

While a section titled "Results" seems quite self-explanatory as to its content, there are certain conventions and guidelines for preparing it. A Results section is the objective presentation of the findings of the study—and nothing more. We focus here primarily on quantitative research; *Chapter 13* provides further details on presentation of qualitative research findings.

When presenting your results, the most important characteristic is truthfulness. Report the observations you made during your data collection without bias. Do not adjust, add, or fabricate data to alter how your research appears or to provide the results you think you should have gotten. The consequences in the scientific community of being accused of this type of behavior, much less having it proven, are severe. So, just "tell it like it is," and retain your integrity in the process.

The second important feature of a good Results section—which includes both written and graphic (tables, charts, graphs, photographs, illustrations) content—is clarity. If results are not clear, the reader will understand neither the value of your research nor how it applies to the field. Results are the essential numbers, statistics, responses, and reports that show the relative success of your project.

It is usual in this section to refer to accompanying tables or other nonnarrative material. For instance, use a sentence like this: "As shown in Table 1, hip-hop dancers reported more injuries, more days lost from dancing, and more physician visits than did modern dancers." Or, for a qualitative study: "Hip-hop dancers' responses to the interview questions revealed much stronger ties to their community and deeper emotions when they were injured than modern dancers (Table 2)." Table 1 or Table 2, then, would display data that substantiate these statements. Do not repeat data in your narrative that is included in a table or graph. Journals place limits on the number of tables and graphs, so strike a good balance between words and the other forms of communicating your results. Make general statements narratively and refer your reader to a table, graph, or other illustrative element for specifics.

Discussion

In this section you remind the reader of your study's purpose and explain your results. Generally, start by restating the purpose as you wrote it at the end of your Introduction. There are seven common components of most discussion sections that we discuss below.

Primary Findings. State your primary findings; that is, give the main points of your results in a plain language form—so that the readers can clearly note what you found after you conducted the study.

Comparisons and Contrasts with Prior Literature. Then proceed to compare and contrast your results in the context of prior research. Here you should explain how your data support or refute those of other investigators and the implications of these comparisons. You may be able to show similarities or differences between your work and other prior work so that readers can answer the question, "Why is this study important as a contribution to the field of dance science?"

Possible Explanations for Hypotheses or Positions Supported or Refuted. In this portion of the discussion explain why you believe that your hypotheses (quantitative studies), interpreted representation (qualitative studies), or position (historical or theoretical work) were supported, or how and why they were not supported. You also will explain (or speculate about, if that is more appropriate) the reasons for any conclusions in your study that differ from prior research. Here you are aiming to explain your results. This portion is often combined with the preceding comparisons mentioned, but you are adding your results to the overall body of knowledge as well.

As a general principle, the way you describe how your research contributes to scientific knowledge is important. Use phrases such as "our findings suggest . . ." or "we surmise from these results that . . ." rather than "our data prove . . ." or "it is indisputable from our results that. . . ." The quantitative approach is such that studies should be conducted multiple times to increase researchers' confidence in their veracity before absolute statements about "proof" or "indisputability" are warranted, whereas the qualitative approach embraces uniqueness and takes steps to ensure authenticity. What if another researcher repeated your study and found a different result? Clearly, with either paradigm of inquiry, further research is needed! Thus, when reporting scientific research, it is best to take a moderate path—leaving room for critical analyses rather than being overly authoritative.

Limitations. In the next portion you should discuss your study's limitations. When discussing results, remember that all studies have limitations. As a researcher, you are obligated to identify any important way that your study was limited in its execution. Such explanations help readers interpret results alongside areas of potential weakness or bias that may affect those results. Identify only limitations that are truly plausible and indicate their potential effects on the study.

Most new researchers' first experience with study limitations occurs during the preparation and defense of a thesis or dissertation project in an academic setting. Such projects have at least one purpose that is different from what is expected in professional research: demonstration of the student's analytic competencies. An extended discussion of limitations in this context may provide insight into a student's understanding of research methodology. This is an important signal of scholarly maturity to their committee members.

However, a heavy focus on limitations may be less important for experienced researchers from whom a less structured mention is preferred. In journal articles the emphasis should remain on helping readers understand the results and their implications. Striking the right balance may require some guidance and can be improved with feedback.

Recommendations for Future Research. Next you should present recommendations for future research. Here you describe how the topic should be examined henceforth when framed against the backdrop of prior work and your study. What are some paths forward and where should researchers expend effort to continue further exploration? These highlight your research as a valuable step in an ongoing process of gathering knowledge.

Implications or Applications of Your Findings. Then describe the practical implications or clinical applications of your study. Some study results may have direct implications or applications (e.g., in our example study about hip-hop dancers, clinicians can start implementing upper body strengthening programs in an attempt to reduce dancers' injury risk). Other study results may be indirect, with no immediate application, but may provide foundational information. Overall, in this sub-section you should provide specific information to readers about the ways your study helps advance the field.

Conclusions. The last portion of all research reports is the Conclusion. No new information should be presented here, and it is atypical to have any statements with citations. Rather, the focus is on summarizing your project and leaving the reader with a final impression of the value of the work and what new, useful knowledge it provides. One helpful method for deciding what to write in this final paragraph is to answer the question, "If readers read only my conclusions, what would I want them to know about this research study?"

Special Research Reporting Situations

Journal Articles

The singularly important step for properly reporting your research as a journal article is to obtain the journal author guidelines and then follow them explicitly. While many guidelines are similar across journals, they are rarely identical. The goal in approaching a journal for publication is to impress the editors and the peer reviewers who will judge your article's merit. Therefore, adhering to what they require is imperative and cannot be over-emphasized. Table 20.2 shows the typical information required by most journals.

One area of potential confusion that requires attention to detail is creating a correctly formatted reference list. *Chapter 7* offers information about software to make the collection and formatting of references as straightforward as possible. These tools help create a reference list by matching citation locations in your document with an information source contained in the reference software. Author guidelines explain each journal's specific requirements, so consult the one for your target journal.

Presentations at Professional Conferences

The two main methods of reporting research at a professional conference are orally and via poster. An oral presentation is a dynamic method of reporting with an author speaking about the research while showing a series of slides.

Table 20.2. Types of Useful Information Found in a Journal's Authors' Guidelines

Mechanical Preparation Information
Statements on who qualifies as an author and how plagiarism is handled
Type style, type size, margin size, and line spacing
Line numbering requirement (or not)
Article word or page count requirement
Abstract formatting, including word count, titles, subtitles, and key words
Number of tables and figures allowed
Formatting and placement of tables and figures within the manuscript
Reference formatting
Submission Information
Accompanying cover letter requirements
Method of submission
How the peer review and editorial processes are managed

A poster presentation (either a large, printed document displayed on a board or an electronic document displayed on a monitor) is usually a static presentation containing one's research for others to view at their leisure. Poster sessions also may contain a dynamic component if a conference requires presenters to stand alongside their display for a specified period to discuss their research with attendees. While detailed instructions about methods and caveats for each of these is beyond this chapter's scope, we offer a few general concepts to help report your work.

Oral Presentations

Oral presenters are allotted a specific time length to present. The crucial factors to consider in confining oneself to this allocation are the number of slides and the amount of information on each slide. Virtually everyone who has attended a professional symposium is acquainted with presenters who did not manage these well and went overtime. It is embarrassing to the speaker when a moderator interrupts their talk if the speaker goes over time; but more important, delivering a longer-than-allotted presentation is inconsiderate to the audience, the speakers who follow, and the conference organizers who have carefully planned their schedule.

To minimize this occurrence, plan your presentation carefully. A good rule of thumb is to prepare one slide every 40 seconds and place no more than five lines of easy-to-read bullet points (rather than complete sentences) per slide. To calculate the recommended slide count, divide your time quota by 0.67. For example, in a 10-minute talk you should have no more than 10 ÷ 0.67 = 15 slides. Disregard the title and final "Thank You" or "Questions?" slide in your count. Keep in mind that when you use the suggested maximum, you must average 40 seconds of speaking per slide. While initially you may find this to be challenging, practice giving your presentation in the allotted timeframe once you complete the slides. Adjust your slides and speaking points as necessary until you can complete the presentation on schedule.

We recommend not using a "References" slide at the end of your oral presentation. If you have ever noticed these at a conference, they show a long list of references in tiny type that is on the screen for no longer than it takes the presenter to say, "Here are my references." Thus such a slide is useless. Rather, appropriately cite information in slides where this is required, so that the audience will know that you used references.

Poster Presentations

For poster presentations, usually the organizers will provide instructions; follow them precisely. Determine the conference's required poster size and whether the poster should be arranged in landscape (horizontal) or portrait (vertical) mode. Depending on the output method (typically some sort of large-format printer) and the requirements of whoever does the printing, a poster can be designed in Microsoft PowerPoint, Apple Keynote, Adobe Illustrator, Canva, or other software that allows exporting to printable formats such as PDF (Portable Document Format) files.

Your poster's design and primary message must attract attendees' attention sufficiently that they want to spend time reading about your research. This means that the title must be large and clearly visible while also accurately communicating the project's essence. Text in the body must be easily read from a distance of at least six feet. The reference list can be smaller but still be readable. Choose a simple, clean font for easy readability. Figure 20.2 provides type sizing guidance in relative sizes.

Use contrasting colors that will allow your text to stand out and be legible even from a distance. Do not cram a lot of information into a poster's confined space because it detracts from the poster's appearance, readability, and, ultimately, its ability to communicate your investigative efforts. Use bullet points rather than narrative sentence blocks. Design tables and graphs so that readers can easily understand your highlighted main points.

Poster Section	**Suggested Minimum Type Size (points)**
Title	96
Main Headers, Authors	48
Subheaders	32
Text	24
References	18

Figure 20.2. Suggested Minimum Font Point Sizes for Poster Presentations.

Remember, your poster presentation is a way to engage your audience in a discussion. So construct a poster that enhances conversation rather than overwhelms its viewers with excessive information. Finally, place your contact details on the poster so that viewers may contact you afterward about your research. These references will hone your skills in both oral (Hoffman & Mittelman, 2004; McConnell, 2009; Pierce, 2020) and poster (Ferreira & Patino, 2019; Hess et al., 2009) presentations.

Summary

Overall, when reporting research, keep your audience at the forefront. Whether your readers are research experts or non-researchers, they should be able to read and understand your research. Thus, provide sufficient detail, be specific, and use the easiest and simplest words possible when communicating your work, so that your readers can use it to improve their own work in the dance sciences.

References

Cuschieri, S., Grech, V., & Savona-Ventura, C. (2019). WASP (Write a Scientific Paper): Structuring a scientific paper. *Early Human Development, 128,* 114–117.

Ferreira, J. C., & Patino, C. M. (2019). How to prepare and present a poster at a conference and communicate your research findings effectively. *Jornal Brasileiro de Pneumologia, 45*(3), e20190167.

Hess, G. R., Tosney, K. W., & Liegel, L. H. (2009). Creating effective poster presentations. AMEE Guide no. 40. *Medical Teacher, 31*(4), 319–321.

Hoffman, M., & Mittelman, M. (2004). Presentations at professional meetings: Notes, suggestions and tips for speakers. *European Journal of Internal Medicine, 15*(6), 358–363.

International Association for Dance Medicine and Science. (2021). *Journal of Dance Medicine and Science Guidelines for Authors.* Retrieved March 17, 2022, from https://iadms.org/media/2935/jdms-guidelines-for-authors.pdf

McConnell, C. R. (2009). Effective oral presentations: speaking before groups as part of your job. *Health Care Manager, 28*(3), 264–272.

Performing Arts Medicine Association. (2022). *Information for Prospective Authors.* Retrieved March 17, 2022, from https://www.sciandmed.com/mppa-manuscript-submission

Pierce, L. L. (2021). Professional development for inexperienced speakers: Making effective oral presentations and audience connections. *Rehabilitation Nursing, 46*(4), 197–202.

21

Applying Research Findings to Practice

DONNA H. KRASNOW AND VIRGINIA WILMERDING

Questions to Stimulate Thinking

- How can researchers disseminate their findings and determine what are the best venues for their particular subject?
- Is it possible to generalize research to a broader population than those participants in a particular study Why or why not?
- What are the barriers that prevent educators and medical practitioners from applying research in their school or clinic?
- How can researchers and practitioners address the gap between research and practical application?

Various dance and sport science articles identify a gap between research findings and their application to practical settings, including educational and clinical (Haines & Donald, 1998; Kristensen et al., 2016). We address how to disseminate the results of research, and how to tailor presentations and informational workshops to specific populations. We explore ways to refine the approach to the research depending on settings and available resources. Finally, we examine the gap between research and application by describing the results of studies from the literature that have not been translated into practice and make recommendations to bridge this gap.

Your primary role as a researcher is to design and implement research that has methodological rigor and answers specific questions that are deemed important to the field. After the research is complete, you have an equally imperative role in disseminating the findings and addressing applications to various populations, such as educators, medical practitioners, and dancers. The importance of application is to ensure that training provides the best

Table 21.1. Disseminating Research Findings

Organizations	Journals	Resources	Places to disseminate work
International Association for Dance Medicine & Science (IADMS)	*Journal of Dance Medicine & Science*	IADMS resource papers	Conferences: lectures, posters, workshops, networking
Performing Arts Medicine Association (PAMA)	*Medical Problems of Performing Artists*	PAMA online bibliography	Local organizations offering workshops
Healthy Dancer Canada (HDC)		HDC resource papers	
Ausdance (Australia)		Ausdance factsheets and reports	
British Association for Performing Arts Medicine			
Dance UK			
Dance USA			
International Dance Association (Italy)			
Médecine des Arts (France)			
National Dance Education Organization (USA) (NDEO)			
Tanzmedizin Deutschland (tamed) (Germany)			

possible skill acquisition with the least potential for injury. Several strategies and methods can facilitate achieving this goal. A major method of disseminating research is through publications. Second, there are local, national, and international conferences that provide venues for sharing research. Third, you can attend or organize informal workshops and presentations. Table 21.1 presents examples of these opportunities.

Why Is Applying Research Important?

People have been conducting research in the sport and dance sciences for decades, but there exists a gap between the research findings and their applications to the field. As far back as 1982, Eddy (as cited in Bero et al., 1998) commented that while money and resources have gone into research, nothing is ensuring that the findings are being translated into practice. Haines and Donald (1998) added that barriers to application include social,

Table 21.2. Steps to Encourage the Application of Research Findings

1. After identifying the main findings from your research, describe the strategies you will use to address the gap between research and application.
2. Determine how to involve the main personnel who are in a position to effect the necessary changes.
3. Recognize the barriers and consider how they might be confronted.
4. Choose targeted strategies and identify the means that can promote your goals.
5. Establish ways to assess whether the research interventions have had the desired effect and how to proceed going forward.

Note: Adapted from Haines & Donald, 1998.

organizational, and institutional barriers. Social barriers are the climate or belief systems of the dancers and their support systems. Organizational barriers are the hierarchy embedded in a particular working group and the power of the traditions that these people embrace. Institutional barriers are in the broader structure that might be in national or international systems that support a given medical protocol or educational syllabus. Hewitt-Taylor and colleagues (2012) identified lack of time and a heavy workload, which can prevent practitioners from gaining the needed education, as additional barriers to incorporating current research. Kristensen and colleagues (2016) stated that it can take more than a decade for the results of research to be implemented in medical settings.

Researchers are often drawn to conducting studies because they inherently respect the scholarship that research requires and the authority it gives to content (Arnold et al., 2005). Further, it develops critical thinking skills and the ability to accomplish logical observation. However, without applying the research findings, potentially useful interventions may remain shelved in libraries instead of being incorporated into practice where they can benefit dancers. Application of research findings helps to improve the techniques of medical and educational professions, and thereby enhances the lives of dancers. However, applied research must reveal information that practitioners find useful to their practice. Walley and colleagues (2007) suggested that researchers talk to practitioners before designing their studies so that researchers can understand the barriers to implementation from the practitioners' point of view and take into consideration constraints within various settings. The importance of involving those experienced with dance in research planning to enhance relevance and acceptability of interventions was emphasized in *Chapter 1.* Table 21.2 presents a pragmatic and sequential way to determine how to best achieve the goal of applying research to practice.

Why Is It Important to Find a Variety of Ways to Disseminate Research Findings?

Haines and Donald (1998) noted that because of the barriers to application, it might take a combination of strategies to overcome any resistance to change or to reach different audiences. It is therefore imperative that you disseminate your research in several formats and environments. Publishing your research in a journal might not be sufficient to reach the greatest number of people who could benefit from your findings. Further, reaching out to dancers and dance educators presents different challenges than those presented when speaking to the medical practitioners (Krasnow & Wilmerding, 2018). Due to the multiple fields within the dance sciences, and the variety of practitioners, researchers must find ways to translate the knowledge into many perspectives and use different terminology.

Translating Research to Speak Directly to Dancers and Dance Educators

Dancers and dance educators might be curious about the research results or the theoretical implications, but they are more interested in the immediacy and practicality of approaches they can use in the studio. Results can be presented as direct strategies and teaching methods. It is important to make it clear that you are not suggesting a complete overhaul of the art form, but rather you are recommending refinements that can enhance practice. For example, McCormack and colleagues (2004) suggested that selecting dancers who are hypermobile and using forced stretching techniques can create an injury-prone population. Rather than eliminating extreme stretching regimens or telling teachers to avoid working with hypermobile dancers, who are often considered aesthetically pleasing, emphasize stretching progressively over time, attending to good alignment and developing stability and proprioception (Redding & Geber Handman, 2017).

Another means of making research accessible to teachers and dancers is to use language familiar to their daily practice. Often researchers use terminology that is highly academic, including statistical or anatomical references, and these may not be well-known to the group being addressed. This use of language can suggest that research is only for academia, and it may alienate educators and dancers (Hewitt-Taylor et al., 2012). If there is a need to include statistics, a simple description of what the term *significance* means, for example, should be included along with the definitions of terminology important to the work presented. Unfamiliar anatomical terms can be translated into language known to dancers, such as "thigh bone" for "femur," or you can use both. Additionally, it is recommended that in any context, you should allow for practitioner discussion of the importance of knowledge based on personal experience (Hewitt-Taylor et al., 2012), and consider how this might inform the research and its application.

Translating Research to Speak Directly to Medical Practitioners and Provide Suggestions for Communicating with Dancers

When presenting your research findings to a clinical audience, there may be aspects of the dance profession that are not well understood. While medical practitioners certainly have the knowledge and language to understand your research, other difficulties exist in presenting your material in a way that is useful in the clinical setting. First, allied health practitioners may not fully understand the aesthetic and challenging demands of the dance profession. Even if their recommendations are valid in terms of best practice for dancers' health, it may be that the suggestions are radically different from accepted, traditional dance practice. The result may be resistance to change in the dance setting. If you are aware that your research presents material that may diverge from accepted norms, you will need to address this conflict in the discussion of your recommendations and offer possible ways to use the information that are non-threatening to the users. There are always ways to modify and temper your suggestions so that they are less alienating to the dance population.

A second area that may be unfamiliar to medical personnel is the extreme pressure on dancers related to time off from class and rehearsal for rest and injury recovery. In general, there is a climate of denial in dance and youth culture, which assumes that they can dance for unlimited hours, and heal injuries without altering their practice. Before medical personnel can make recommendations for training modification or even complete rest, they need to be aware of these persistent attitudes. As a researcher, you can include this information in your presentation to the medical community.

Finally, it is not uncommon for dancers to receive conflicting information from dance teachers and choreographers after receiving advice from a medical practitioner. As the researcher, you can suggest that the medical professionals, educational experts, and parents (if dealing with children) become a support team in dealing with the injured dancer (Mainwaring et al., 2003). In this way the team can discuss possible treatment strategies and agree on a collective effort that may increase compliance.

What Is the Practitioner's Role? Examining Ways the Practitioner Can Implement Research Findings

It is the practitioner's role, once exposed to new research findings, to implement any desired changes in the dance school, company, or medical clinic. The first step is to evaluate what is applicable to the particular environment. For example, the research may have addressed the needs of adult professional ballet dancers, but the dance director may be running a school for young children. Dance material should always be age appropriate, as should the parameters of practice, such as hours per week and length of class (Krasnow & Wilmerding, 2015). As the researcher, you must be very clear about the population you studied, and whether or not you think the results can be generalized to other groups. This clarification can assist the practitioner in applying the results of the research.

Second, the practitioner will have constraints in both time and finances and may not be able to implement your suggestions in their entirety. For example, while it may be optimal to do one-on-one training with certain recommendations, this format may be impossible in an educational setting but entirely conceivable and even expected in a medical setting. As a researcher, the more specific you can be in your discussion, the more useful your work will be to practitioners.

Third, it is often the case that the director of a school, company, or clinic might attend a conference, but the staff may not be able to do so, due to time and finances. The director will need to address the changes that need to be implemented, and how to pass this information along to staff. You might suggest handouts they can distribute, or group discussions at their school or clinic. In a more involved format, you could suggest that the director brings in experts to give informal workshops to the staff. There are a variety of strategies to implement new ideas that can be suggested, and, as the researcher, you can be a source to offer solutions.

Determining Relevance and Adapting Findings to Specific Contexts and Populations

What is meant in research by generalizability? This term refers to the applicability of research findings from a study on a particular sample of dancers to the population of dancers that the sample was intended to represent (see *Chapter 8*). It may be the case that your findings can be generalized to a larger group than the sample you studied. It is important to avoid making assumptions that may not be supported by your research. However, the ultimate test of the generalizability of recommendations that grow from a research study is to try them in a new context and see if they work. Ideally, the test should be done as a subsequent experiment. However, practitioners who remain open to looking objectively can make their own, informal determination of applicability until they or others have an opportunity to conduct an experimental evaluation.

In presenting your research, it is crucial that you describe the participants in your study with specific details about age, gender, height, weight, years of experience, amount of daily practice time, level of proficiency, type of training, and any other relevant data. In this way your audience can compare your participants to their students or clients. If you included an intervention in your study, you need to describe the process explicitly. Whereas research conducted within the traditional scientific method (see *Chapter 10*) aims to be generalizable to broader groups than the one studied, the results may actually apply only to your target population. For example, extensive research over decades has indicated a high level of injuries in the lower limbs of dancers, but much of the research was conducted on ballet dancers. When modern/contemporary, jazz, or urban dancers are studied, the injury rates are different, and can occur in the spine and upper extremities to a greater extent. As a researcher, you may choose to conduct your study within the guidelines of the traditional scientific method and through generalization, inferring that your findings apply to most dancers, but it is important not to assume that they do. In contrast, research findings emanating from a qualitative research methodology or from studies with limited sample sizes are not meant to be generalizable but may have particular relevance and utility for selected circumstances.

Clarifying How the Researcher's Background Informs the Methodology and Perspective

The scientific process assumes that you will be conducting your research with integrity, rigor, and transparency. Traditional science dictates that the researcher is objective and detached from the research and especially the participants, but in reality, this is not entirely possible. Researchers must make every effort not to skew or bias their results intentionally, but everyone brings a particular set of experiences and point of view to the work. This is not necessarily detrimental, but it needs to be recognized and acknowledged. Your perspective will affect how you design your methodology and what questions interest you. It is important when you present your research to be transparent about these personal points of view and how they influenced your choices, but this does not negate the value of your work. For example, a motor learning expert will take a different approach to a particular movement question than a biomechanist.

Tailoring How the Research Is Presented to a Particular Audience

It is possible to present the same research material in different ways to different audiences. Even within the educational setting, describing your work to studio teachers presents challenges that differ from presenting to educators in a university setting. And you may need to adapt your emphasis when talking about the results of injury mechanisms to surgeons versus physical therapists. Consider your audience members and what their backgrounds and needs are. The more you can tailor your message to your audience members' interests, needs, and ways of working, the more useful your research will be.

Assessing Effects in Your Context: Refining the Approach

After your research is complete, you and others may be involved in the task of implementing and assessing your recommendations. It is essential that these efforts be evaluated to determine efficacy and practicality in order to make future improvements. A number of factors can influence this assessment process, such as the variety of settings, available resources, and how much time you or the practitioner may have to work with a given population.

Variety of Settings (Educational, Clinical, Conditioning)

Assessing intervention results will have different challenges in different settings. In an educational context, the teacher is usually dealing with groups of dancers, rather than working in one-on-one situations, and this format makes it difficult to meet with individuals. However, the teacher has the advantage of seeing the dancers perform in class. In contrast, medical practitioners have individual sessions with the dancers, but there may be little exposure to a dance context. In the conditioning or fitness training setting, the teacher or trainer can easily identify problem issues that need to be addressed, but again, may not see how these relate to actual dance practice. When evaluating a new intervention, it is ideal to assess effectiveness in more than one context.

Available Resources (Time, Funding, Culture, Skills, Age, and Gender)

Time and funding are probably the biggest limiting factors for appraising the effects of applying research findings. Once the practitioners at a medical clinic or dance studio have added new strategies to their usual treatment protocol or studio classes, time to evaluate the effects formally may not be available, especially if it is critical to assess the dancers individually. For the teachers, the time factor is more problematic as it can add hours to their schedules. For the medical personnel, access to funding can be an impediment, as they are usually required to bill for any additional time spent with a patient.

There may be resistance in a given environment to taking on this time-consuming process. For example, in some dance settings, such as public school, dance may not be viewed as a strenuous, athletic endeavor. Devoting additional time to conducting a formal evaluation may be deemed unnecessary. In a medical clinic, there is the question of who will absorb the cost of conducting an evaluation, the clinic or the client. These timing and funding dilemmas can create resistance to what might otherwise be an accepted and responsible practice; that is, evaluating the effectiveness of new procedures (see Royse et al., 2016, for a discussion of program evaluation).

While appraising new protocols might be seen as advantageous, there may be few people in a given environment who have the skills needed to conduct a formal evaluation. Determining changes in dancers over time requires a particular skill set, and many teachers or medical practitioners may not have the time to learn those skills. Before implementing new protocols, the clinic or school might want to consider how to proceed to the next stage, and the researcher might be able to provide guidance about how to conduct an evaluation.

Other factors, such as age and gender, can influence the process of assessing results. It may be simple to add new strategies to the dance class of children, but quite demanding to involve young ones in extra time and effort outside of class. There may be gender differences in how dancers react to new interventions and evaluation procedures. Practitioners need to take all the possible factors into account when considering new ideas and strategies.

Reassessing Effects Over Time

A new approach to training might initially demonstrate a beneficial effect, but over time, dancers may revert to their previous habitual methods or lose newly established ones. It is important to reassess the dancers' capabilities over time. Even a month may not be sufficient to determine the long-term outcome of new interventions. To guarantee comparable results, establishing identical conditions for each assessment is beneficial.

Conducting and Reporting Research to Contribute to the Knowledge Base

What would it mean to say that a research study has contributed to the knowledge base? Of course, all rigorous research adds to the general breadth of what is known about a given subject. Applied research should impact the pragmatic and general methods that are employed in the field. The gap between research and practical application has already been described. Looking at some examples from the literature can shed light on the difficulties of implementation and give new researchers ideas addressing those difficulties.

Application of Research Findings from the Literature to Practice

To clarify and exemplify the issue of applying research findings to practice, we examine three areas of research in the literature. The first is having dancers observe class when they are injured. The second is believing that alignment should be the same in all conditions. The third is the development of conditioning programs for dance.

Class Observations for Injured Dancers. It has long been common for teachers to recommend that injured dancers observe class and rehearsal when they are unable to participate. The reasoning is that dancers can learn through observation and will be more prepared when they return to class. What has not been questioned is the effectiveness of observing class, and if there is a time limit to any potential benefits.

Mainwaring and colleagues (2003) considered these issues and found that while there is benefit to observation, there are only short-term gains. When the time required to observe others was extended, the psychological detriments outweighed the learning benefits. The dancers became isolated and despondent, and some took part in self-destructive behaviors, such as severe dietary restrictions and self-mutilation. The researchers made several positive recommendations to replace observation, including partaking in conditioning and rehabilitation programs, and assisting the teacher or choreographer. They also suggested that social activities such as field trips and film viewings for the whole group be included, to avoid the feelings of isolation and abandonment of the injured dancer. It is important to recognize that other dancers can be negatively affected by a close colleague being injured, and often unwittingly avoid talking to the dancer because they are not sure what to say or whether it is appropriate to mention the injury. These social events can ease this discomfort. In spite of this study, and dance psychologists making similar recommendations, the practice of insisting that injured dancers observe classes and rehearsals for extended periods of time

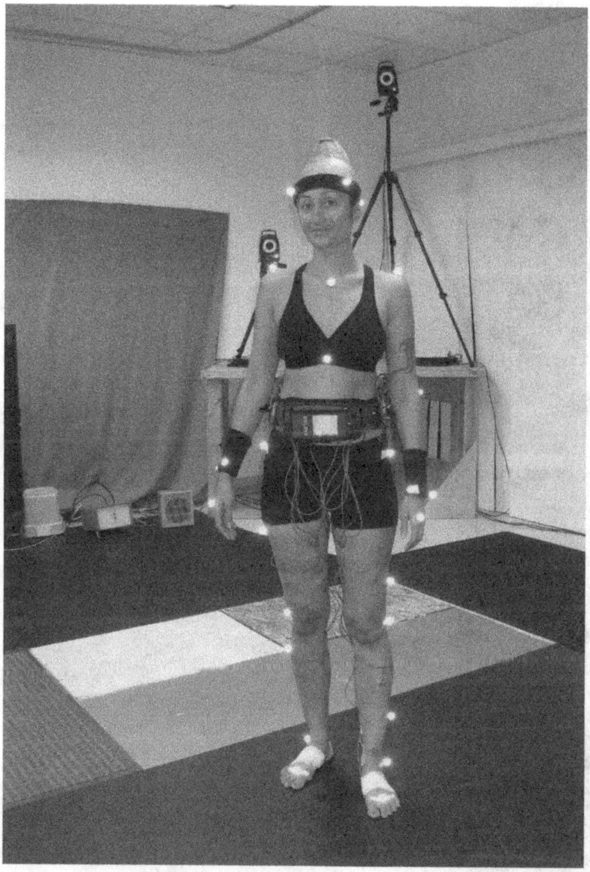

Figure 21.1. Dancer Prepared for Data Collection. Dancer: Rika Traxler. Photo: Shane Stecyk.

continues. More research is recommended to reinforce the findings and recommendations from the Mainwaring study.

Alignment in Various Conditions Used in Dance Practice. Another assumption in dance is that dancers maintain the same alignment in all stances and traveling work throughout practice. Kinematics is an excellent method for exploring the organization of the body in motion. Krasnow and colleagues (2012) studied 40 female dancers—beginner, intermediate, and advanced/professional levels, ages 17–43—executing grand battement devant in three conditions: at the barre, in the center, and traveling. The dancer in Figure 21.1 has been prepared for data collection with reflective markers (white balls), which record movement of limbs relative to joint action.

The focus of the data analysis was the use of weight transfer during the battement, and there were significant differences between the three conditions, directly related to how the dancers shifted the pelvis during different phases of the movement. Contrary to expectations, the advanced/

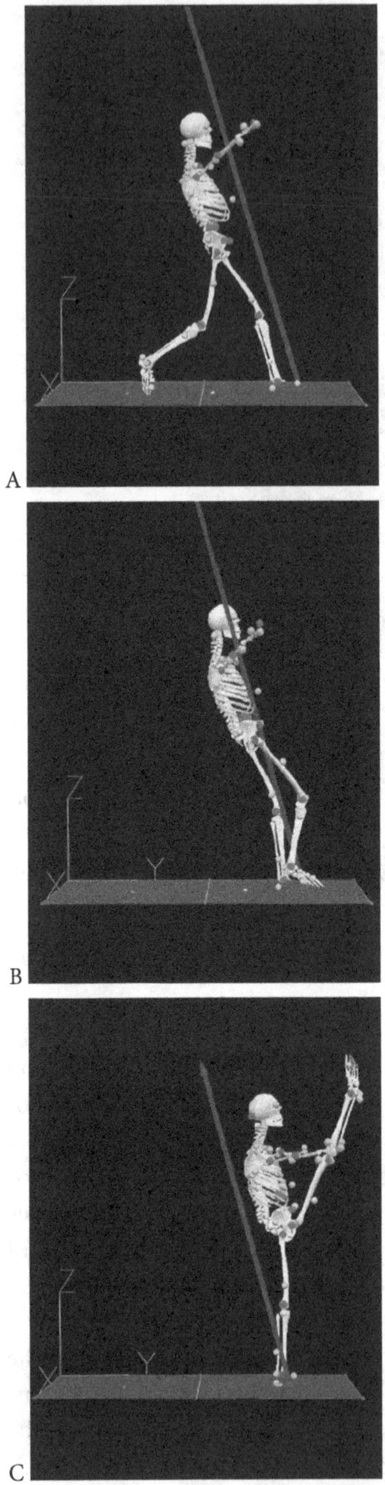

Figure 21.2. Analysis of Advanced Dancer Performing Grand Battement Devant While Traveling Through Space. (*A*) Weight transfer onto left standing leg; (*B*) initiation of the battement, right leg; (*C*) peak of the battement, right leg.

professionals relied just as much on the barre as the beginning and intermediate dancers, rather than shift the weight completely onto the supporting leg. The strategy for weight transfer during traveling was more complex than the other two conditions, but similar across levels. Further, there were surprising differences in the dancers' alignment, most extreme in the traveling condition, showing the dancers leaning far behind the plumb line in executing a traveling battement. Figure 21.2 shows the images derived from kinematics of stance (when weight transferred onto the standing leg), initiation of the battement, and peak of the battement.

In these two examples, certain assumptions have been questioned, and data have provided evidence to suggest that changes can be made to current practice to make it more effective. One might also extend these conclusions to floorwork; that is, dancers use different strategies executing movement on the floor as opposed to standing and traveling. While not suggesting that barre and floorwork should be eliminated, teachers might consider giving more class time to standing and traveling conditions. It is often the case, particularly with beginners, that a large portion of the class is dedicated to the barre or floorwork. The recommendation is to employ the third/third/third rule, giving one third of the class time to barre or floor, one third to center, and one third to traveling.

Although these studies are not recent, the recommendations have been limited in implementation in dance schools and companies. These examples help to illustrate the gap between research and application described previously. The traditional method of structuring dance classes, and how time is managed, particularly for beginners, has been entrenched in dance training for decades. These traditions have been passed down from generation to generation, and it has become quite difficult for teachers to consider major changes. Similarly, medical position statements should change as new evidence is discovered (Kolt, 2009). To understand fully how to address the gap between research and application, let's examine a successful case.

The Development of Conditioning Programs in Dance Training. The considered use of supplementary conditioning programs for dancers has been developing for at least the past forty years, and there are thousands of articles and texts in Google Scholar covering research in a variety of aspects of conditioning: cardio-respiratory, muscular strength, flexibility, proprioception, balance, motor control and applications of imagery (Krasnow et al. 1997). It has benefited dancers internationally that conditioning programs and classes have been implemented in professional schools, colleges and universities, and dance studios. Many schools and institutions provide workshops and lectures on how conditioning can enhance dance technique and prevent injury, major goals of all training programs. The success of this research application to practice can be attributed to a number of factors, such as the dissemination of research findings through numerous journals, newsletters, and conferences related to dance and sport sciences and dance education. Finally, the massive quantity of research in this area has certainly helped to propel it to acceptance, and this factor may be the key difference between this area of research and the studies examined in the previous section.

Summary

We began this chapter with a discussion of a variety of avenues for disseminating research findings in terms of applications to various populations, including publishing sources, conferences, and informal contexts. We described the importance of practical applications of research, and how it can be translated to communicate effectively with dancers, educators, and medical practitioners. We explored ways practitioners can apply research findings in the settings where they work, and how researchers can assist with the process. We discussed how to adjust research findings to fit a specific context, the idea of generalizability, how the researcher's background can inform the work, and how to tailor a presentation to a particular audience. We examined ways to refine the approach to the research depending on settings (e.g., educational, clinical), and available resources (e.g., time, funding, culture). Finally, we considered the gap between research and application by describing examples from the literature, and we made recommendations to bridge this gap.

Highlights

- Communicating research findings in practical information for educational and medical practitioners is an important aspect of the research process.
- It is important for researchers to recognize barriers to implementation of research findings and consider methods to disseminate the new knowledge, such as journals, conferences, and workshops.
- Once new strategies have been implemented, it is important for researchers to assist practitioners in evaluating the effects of the new protocols and determine future research endeavors.
- The current research literature can be examined to determine what has previously been uncovered and whether it is being applied in practice.

References

Arnold, B. L., Gansneder, B. M., & Perin, D. H. (2005). Introduction to research in athletic training. *Research Methods in Athletic Training* (pp. 3–11). F. A. Davis.

Bero, L. A., Grilli, R., Grimshaw, J. M., Harvey, E., Oxman, A. D., & Thomson, M. A (1998). Closing the gap between research and practice: An overview of systematic reviews of interventions to promote the implementation of research findings. *British Medical Journal, 317,* 465–468.

Haines, A., & Donald, A. (1998). Making better use of research findings. *British Medical Journal 317,* 72–75.

Hewitt-Taylor, J., Heaslip, V., & Rowe, N. E. (2012). Applying research to practice: Exploring the barriers. *British Journal of Nursing, 21*(6), 356–359.

Kolt, G. S. (2009). Practical applications of research findings. *Journal of Science and Medicine in Sport, 12,* 251.

Krasnow, D. H., Chatfield, S. J., Barr, S., Jensen, J. L., & Dufek, J. S. (1997). Imagery and conditioning practices for dancers. *Dance Research Journal, 29*(1), 43–64.

Krasnow, D. H., & Wilmerding, M. V. (2015). Development of locomotor skills. *Motor Learning and Control for Dance: Principles and Practices for Performers and Teachers* (pp. 44–60). Human Kinetics.

Krasnow, D. H., & Wilmerding, M. V. (2018, September 28–30). Practical steps in communication and collaboration among dancers, educators, researchers, and medical practitioners [Conference presentation]. Healthy Dancer Canada 10th Annual Meeting, Toronto, Canada.

Krasnow, D., Wilmerding, V., Stecyk, S., Wyon, M., & Koutedakis, Y. (2012). Examination of weight transfer strategies during the execution of grand battement devant at the barre, in the center, and traveling. *Medical Problems of Performing Artists, 27*(2), 74–84.

Kristensen, N., Nymann, C., & Konradsen, H. (2016). Implementing research results in clinical practice: The experience of healthcare professionals. *BMC Healthcare Services Research, 16*(48).

Mainwaring, L., Krasnow, D., & Young, L. (2003). A teacher's guide to helping young dancers cope with psychological aspects of hip injuries. *Journal of Dance Education, 3*(2), 57–64.

McCormack, M., Briggs, J., Hakim, A., & Grahame, R. (2004). Joint laxity and the benign joint hypermobility syndrome in student and professional ballet dancers. *Journal of Rheumatology, 31*(1), 173–178.

Redding, E., & Geber Handman, P. (2017). Cross-training and conditioning. In M. V. Wilmerding and D. H. Krasnow (eds.), *Dancer Wellness* (pp. 37–53). Human Kinetics.

Royse, D. D., Thyer, B. A., & Padgett, D. (2016). *Program Evaluation: An Introduction to an Evidence-Based Approach* (6th ed.). Cengage Learning.

Walley, J., Khan, M. A., Shah, S. K., Witter, S., & Wei, X. (2007). How to get research into practice: First get practice into research. *Bulletin of the World Health Organization, 85*(6), 424–425.

VI

Pursuing a Career in the Dance Sciences

22

Pursuing Graduate Studies

MATTHEW WYON AND YIANNIS KOUTEDAKIS

Editors' Note—The authors of this chapter created the first PhD program devoted to dance science. Here they share insights into graduate studies in our field.

Questions to Stimulate Thinking

- How can graduate studies help you become a dance science researcher?
- What networking opportunities might lead to research collaborations?
- What is the social significance of your research interests?
- Who are some stakeholders that might support your research activities?

In Greece in the fourth century BC, the polymath Aristotle initiated the techniques of logic (for which he is particularly known), observation, inquiry, and demonstration. These would later shape Western philosophical, scientific, and research culture, which is why Aristotle is regarded as the "father of scientific methodology" or the "first scientist" (Presti, 2014). Influenced by two of the greatest ancient philosophers, Plato and Socrates, Aristotle also introduced the concept of causality—that there is more than one "reason" that can explain why things happen in a certain way. His four concepts of causality were material cause, efficient cause, formal cause, and final cause.

Research can be defined as the systematic investigation of materials and sources in order to establish facts and reach new conclusions, thus expanding the frontiers of knowledge in a given area. According to the *American Journal of Public Health:*

> Research is a planful and systematic activity directed at the discovery of new facts or at the identification of relationships among facts. It attempts to achieve these objectives in a manner assuring that the data obtained can be generalized (i.e., can be applied) to a wider range of situations than the one studied, and that the data are not influenced by the conscious or unconscious bias of the investigator or by other extraneous factors. (Rosenstock & Hochbaum, 1961, p. 266)

Dance science research spans multiple paradigms and types of inquiry. It consists of several different but interrelated features, which derive from such diverse areas as material science, body science and medicine, and even space science. Dance science is a complex entity with links to a large number of technical, medical, psychological, nutritional, physiological, economic, environmental, and other features (Koutedakis & Sharp, 1999). However, despite the fact that dance itself has a long history that can be traced back to ancient Egypt and classical Greece, dance science has a rather shorter life. The first dance science research publications appeared in the 1960s and 1970s, and they were observational in nature with a medical focus. The first intervention studies appeared about 20 years later. In general, a dance science research program is a means to improve existing knowledge or generate new insights on any aspect of dance health, performance, materials, teaching techniques, economics, nutrition, management, or other elements.

To the best of our knowledge, the first large-scale dance science study involving more than 600 dancers was conducted in early 1990s, and it was the result of a collaboration between two UK institutions, Dance UK and the University of Wolverhampton. This collaboration was financed by the late Peter Brinson, who was perhaps one of the first dance professionals who fully appreciated the association between a prolonged dance career, injuries, physical fitness, and nutrition. The results of the study appeared in the *Fit to Dance?* report, which was published in 1996 (Brinson & Dick 1996). Since then, numerous research studies have been conducted worldwide, and relevant reports have been published in a number of dance-, sport-, and medicine-related journals.

Unlike sport science, where the sophistication of research methodologies (research design, statistical testing, etc.) have steadily increased during the last few decades, research methodologies that are applied to dance have not experienced the same development (Amorim et al., 2015). This is partly due to the fact that scientific research has never been among the profession's main concerns, as dancers and their teachers focus almost exclusively on aspects of technique and performance. Further, the curriculums of most university dance departments offer little training in how to design and conduct dance science research. Therefore the aim of this chapter is to outline essential steps that graduate student researchers can undertake while designing and conducting a research project linked to the dance sciences. We begin by briefly describing the qualities of competent researchers.

Researcher Qualities

Research is more than just learning more about dance science. It is also a great opportunity to learn about yourself and the world around you. While training as a researcher, you will learn skills related to, *inter alia*, achieving goals, time management, generating new research ideas and addressing old ones, laboratory techniques, data analyses and interpretation, and conducting literature searches. You will further learn to appreciate the importance of teamwork, where different team members have different roles in large projects. Effective researchers possess or develop a number of personal qualities (listed roughly in alphabetical order):

- **Analytical mind** and plenty of **common sense**. The obvious answer is not necessarily the best, and "analysis" is the opposite of "synthesis."
- **Able to stay calm,** especially when the unexpected (e.g., apparatus failure) occurs.
- **Ability to communicate clearly,** both in writing and when speaking.
- **Curiosity** about how things work and the **drive** to dig deeper to reveal insights.
- **Dedication** to becoming a researcher is a demanding exercise.
- **Attentive to details.** Check, re-check and check again at all stages of the process.
- **Compulsive about planning** in research and personal matters.
- **Flexible** enough to re-think and re-focus as the interactive research process unfolds.
- **Friendly** with associates and people in general, especially toward human volunteers.
- **Able to listen carefully** to others whose strengths are different from your own.
- **Open-minded** and **free from prejudice** about what you can do in your research project, consistent with ethical guidelines.
- **Rigorous** and **patient** in approaching the laborious and hard work required to conduct research.

- **Systematic** in approaching research activities, including daily record keeping, meetings, and discussions.
- **Truthful** in reporting research findings accurately and never manipulating data, thus avoiding actions that can end a researcher's career in a rather unceremonious way.

Most research in dance science is conducted in conjunction with masters or doctoral degree programs. However, roughly 25% of masters and doctoral students drop out before finishing their degrees, with numbers varying across programs and from country to country. The hardest hit are students who opt for part-time study, which can take twice as long as full-time study. The challenge is even greater when human volunteers serve as participants in a given research project, as is the case with most dance science projects. For example, accommodating the tightly programmed duties that professional dancers have, can be particularly difficult for a part-time researcher. Doctoral researchers can spend 55 to 70 hours a week on a research project, principally during periods of data collection.

Before Starting a Research Project

The planning and development phase of a research project is critical for the successful completion of a graduate degree. Becoming a student of research is similar to an apprenticeship; you are learning the skills under the guidance of a more experienced mentor. It is critical—especially at postgraduate level—to choose the right institution and the right research supervisor. Every research supervisor has an area of expertise and a different philosophical approach to research. As your degree becomes more advanced, finding a compatible program and supervisor becomes strategically important.

With assistance from the supervisor, the student researcher next needs to identify an under-researched area that deserves further scientific attention. The process can be tedious but it is absolutely necessary for a smooth progression. The goal at this stage is to be able to explain the significance of your research idea and how you plan to address its components.

Dance science research is a journey that never ends, as new questions often need new methodological solutions. As an apprentice researcher it is advisable to experience different (quantitative and qualitative, large group and individualized) methodological approaches encompassing interdisciplinary and mixed methods paradigms.

Choosing an Institution and Supervisors

Taught Degrees

It can take a while to cast your net wide and review the full faculty of about 20 or so top institutions offering dance science programs at the undergraduate and graduate levels. *Taught Degrees* are programs where the majority of the credits are earned by taking courses taught as modules (MSc, MS, MA, and some MFA programs). Your goal is to find an institution that offers a program or curriculum as close as possible to your area of interest. Once you have created a short list, an initial contact via email is recommended to get more detail about the curriculum and the institution. You will want to visit the institutions that remain on your short list, as each of them might have a certain focus within their course work that is based mainly on the expertise of the teachers and on the facilities. However, matters may be different when pursuing a degree with a research emphasis.

Research Degrees

In most *Research Degree* programs, the majority of the credits are earned not by taking taught courses but by engaging intensively in the execution of research (MRes, MPhil, Professional Doctorates, and PhDs). Learning to conduct research competently requires expert mentoring, which makes choosing who to study with an important consideration. Every established researcher has developed an area of expertise and a methodological and philosophical approach that may be unique in the field. A search of the published research literature will provide an initial indication of these aspects for potential mentors. You might also look for their online profile, to see how many research students they have successfully supervised (often referred to as completions) and how many students they are currently supervising.

Experienced researchers may receive 5 to 10 inquiries a month from prospective students. They will normally accept students who are likely to pursue a line of research that is parallel or complementary to their own research. Consequently, an initial email followed by telephone or online communication may provide an indication of your compatibility with a potential supervisor. This initial contact should be accompanied by a well-written summary (proposal) of your own research interests and an explanation of why you believe these interests warrant further investigation as part of a masters or doctoral program. Highlighting how your energy and interests might support your prospective supervisor's research efforts may increase your chances of being considered for a position. Not hearing back or receiving a rejection does not necessarily mean that your idea has no merit.

Doctoral students will have a team of supervisors who do not all need to be at the same institution, but they should complement each other. At least two of the three to five committee members should have evidence of supervising a number of research students to completion. You will have a lead supervisor, the Director of Studies (or the Dissertation or Thesis Committee Chair). The remaining members of the supervisory team should be agreeable to working with you before the Director of Studies sends them an invitation to serve on your team.

Your primary institution will need to provide access to the equipment and resources you will need to research your topic. This may include space for you to study, library resources, short courses on research methods (especially for data collection and data analyses), a research community, and specific equipment to support your project. Another important consideration is the ability to gain access to participants so you will be able to observe, measure, train, and/or treat them. Ask questions such as: Can I get the equipment to the participants? Are specific permissions required, such as parental or school consent?

Identifying the Focus of Research

Depending on the level of the degree sought, there are often specific requirements that need to be met. At the masters level, it is often beneficial to focus on developing methodological skills; replicating a previous study is a good way of achieving these skills. A replication is when a study's methods are repeated exactly as in a prior study but using a different sample of participants. As explained in *Chapter 3*, replication is a fundamental scientific tool that is too often neglected in science. Each time a study is replicated, and the results are similar to those in a previous study, it increases the validity of the original findings and the confidence that they are real and not a mere artifact of unexamined details. However, it should be stressed that doctoral research (PhD, Ed. Doc, Prof. Doc) is expected to make an original contribution to the existing body of knowledge, not necessarily in a spectacular way. Please remember: "Less is more, simple is better" (Cohen, 1990).

During the Research Project

A research project rarely follows a pre-planned path with precision, so you need to be ready for changes. For instance, new evidence might come to light, or the results of a previous study may reveal unexpected outcomes, which may make it necessary to introduce certain modifications regarding the direction of your project. It is, therefore, important in a multi-study project for the research team to analyze and discuss the results from the previous study before moving onto the next; otherwise, the entire project may never reach the intended outcomes. For example, the present authors experienced a setback in a project examining the effect of stretching on muscle inflammation and recovery, simply because an insufficiently sensitive inflammation marker was used for a study based on limited analysis of data from a previous study. There will always be a need for course correction in research, but through planning and discussion one can minimize such distractions. If you do make a procedural mistake that has implications for interpreting your results, describe the limitations honestly in your report.

Serving a Particular Population or Contributing to Theory?

When thinking about a program of research, you may have two priorities: (1) Serving a particular population, or (2) Contributing to theory. Although these two priorities are not exclusive, one will often dominate. You may plan to carry out a fixed number of studies on a particular theory or population and then change your focus. For instance, Professor J. Duda has focused on motivational climate psychology across sport and dance populations (e.g., Reinboth & Duda, 2004) whereas others studied the physiological elements of high intensity intermittent exercise (e.g., Bangsbo et al., 1993;), and others examined elements of osteoporosis in dancers (e.g., Amorim et al., 2017; Amorim et al., 2019). An alternative route is to establish a collaboration with "stakeholders" such as dancers, dance teachers, or medical support teams.

When collaborating with professionals who are not scientists, establish at the outset that it can take some time to answer even a simple question conclusively. This is exactly what happened when our research group was approached by the medical team of a professional ballet company that was worried about vitamin D levels in their dancers. To address their concerns, we collaborated for six consecutive years on a series of studies designed to answer the question: "Should dancers receive vitamin D supplements?"

The first study employed an observational design to see how vitamin D (serum 25[OH]D3) levels varied between winter and summer in an elite ballet population; it was found that the majority of dancers had serum [OH]D3 insufficiency regardless of season (Wolman et al., 2013). This led the research group to develop the theory, based on available data on other populations, that vitamin D supplementation would improve muscle function, which in turn would have an effect on injury incidence. A randomized controlled trial (RCT) study was therefore planned using vitamin D3 supplementation and the dependent variables of muscle function and injury incidence. However, since the previous study revealed high levels of deficiency and insufficiency, it was deemed clinically prudent not to have 1:1 intervention to control allocation but to ask for volunteers (this is an example of how research needs to adapt when working with dance companies).

Thus, the originally planned 4-month RCT became a controlled study, which revealed significant improvements in muscle function and decrease in injury incidence in those dancers who received appropriate supplementation (Wyon et al., 2014). This was also the case when the supplementation changed to a single high dosage bolus and measured muscle function improvements over a 1-week period using a double-blind RCT protocol (Wyon et al., 2016); results indicated significant improvements in muscular strength. The final study was characterized by a longitudinal muscle function and injury incidence design, this time implementing an RCT protocol on elite pre-professional dancers (Wyon et al., 2018).

Collaboration as a Means of Developing Research Skills

Researchers continually improve and develop their skills, whether in writing, methodological design, or understanding the theoretical underpinnings of their area of study. One of the major benefits of scientific research is that most investigations in the dance sciences are conducted by teams, with different members bringing different skills and knowledge to the research process. Training to be a researcher is an apprenticeship. You learn how to be a researcher by working closely with more advanced researchers. Dance science research teams often consists of three to four people, and a good team dynamic is indispensable.

There is usually more than one good way to answer a question, and the path chosen is often determined by the skills and experience of the team as a whole. Time spent planning and discussing how the question is to be addressed, previous research, comparable methodologies, and who is going to do what are vital preliminary conversations for a team. These discussions should be an open forum where everyone can contribute and be heard. The principal investigator is expected to have more research experience, but each decision should be clearly explained, so that every team member understands the reasoning behind it. Speaking out might seem scary at first, but good researchers are, by nature, open to considering new ideas when convincing evidence is offered. They are also ready to accept criticism.

It is important for researchers to avoid working in silos. Be open with colleagues about what you are doing, as they might be able to provide—often quite accidentally—an alternative perspective that might be useful for your work, or to develop a study that complements your own. For instance, researchers studying the biomechanics of pirouette might invite psychologists to interview participants about their attentional focus during the movement or invite coaches or teachers to document the essential features of how the participants were taught.

Such seemingly unrelated collaborations may sometimes lead to the development of new perspectives and new theories, both of which are the essence of science. For instance, a collaboration between dance scientists and experts in robotics opened the door to a new product (home robot) which was able to support the rehabilitation of stroke sufferers. Based on information obtained via biomechanical analyses of simple dance movements, this home robot was programmed to "remind" patients of how to execute certain limb movements with the aim of increasing mobility.

When studying at undergraduate and masters level it is good practice to collaborate with a student who is working on a more advanced degree by helping with data collection, literature reviews, data coding, or participant recruitment, or by preparing research ethics applications. Be an active learner; ask questions about why things are being done in a certain manner (but never in front of participants); sit with the statistician when doing the numbers; and offer constructive suggestions if you can do so without distracting your colleagues from the primary purpose of their activity.

After Completing Your Research

Your research efforts may serve as a launch pad for answering a series of questions arising from each study you complete. Building on your work as a graduate student will enable you gradually to develop your own "line of research." Of course, publishing your research findings is essential, not only for building a reputation as a researcher but also for contributing to the body of knowledge in the field. As your work becomes known (through citations of your published material), more opportunities will become available, including grants to fund future research. Presenting your work at conferences is a great way to get feedback from professionals in the field, but publishing your research will give it a broader reach. It will also help you become competitive in the academic environment and will provide critical support for a career in academia. As a third step, it is also good practice to "translate" your research into a format that will communicate effectively with the dance community, as this will help improve the impact of your research. This may be in the form of magazine articles, infographics, blogs/vlogs, and podcasts.

Summary

In this chapter we addressed key aspects of pursuing graduate studies in the dance sciences. We explained the importance of selecting an institution, a program of study, and a supervisor to provide the resources and mentorship you will need to succeed. We identified a number of personal qualities that can help you focus your research and develop into an independent researcher with the potential to make unique contributions to the field. We recommended collaboration and participation on research teams as strategies for developing research skills. To advance in a career as an independent researcher, you will need to focus on publishing your research and acquiring grants to support your research.

The journey from student to dance scientist is an exciting and multi-faceted form of art on its own. We wish you much success and hope the ideas we have shared help you launch a promising and satisfying career in dance medicine and science.

Highlights

- Becoming a research student is an apprenticeship.
- Replication is a fundamental tool of science.
- No one knows everything, so work in teams to achieve your goals.

Application Exercises

- Identify an institution and a mentor where you will be likely to get the support you need to develop into the type of researcher you want to become.
- Identify at least two journals that might be interested in publishing your articles addressing a research question that interests you.

- Examine a conference program and identify two researchers who are doing work that relates to your interests. What is their professional affiliation?
- Search the current literature to find a research article related to your topic of interest by one of the presenters at the International Association of Dance Medicine and Science conference.

References

Amorim, T., Koutedakis, Y., Nevill, A., Wyon, M., Maia, J., Machado, J. C., Marques, F., Metsios, G. S., Flouris, A. D., Adubeiro, N., Nogueira, L., & Dimitriou, L. (2017). Bone mineral density in vocational and professional ballet dancers. *Osteoporosis International, 28*(10), 2903–2912.

Amorim, T., Metsios, G.S., Flouris, A.D., Nevill, A., Gomes, T.N., Wyon, M., Marques, F., Nogueira, L., Adubeiro, N., Jamurtas, A.Z., Maia, J., & Koutedakis, Y. (2019). Endocrine parameters in association with bone mineral accrual in young female vocational ballet dancers. *Archives of Osteoporosis, 14,* 46.

Amorim, T., Wyon, M., Maia, J., Machado, J., Marques, F., Metsios, G., Flouris, A., & Koutedakis, Y. (2015). Prevalence of low bone mineral density in female dancers: A systematic review. *Sports Medicine, 45*(2), 257–268.

Bangsbo, J., Johansen, L., Graham, T., & Saltin, B. (1993). Lactate and H+ effluxes from human skeletal muscles during intense, dynamic exercise. *Journal of Physiology, 462*(1), 115–133.

Brinson, P., & Dick, F. (1996). *Fit to Dance?* Calouste Gulbenkian Foundation.

Cohen, J. (1990). Things I have learned (so far). *American Psychologist Journal, 45*(12), 1304–1312.

Koutedakis, Y., & Sharp, N. C. C. (1999). *The Fit and Healthy Dancer.* John Wiley & Sons.

Milton, D., Appleton, P. R., Bryant, A., & Duda, J. L. (2018). Initial validation of the Teacher-Created Empowering and Disempowering Motivational Climate Questionnaire in Physical Education. *Journal of Teaching in Physical Education, 37*(4), 340–351.

Presti, R. L. (2014). History of science: The first scientist. *Nature, 512*(7514), 250.

Reinboth, M., & Duda, J. L. (2004). The motivational climate, perceived ability, and athletes' psychological and physical well-being. *Sport Psychologist, 18*(3), 237–251.

Rosenstock, I. M., & Hochbaum, G. M. (1961). Some principles of research design in public health. *American Journal of Public Health and the Nation's Health, 51*(2), 266–277.

Wolman, R., Wyon, M., Allen, N., Nevill, A., Eastell, R., & Koutedakis, Y. (2013). The vitamin D status of professional dancers in the winter and in the summer. *Journal of Science and Medicine in Sport, 16,* 388–391.

Wyon, M., Koutedakis, Y., Wolman, R., Nevill, A., & Allen, N. (2014). The influence of winter vitamin D supplementation on muscle function and injury occurrence in elite ballet dancers: A controlled study. *Journal of Science and Medicine in Sport, 17*(1), 8–12.

Wyon, M., Nevill, A., Cloak, R., Metsios, G., Gould, D., Ingham, A., Wolman, R., & Koutedakis, Y. (2016). Acute effects of vitamin D3 supplementation on muscle strength in judoka athletes: A randomised placebo-controlled, double-blind trial. *Clinical Journal of Sport Medicine, 26*(4), 279–284.

Wyon, M. A., Wolman, R., Kolokythas, N., Sheriff, K., Galloway, S., & Mattiussi, A. (2018). The effect of vitamin D supplementation in elite adolescent dancers on muscle function and injury incidence: A randomised double-blind study. *International Journal of Sports Physiology and Performance, 14*(1), 1–15.

Additional Resources

Edwards, N. (2018). *Developing a Program of Research: An Essential Process for a Successful Research Career.* CHNET Press.

Feibelman, P. J. (2011). *A PhD Is Not Enough!: A Guide to Survival in Science.* Basic Books.

23

Building a Researcher's Repertoire

K. MICHAEL ROWLEY

Questions to Stimulate Thinking

- What are the practical skills you need to develop to be a successful researcher in your area of interest?
- What are some research communities globally and locally with which you could connect?
- Who are some mentors or collaborators with whom you could connect?

Reading this text has provided you with an overview of the activities, skills, and approaches involved in conducting research with dancers. As is true for learning to dance, reading and hearing about research is not enough, by itself, to make one a skillful researcher. Dancing employs a complex and sophisticated collection of skills taking years to master. Becoming a capable researcher is likely to follow a similar path. Building your research repertoire is in many ways like building your dance repertoire. In this chapter we offer suggestions on how to expand the fundamental research knowledge that you have been building.

Rather than offer only one perspective from a single author, we "interviewed" a number of experienced research mentors in the dance sciences, so as to provide an array of perspectives on how you might continue to develop your research skills (they are listed at the end of the chapter and profiled at the back of the book). You may find one approach that is well matched to the way you like to work, or you may want to choose and combine elements from several approaches. Incorporating ideas from your own life experiences will help you build a research-skill-development strategy that is tailored to your own learning strategies, work habits, passions, and skills.

Finding a Research Community

The first step in building your repertoire as a researcher will be to connect with a research community, an environment where you can learn and practice the necessary skills and get support from others at different stages of the process. At the broadest level geographically, you can use social media to find and follow dance science groups and influencers. There are online communities of dancers, therapists, doctors, and researchers who are interested in the same things you are. Connect and join the conversation as a means of rehearsing and further developing your ability to think and communicate like a researcher.

You could also attend conferences where you can network with other professionals, such as the International Association for Dance Medicine & Science (IADMS) annual conference. At conferences, you will be able to attend presentations to understand the current state of a specific field that interests you and participate in networking groups that support researchers new to the field.

More locally, you may also look for a research group to join. For example, connecting with like-minded students in a student research group, sitting on a research ethics committee as a student representative, volunteering in an established lab, joining a journal club or volunteering in a research study are a few ways that you can learn about the research process through participation.

Given that the field of dance research is still growing, you may have trouble finding a local community with whom to connect. In this case one of our contributors, Tom Welsh, a dance scientist, educator, and research mentor, offers this valuable advice on how to build your own dance research community by connecting with broader research groups:

> Particularly in fields that are new to research, like dance, jobs that have research as a primary assignment are uncommon; we will have to build them. Look for researchers in the area that are working with other performing artists or athletes. Also consider asking researchers outside your area if you may participate remotely in their research group meetings. If you find a group where you can continue to exercise your research skills, remember that you are a guest. Start as a fly-on-the-wall, offering your opinion only when invited. If your contributions are informed by what you have been able to absorb by listening, invitations will become more frequent. Eventually, you may be accepted as

> an established member of the group with full participation privileges, earning the trust of those in your group.—TW

Building and Expanding Your Conceptual Foundation

After reading this textbook, you will have a grasp of research methods commonly used in our field. You can expand this foundational understanding through taking courses such as research design, research methods, quantitative analyses, and qualitative analyses. You might volunteer in a research lab or acquire a position as a research assistant working with a faculty member. Learning the process of research within a research community is beneficial and provides students with experiences that they would not otherwise have. As the sport, performance, and dance scientist and psychologist Lynda Mainwaring suggests:

> Undergraduate volunteer experiences have kick-started academic career paths. Those pursuing a research career will want to stay involved with research even if early career opportunities are not research positions. Keep up with the current literature and collaborate with others on projects.—LM

Exposure to research through collaboration with others provides opportunities for reinforcing what you have already learned and advancing your knowledge through practice and seeing how research studies are run. You will be encouraged to read in the area of the research topic and will have the opportunity to discuss scholarly articles with others.

Reading research books and current peer-reviewed articles in scientific journals can provide sound advice and models for research methods. Our contributors suggest reading dance research articles in the *Journal of Dance Medicine & Science* and *Medical Problems of Performing Artists* as a way to exercise your ability to think about writing a clear research question, analyze data, and write for a scientific publication. Reading an abundance of research studies to learn what other researchers have done and how they have conducted their research is an important strategy for developing your abilities.

If you have the opportunity to engage in full-time student life, this can afford rich and varied opportunities, such as access to information through university libraries, interactions with other student researchers, faculty mentorship, and exposure to a diverse range of learning experiences. Sarah Kenny, dance scientist, injury epidemiology researcher, and contemporary dancer, provides the following advice:

> Study at the post-secondary level can be quite intense for some. If the opportunity allows, pursue studies full-time, rather than part-time. Being full-time can lead to an immersive experience that allows you to take advantage of different circumstances and opportunities that may not otherwise happen.—SK

For readers returning to school after a performing or teaching career, it can be difficult or intimidating to feel behind when it comes to building a conceptual foundation in your dance research area. If you are still working and can only engage in student life part-time, it can feel as though you are at a disadvantage. But recognizing how to leverage the skills, connections, and experience you already have from your former career can actually be an advantage that sets you apart as a valuable member of a lab group or research team:

> Having significant experience in dance is an ideal background for a dance science researcher. It allows researchers to understand, from their own experience, what is relevant and what is important to dancers. It also allows researchers to know how to approach dancers and what types of interventions are likely to be acceptable and sustainable in the world of dance.—TW

Developing Practical Skills

Just as it is impossible to become a competent dancer solely by watching videos of dancing, it is impossible to become a competent researcher solely by reading a book about research. Being a researcher involves the use of a collection of skills that must be developed and refined. The sooner new researchers begin to exercise their research skills, the sooner they will develop competence as researchers. Here are some strategies to develop and practice those skills.

Collaboration

Engaging in collaborative research is a great way to learn the process of research. It also provides opportunity to build on combined efforts, learn from others, and experience the synergy of teamwork. The great thing about collaboration is that each research partner gets to multiply the research team's conceptual and research knowledge instantaneously. Instead of pursuing a second graduate degree, you could collaborate with a researcher in another field to broaden the scope and impact of your research. For a student or early-career researcher, a research collaboration could be a way to offer your own expertise, perhaps as a dancer or dance educator.

Assisting other researchers will help you develop a solid foundation for your own research studies. These experiences might include assisting or learning to use a specific piece of equipment, note-taking while observing an interview process, interacting with research participants between data collection times, and scribing data during complex data collection phases. Even if the role or tasks offered appear mundane at first, what you learn from the experience is likely to be valuable when you initiate your own research. As a student or dancer, you may be able to engage in ongoing projects as a participant in the research. Seeing the research process from many perspectives will help you become a well-rounded and, likely, a more effective dance science researcher.

Each of these research experiences provides opportunities to learn, develop, and enhance your skills, which you can then document on your curriculum vitae. A vitae that includes extensive practical research experiences and letters of reference that articulate your work in areas that align with developing research skills are often considered with high regard.

Searching, Reading, and Synthesizing Research Publications

A crucial skill for researchers to add to their repertoire is the ability to find, comprehend, critically evaluate, and synthesize research publications. These skills are discussed in *Chapters 4* and *7*. Here are suggestions from our contributors for developing these skills further.

A typical place to begin searching for articles is through using databases like PubMed, Web of Science, or GoogleScholar. For dance-specific research, you may find success starting with popular press articles and tracking down references or citations. Margaret Wilson, a dance, Pilates, history, and kinesiology teacher and dance scientist, emphasizes:

> While peer-reviewed studies are the gold standard, orienting yourself to the area of interest includes reading popular press articles to see what the field is referencing. For example, *Dance Magazine* often does features on health and wellness, injury and injury prevention, burnout and motivation. Reviewing the reference list at the end of the article will point you in the direction of other studies related to your question.—MW

As you read articles as part of a review project or on your own, begin adding them to a database or citation manager (see *Chapter 4* for a list of options). Summarize the information in the article in your own words and keep track of findings that could be useful for your work.

> I always learn something new when I try to summarize through my own lens what someone else has done. When I begin to feel that I am too busy to take time to write a research article summary, I remind myself that I have just invested an hour or more in reading the article, and it will never again be as fresh in my memory. Taking a little additional time to write a summary now will offer important, long-term benefits.—TW

Comprehending research publications and critically analyzing them are skills that can be practiced. Unfortunately, they can also deteriorate with disuse. Start a practice of reading research articles regularly, perhaps weekly or biweekly, to build and maintain this skill. An informal reading group or journal club with peers can be a strong motivator to keep you on track.

Communicating and Translating Your Research

All researchers need to develop their scientific writing and communicating skills. Scientific writing can be quite different from the styles used in fiction, art critique, journalism, or other media. Jeff Russell, dance, theater, and sport scientist and athletic trainer, has some words of wisdom about how to develop these skills:

> The two best academic decisions I ever made were to take a touch-typing class in high school and a technical writing class in college. I tell my students, "Never say that you are a terrible writer or a terrible speller. Those are lies and rip-offs!" Yes, of course some people are more gifted in these areas than other people are, but everyone has the ability to learn these key skills. Pursue classes and seminars, identify the wealth of books and journal articles about scientific writing, and use a dictionary and thesaurus whenever you write. Above all, though, you must write! Practicing writing and getting feedback from trusted colleagues are essential. The process is not any different than learning new choreography.—JR

Being willing to receive feedback and knowing where and when to ask for it is a skill that can be developed and practiced as well. Start by asking peers to review your writing, and work your way up to mentors, supervisors, or other writers you respect or admire.

Research Presentations

One of the most important ways research is communicated is through conference presentations in short oral, poster, or other formats. Many organizations publish guidelines for presenters, which outline how these presentations should be constructed and what researchers should prioritize when communicating their research. Our contributors offer presentation tips that reinforce what was outlined in *Chapter 20:*

> Perhaps the most frequent mistake presenters make is using text, either in slides or on a poster, that is much too small to be read. Try projecting your slides in a large room or printing text at various sizes on paper and hanging it on the wall like a poster. Can you read it? If not, cut your words back to only the most necessary ones and make the text size bigger. You also can critique other presenters' slides and posters. Make notes about what works and what doesn't, so that you can fine-tune your own presentations.—JR

Watching research presentations will provide guidance and inspiration. You will likely identify presentation strategies you like and ones you find don't work well. Conference presentations are often archived, and you may be able to view them digitally. Review other researchers' posters as well, perhaps previous posters created from your lab group or department. Does the format communicate the question, methods, findings, and application clearly and have an aesthetic appeal that attracts viewers?

Once your presentation is constructed, it is important to practice by delivering it to peers, relatives, roommates, etcetera. Encourage them to be critical and ask hard questions. Preparing for difficult questions with people you know and trust will make it easier to engage with professionals in a conference setting.

Social Media

Another increasingly important means of communicating research activity is through social media. Many research labs, research journals, and individual researchers maintain active social media accounts to promote their work. Promoting research activities on social media can allow diverse and interdisciplinary readers to engage with both the material and the researcher in ways not possible in academic journals or conferences. Sharing your findings on social media can also expand the reach and impact of your work. Before sharing research findings, however, consider your plans for publishing the findings in an academic journal and the copyright rules of that journal. Some journals will not publish work that has already been presented in a public forum. Most will allow you to post brief summaries of your research after a manuscript is published, and in fact some journals will promote your work on their own channels.

Fully engaging in the social media sphere can be a challenging task and is a whole area of study on its own. There are resources and trainings available to help you optimize post aesthetics and content, use analytics to maximize reach, and understand how to connect and interact with other users. This is why social media managers are now commonly full-time paid positions at many organizations. As a researcher, you may be able to support a lab member who has an interest in social media promotion—perhaps an undergraduate student or a student dancer—by funding the student to attend trainings or workshops. Or if you yourself have these skills, you might propose offering your expertise to a researcher or research group.

Accessing Resources to Develop New Skills

As you build your researcher repertoire, you will certainly need to find other resources to answer questions you have or to acquire the knowledge and skills discussed so far. Many universities or local libraries have workshops useful for new researchers. These might include sessions on using scientific search engines, referencing software, and other tools. Large libraries often have librarians assigned to specific subject areas. These skilled professionals can help you sort through field-specific content.

In addition to library resources, our contributors suggest joining or creating a journal club as a good resource for early-career researchers:

> Journal clubs often meet monthly to discuss pre-assigned articles. When it is your turn, recommend a dance research article that is likely to have broad appeal across the membership of your journal club. When the group reads the article, it will be your responsibility to present that article, highlighting its essential details, summarizing its findings, and initiating a discussion of the implications.—TW

Of course, it is not possible to foresee all your needs at the outset of this journey. Therefore, building a support network is a good option:

> Early on, put effort into cultivating your academic support network. This includes making friends with the librarian, the biostatistician, like-minded peers, colleagues, and fellow research students.—SK

Mentorship

One thing is true of all experienced researchers—they did not become successful researchers without mentorship and support from others in their field. Finding a research or career mentor is a crucial early step in moving forward in your research career. Jeff Russell and Jennifer Deckert, ballet and kinesiology teacher and dance scientist, have the following to say:

> There are plenty of mentors who really enjoy working with students and young professionals. On the other hand, there also are plenty of people who do not enjoy mentoring or are not good at it. When you find a good mentor—and it may take you a few tries—that person will be like gold to you.—JR

> Many of my mentoring meetings consist of listening and asking questions, suggesting additional reading, and helping the student researchers to find their own path. If others on your research team begin to override your process, perhaps they are not the best fit for your project. As we have discussed throughout the chapters in this book, research takes time, so make sure your mentor and collaborators are in it not just for the results, but for the adventure.—JD

Given the time mentors and mentees commit to one another, it is very important you consider your mentor's interests, expertise, and personal mentoring style when choosing a mentor. Research requires the use of a complex and nuanced set of skills, often uniquely developed from each individual's experiences and what they need to be successful in their discipline. Selecting a mentor whose expertise and experience align with your needs is critical for the success of your research efforts. Further, your "beginner's mind" can bring clarity and new understanding to your mentor's own work as well, a mutually reinforcing relationship.

> One of the most important pieces of advice that was given to me was to align myself with a supervisor with whom I got along. First and foremost, a good supervisor will guide what you do and influence the quality of what you contribute to the field (i.e., your overall success). Be sure to seek out a mentor (or multiples of them!) whose interests align with and complement your own. If all are going to benefit from the experience of working together, then the success of the project is highly likely.—SK

How do you go about initiating a mentor-mentee relationship? First, if you meet this potential mentor in a conference setting, consider asking a sincere and interesting follow-up question to their presentation. Showing your engagement with the material will leave a good first impression. Next, you can reach out to explore mutual interests and get a sense of the rapport and relationship that may be built between you. Mentors may have limited time, so be prepared and efficient with your meetings. You can also join journal clubs or lab meetings organized by your mentor. Here you can read interesting papers, discuss ideas and research, and collaborate with others in the group if the opportunity arises.

Engaging in an Independent Research Project as a Trainee

Depending on your path into the research community, you may have the opportunity to move from engaging in research as a team member to engaging as a project leader. Independent research projects such as capstone projects, theses, or dissertations are formalized research projects usually done under the guidance of a research mentor, reviewed by a panel or committee, and published in a university library. Often, students completing these projects will need to defend their findings and interpretations to a committee of faculty. Here are some of our contributors' thoughts on completing a major student research project:

> This project is often the most involved, self-directed project a new researcher has completed in their lifetime. There is a tendency to feel obligated to change the field in some major way, to perform a magnum opus. Of course, that would be like a beginning dancer expecting to perform his or her first recital as a soloist on the Kennedy Center main stage. New researchers will do themselves, and their field, a favor by scaling their aspirations to match their evolving skills. The primary purpose of your first research project is to develop your skill as a researcher and to complete a study that makes a humble contribution to the field.—TW

> Make sure you provide enough time to complete the writing and to receive and implement feedback from your mentor. As with all creative activities, your writing can feel quite precious to you, and the editing process can sometimes feel like the baby you have spent months creating is being torn to pieces. Take a deep breath and know that this is also part of the journey. Your editors and mentor are working with you to

make it the best project possible, and the end product will be worth the struggle.—JD

Integrating Research with Other Professional Responsibilities

For readers aiming to incorporate research into an already-established career and professional responsibilities, we applaud you! This is not an easy task, but the research community needs your hands-on experience and expertise to be able to conduct research impactful for dancers. We hope these comments from our contributors can offer useful recommendations, support, or inspiration:

> There is an incredible advantage to be had when the researcher is also the knowledge user of the research findings. For example, if you are a dancer or a dance teacher who is pursuing a research degree, then your research questions can come from a place of knowing how dancers' bodies typically function, of knowing how student dancers typically respond, of knowing the culture with which you want to engage.—SK

> There is never an ideal time to conduct research; most researchers in the dance science community have several other responsibilities, and we would not have it any other way. Staying active in the world of dance is what keeps dance science relevant and spurs questions to inform our research better. The truth is that research only comes to fruition once it is moved to the front burner, but allowing it to simmer on the back burner until you are ready to dive in lets the questions continue to develop and provides a richer experience once you are able to focus your attention.—JD

> Some forms of research—case studies and single-case experiments, for example—can be conducted as an extension of clinical practice. Conducting applied research related to your own professional responsibilities will inform the work you do, increase your motivation for doing it, and let you contribute new knowledge, however humble, to the field as you continue developing your research skills. We know of practitioners who have convinced their supervisors to add a research component to their job assignments after conducting research their supervisors found relevant and useful.—TW

> If you are working in a clinical setting, observation of trends that need addressing is a good way to start thinking about your interests and what is needed. Once you identify a research question, find a way to pursue the research by talking to administrators about possibilities. Collaborate with others who are conducting research on similar topics in your environment, or investigate if there are opportunities to collaborate with a local university. Invite speakers to do lunchtime talks and open doors for collaboration.—LM

> Research what you love to do. I have had the opportunity to develop a Vertical dance program as a performance opportunity for me and my students. First, I looked to measure the forces acting on the body, then I researched ways of preparing the body for those stresses and developed and tested a conditioning program. My research is helping me refine the way I train and prepare dancers for performance. All of this has made me understand the art form even more, and I am taking care of my students at the same time.—MW

Qualities of Successful Researchers

In the query to our contributors about how those new to dance science research should develop their research repertoire, one theme emerged quite strongly—*persistence.* Following here are perspectives from our contributors about the challenges of research and how to overcome them:

> An important lesson to learn is that good research generates more questions than it answers. Be grateful that you have the chance to learn something small with every project and that it will provide fuel for the next step. No one is a great researcher after their first project. It takes time, and there is an evolutionary process. Enjoy the exploration and keep putting one foot in front of the other. Intrigue lives around every corner.—JR

> Research skills take time to learn. It is like learning the fundamentals of a dance form so that refined and creative choreography can grow from that foundation. Skill develops over time through a process of interacting with people who are conducting research, by reading and re-reading the literature, by keeping a scientific journal or diary, and by taking time to think and reflect on what one has learned.—LM

> Developing expertise in any field, be it dancing or conducting research, requires a persistent effort over time. If you aspire to become an expert researcher in the dance sciences, you should expect to invest as much effort and time as you invested in developing your expertise in dance. As with any long-term project, having interim goals or benchmarks to achieve can inform progress and motivate your efforts.—TW

In addition to persistence, many scientists will tell you that there are other qualities that are important to facilitate a fruitful research career. *Chapter 22* identifies other qualities including an analytical sense, an ability to stay calm, curiosity, paying attention to details, being organized yet flexible, listening carefully, open-mindedness, rigorousness, and being patient, systematic, and truthful. You may have these qualities already, as dance also requires many of them. The research enterprise, like dance, requires you to invest both time and yourself in a process that unfolds gradually and has many twists and turns.

Researcher Wisdom

Finally, here are a few gems of wisdom offered by our contributors from their experiences:

> One of my favorite quotes suggests that there are two ways to get to the top of an oak tree. You can climb or you can sit on an acorn and wait. We gain knowledge in our world because researchers choose to climb. At the bottom of that tree, all we see are branches, leaves, and a trunk—certainly not a beautiful view, and quite intimidating at that. But with every branch onto which we hoist ourselves, the view improves, and the distance to the top of the tree grows shorter. This is true for successfully completing the climb up a single research project as well as for ascending several projects that collectively improve our knowledge in dance science. Concentrate today on the branch within your reach. Read that article, write that first paragraph, email that expert, search for those references, inquire about that grant. Grab the branch and pull yourself up. Then look for the next branch.—JR

> I think one of the biggest misconceptions about dance science research is that you need a biomechanics lab, expensive equipment, and lots of dancers to conduct a meaningful study. While those resources are nice to have, you do not actually need any of them to complete solid,

valid research that can make a difference in dancers' lives. This field is still defining itself, and through your research you can help shape the future of dance science.—JD

Put effort in, be strategic, and be purposeful when it comes to managing your mental and physical health and well-being. Spend time reflecting and thinking about topic areas that are troubling you, that are intriguing to you, that are motivating you.—SK

The journey to develop your research expertise and repertoire will involve many lessons. This text has provided you with some knowledge and tools to continue this journey as a dance science researcher.

Contributor Acknowledgments

This chapter was compiled with contributions from Jennifer Deckert, Sarah J. Kenny, Lynda Mainwaring, Jeffrey A. Russell, Tom Welsh, and Margaret Wilson. Professional biographies can be found at the back of the book in the list of contributing authors and editors (following the glossary).

Highlights

- Building your dance "rep" in many ways is like building your research "rep."
- Just as it is impossible to become a competent dancer solely by watching videos of dancing, it is impossible to become a competent researcher solely by reading a book about research.
- Being a researcher involves the use of a collection of skills that must be developed and refined.
- All of our contributors agreed that persistence is crucial to succeeding in becoming a researcher.

Appendix A

Consent Document—Sample Template

Note—We recommend using the template provided by your local Research Ethics Committee. We offer this template as an example of what you can expect to see. A parental permission consent document will be structured very similarly.

Study Title

Principal Investigator

Contact Information

INTRODUCTION

We invite you to take part in a research study. Please take as much time as you need to read the consent form. You may want to discuss it with your family, friends, or your personal doctor. If you find any of the language difficult to understand, please ask questions. If you decide to participate, you will be asked to sign this form. A copy of the signed form will be provided to you for your records.

KEY INFORMATION

The following is a short summary of this study to help you decide whether or not you should participate. More detailed information is listed later in this form.

- Being in this research study is voluntary–it is your choice.
- You are being asked to take part in this study because *[fill in study-specific details]*. The purpose of this study is *[fill in study-specific details]*. Your participation in this study will last *[fill in how long participation will last]*. Procedures will include *[provide a brief list of the major procedures in 1–2 sentences]*.
- There are risks from participating in this study. The most common risks are *[provide a brief list of the major risks in 1–2 sentences]*. More detailed information about the risks of this study can be found later in this form.

- *[If there are direct benefits for participating:]* The possible benefits to you for taking part in this study may include *[provide a brief statement of the benefits to participating]*.
- *[If there are no direct benefits for participating:]* You may not receive any direct benefit from taking part in this study. However, your participation in this study may help us learn *[provide a brief statement of the indirect benefits]*.
- If you decide not to participate in this research, your other choices may include *[provide a brief statement of the alternatives to participating, such as alternate treatments, alternate course assignments, not participating, etc.]*.

DETAILED INFORMATION

PURPOSE

[Provide in 1–2 sentences a statement of the purpose of your research.]

PROCEDURES

If you decide to take part, this is what will happen. *[Fill in study-specific details.]*

RISKS AND DISCOMFORTS

Possible risks and discomforts you could experience during this study include *[fill in study-specific details]*.

BENEFITS

The potential benefits to you may include *[list the possible direct benefits to participants]*.

[-or-]

There are no direct benefits to you from taking part in this study. However, your participation in this study may help us learn *[list possible indirect benefits to society]*.

PRIVACY/CONFIDENTIALITY

[Your institution may have template language regarding the conditions under which privacy/confidentiality will be maintained.]

ALTERNATIVES

There may be alternative(s) to participating in this study. These include *[list the alternatives]*. The risks and benefits will be explained to you.

[-or-]

An alternative would be not to not participate in the study.

INJURY

[Your institution may have template language regarding how adverse events or injuries will be treated and how treatment will be paid for.]

POTENTIAL CONFLICT OF INTEREST

[Your institution may have template language regarding how conflicts of interest are disclosed.]

VOLUNTARY PARTICIPATION

It is your choice whether or not to participate in this study. If you choose to participate, you may change your mind and leave the study at any time. Refusal to participate or stopping your participation will involve no penalty or loss of benefits to which you are otherwise entitled. *[Your institution may have template language with more specific information to include about how participants can withdraw from the study safely.]*

CONTACT INFORMATION

If you have questions, concerns, complaints, or think the research has hurt you, talk to the investigator at *[insert the principal investigator's contact information]*.

This research has been reviewed by *[your Research Ethics Committee]*. The Research Ethics Committee is a research review board that reviews and monitors research studies to protect the rights and welfare of research participants. Contact the Research Ethics Committee if you have questions about your rights as a research participant or you have complaints about the research. You may contact the Research Ethics Committee at *[insert contact information]*.

STATEMENT OF CONSENT

I have read (or someone has read to me) the information provided above. I have been given a chance to ask questions. All my questions have been answered. By signing this form, I am agreeing to take part in this study.

____
Name of Research Participant

____
Signature

____
Date Signed

Appendix B

Child Assent Document—Sample Template

Note—Child assent is usually required for research participants aged 7–17, though local jurisdictions may have their own age guidelines. We recommend using the template provided by your institution. We offer this template as an example.

Study Title

Principal Investigator

Contact Information

1. *[Name of Principal Investigator]* is doing a research study about *[topic]*.
2. We are asking you to take part in this research study because we want to learn more about *[outline what the study is about]*.
3. If you agree to be in this study, you will be asked to *[describe procedures]*.
4. Sometimes accidents or bad things happen in research studies. Some of the bad things that could happen are *[describe risks]*. Some of these things might happen to you or they might not; or things might happen that we don't know about yet.
5. People also have good things happen to them when they are in research studies. The good things may be *[describe direct and indirect benefits]*.
6. We will ask your parents to give their permission for you to take part in this study. Even if your parents say "yes" you can still say "no." You do not have to join this study; it is up to you. *[Include any alternatives to participating.]*
7. You can say "yes" now and change your mind later. All you have to do is tell us you want to stop. No one will be mad at you if you don't want to be in the study or if you join the study and change your mind later and stop.
8. Before you say yes or no to being in this study, we will answer any questions you have. If you join the study, you can ask questions at

any time; just tell the researcher that you have a question. You can call the researcher at *[contact information]*.

If you put your name below, it means that you agree to take part in this research study.

____
Name of Child

____
Child's Signature

____
Date Signed

____
Name of Person Obtaining Assent

____
Signature

____
Date Signed

Appendix C

Journals Publishing Dance Medicine and Science Research

DANCE/PERFORMING ARTS MEDICINE

Journal of Dance Medicine & Science
Medical Problems of Performing Artists

DANCE EDUCATION

Dance Research Journal
Journal of Dance Education
Journal of Physical Education
Research in Dance Education

ARTS AND HEALTH

American Journal of Dance Therapy
Arts and Health
Body, Movement and Dance in Psychotherapy
Body & Society
Journal of Applied Arts & Health
Journal of Dance & Somatic Practices

EXERCISE SCIENCE AND SPORTS MEDICINE

American Journal of Sports Medicine
British Journal of Sports Medicine
Clinical Rehabilitation
Clinics in Sports Medicine
European Journal of Sport Science
Exercise & Sport Science Reviews
Human Kinetics
Human Movement Sciences

Journal of Applied Sport Psychology
International Journal of Sports Medicine
Journal of Science and Medicine in Sport
Journal of Sport Rehabilitation
Journal of Sports Medicine
Journal of Sports Medicine and Physical Fitness
Journal of Sports Sciences
Medicine & Science in Sports & Exercise
Motor Control
Pediatric Exercise Science
Research in Sports Medicine
Research Quarterly for Exercise and Sport
Scandinavian Journal of Medicine & Science in Sports
Scandinavian Journal of Rehabilitative Medicine
Science and Sports
Sports Health
Sports Medicine
Sports Medicine Clinics
Translational Sports Medicine

ATHLETIC TRAINING, PHYSICAL THERAPY/PHYSIOTHERAPY

International Journal of Athletic Training & Therapy
International Journal of Sport Physical Therapy
Journal of Athletic Enhancement
Journal of Athletic Training
Journal of Orthopedic & Sports Physical Therapy
Journal of Physical Therapy
Physical Therapy
Physical Therapy in Sport
Physical Therapy Reviews
Physiotherapy

STRENGTH AND CONDITIONING AND PHYSIOLOGY

Acta Physiologica Scandinavica
American Journal of Physiology
Annual Review of Physiology
Canadian Journal of Physiology and Pharmacology
Experimental Physiology

International Journal of Sports Physiology and Performance
Journal of Applied Physiology
Journal of Exercise Physiology
Journal of Neurophysiology
Journal of Physiology
Journal of Strength and Conditioning Research
Strength and Conditioning Journal
Science & Sports

MEDICAL SCIENCES

Archives of Physical Medicine and Rehabilitation
Arthroscopy: The Journal of Arthroscopic & Related Surgery
BMC Musculoskeletal Disorders
Clinical Journal of Sport Medicine
Clinical Orthopaedics
Clinical Orthopaedics and Related Research
Journal of Arthroscopic and Related Surgery
Journal of Bone and Joint Surgery
Journal of Bone and Joint Surgery America
Journal of Knee Surgery
Journal of Orthopaedic Research
Journal of the American Academy of Orthopedic Surgeons
Journal of the Southern Orthopedic Association
Knee
Knee Surgery, Sports Traumatology, Arthroscopy
Orthopedic Clinics of North America
Orthopedics
PLOS one
Spine

BIOMECHANICS AND MOTOR CONTROL

Acta of Bioengineering and Biomechanics
Annals of Biomedical Engineering
Clinical Biomechanics
Gait and Posture
Human Movement Science
IEEE Engineering in Medicine and Biology
IEEE Transactions on Biomedical Engineering

International Journal of Sports Biomechanics
Journal of Applied Biomechanics
Journal of Biomechanical Engineering
Journal of Biomechanics
Journal of Electromyography and Kinesiology
Motor Control

NUTRITION

American Journal of Clinical Nutrition
Annual Review of Nutrition
British Journal of Nutrition
European Journal of Clinical Nutrition
International Journal of Sport Nutrition
International Journal of Sport Nutrition & Exercise Metabolism
Journal of Nutrition
Medicine, Exercise, Nutrition, and Health

EPIDEMIOLOGY

Epidemiological Reviews
Injury Prevention: Journal of the International Society for Child and Adolescent Injury Prevention

SOCIAL SCIENCES AND PSYCHOLOGY

Ecological Psychology

NEUROLOGY

Brain
Brain Research
Experimental Brain Research
Journal of Neuroscience Methods
Neuroscience Letters
Neuroscience Research

Glossary

A-B design: See Simple Comparison Design.

Action research: An interactive inquiry process that balances problem-solving actions implemented in a collaborative context with data-driven collaborative analysis or research to understand underlying causes enabling future predictions about personal and organizational change.

AIMRaD format: Refers to a paper that is structured by main sections: Abstract, Introduction, Methods, Results, and Discussion.

Algorithm: A process or set of rules to be followed in calculations or other problem-solving operations.

Alternating Treatments (Multi-Element) Design: A within-subject design that uses repeated measurements of a dependent variable while two or more treatments are applied in an alternating fashion.

Alternative Hypothesis: A statement that is contradictory to Null Hypothesis (H_0) and what we conclude when we reject H_0. Denoted as H_a or H_1.

AMA: American Medical Association.

Analysis of Variance, ANOVA: A collection of statistical models and their associated estimation test procedures (e.g., the "variation" among and between groups) used to analyze the differences among means.

APA: American Psychological Association.

Assessment Protocol: See Observation Protocol.

Beneficence: Act of doing good or being kind, related to the bioethical principle underlying the duty to act in the best interests of the client.

Between-Group Experiments: Explores changes between groups of participants.

Between-Group Variability: Difference between group means.

Bias: A cause to feel or show inclination or prejudice for or against someone or something

Bibliography, Annotated: A list of citations to books, articles, and documents. Each citation is followed by a brief descriptive and evaluative paragraph, the annotation.

Blinding (double): A process in which neither participants nor researchers know which treatment or intervention participants are receiving until the end of the intervention.

Boolean Search Terms: A type of search allowing users to combine keywords with operations (or modifiers) such as AND, NOT, and OR to tailor search results.

Case-based research: A qualitative method that emerges from the identification of a novel observation.

Case Report: A detailed report of the symptoms, signs, diagnosis, treatment, and follow-up of an individual case (e.g., patient, dancer).

Case Series: A research study that tracks subjects with known symptoms, signs, diagnosis, and treatment or examines their medical records for exposure and outcome.

Case Study: A detailed study of a specific subject, such as a person, group, place, event, organization, or phenomenon. A case study often involves qualitative methods but sometimes also uses quantitative methods.

Causality: Influence by which one event, process, state, or object (a cause) contributes to the production of another event, process, state, or object (an effect), where the cause is partly responsible for the effect, and the effect is partly dependent on the cause. Also referred to as causation, or cause and effect.

Central Tendency: Mean, median, or mode.

Chi-square Test: A non-parametric test that examines the discrepancy between the observed and the expected results.

Child Assent: The affirmative expressed agreement of a minor to participate in a research study.

Citation: A quotation from or reference to a book, paper, or author, especially in a scholarly work.

Clinical Practice: A model of practice that involves those activities with and on behalf of clients, especially those activities completed in the client's presence and with the client's collaboration.

Clinimetrics: The science of clinical measurements to examine indexes, rating scales, and other expressions used to describe or measure symptoms, physical signs, and other clinical phenomena.

Cluster sampling: A probability sampling method that involves dividing a population into clusters and then randomly selecting a sample of these clusters.

Coefficient of Determination: The proportion of the variation in the dependent variable that is predictable from the independent variable(s). Denoted as R^2 or r^2 and pronounced as "R squared." See Explained Variance.

Cohort: A group of people assembled on the bases of common characteristics or experimental protocol.

Competition Dance: A sport in which competitors perform dances in any of several permitted dance styles—such as acrobatics, ballet, contemporary, jazz, hip-hop, lyrical, modern, musical theatre, tap, and improvisation—before a common group of judges.

Condition Change Line: A vertical line that marks the point in time when a systematic condition change is applied; sometimes connected in a step-wise fashion across more than one set of axes in a multiple baseline design.

Confidentiality: The duty of an individual (researcher or professional) to refrain from sharing personal information except with the express consent of the individual.

Confirmability: The degree to which findings could be confirmed by other researchers performing the same analysis.

Conflict of Interest: Occurs when personal interests are elevated above public obligations.

Confounding Variable: An uncontrolled variable that has the potential to influence the DV under study, thereby confounding the interpretation of a study's results.

Constructivism: A philosophical position that suggests multiple realities exist and that people construct their own knowledge and realities, which are complex and created in a particular social context.

Convenience sampling: A sample to which the researcher has easy access.

Correlation: A term describing the type of association between two phenomena or variables.

Credibility: Confidence in the interpretation of data.

Critical Realism: A branch of philosophy that distinguishes between the "real" world and thee "observable" world.

Critical Theory: An approach to social philosophy that focuses on reflective assessment and critique of society and culture in order to reveal and challenge power structures.

Cross-sectional sampling: Collection of data at one point in time.

Cross-sectional study: A type of observational study where researchers examine data from a sample, groups, or population at a specific point in time.

Cross-sectional: A particular moment in time that is sampled.

Data: (singular = Datum) are individual facts, statistics, or items of information.

Database: An organized collection of structured information, or data, where the data can then be easily accessed, managed, modified, updated, controlled, and organized.

Debriefing: The communication of information related to the research after the study has been completed. Researchers may share the findings or explain procedures further if they were not entirely disclosed in the initial research communication because of sound scientific reasons.

Deductive Reasoning: The process of reasoning from one or more statements to reach a logical conclusion. This process goes in the same direction as that of the conditions and links premises with conclusions.

Delta: The upper case letter Delta (Δ) is a Greek symbol used to denote difference or change of a quantity in mathematics and the sciences.

Demographic Information: The description of a population based on factors such as age, race, sex, and others.

Dependability: The stability of findings over time through clarity of research question, methods, and analysis.

Dependent Variable (DV): The variable that is being measured or tested in a study. Its value may change depending on the value of another variable, called the independent variable.

Descriptive Statistics: Information that summarizes a given data set and that may include measures of central tendency (e.g., means) and data dispersion (e.g., standard deviations).

Developmental Research: A process that examines growth and development over the lifespan.

Direct Observation: Observing a phenomenon directly, such as seeing if a behavior occurs.

Direct Replication: The practice of repeating an existing study and keeping protocols as close to identical as possible to the original study parameters.

Dissemination: One of the four fundamental aspects of knowledge translation that involves relating information or findings from research to appropriate audiences.

Ecological Validity: Refers to the judgment of whether a given study's variables and conclusions are sufficiently relevant to its population.

Effect Size: A quantitative measure of the magnitude of the experimental effect. The larger the effect size, the stronger the relationship between two variables.

Emergent: The process of developing a research design or analysis.

Empirical: Based on, concerned with, or verifiable by observation or experience rather than theory or pure logic.

Epistemology: Fundamental assumptions about the nature of knowledge.

Equivalence Test: A statistical inference test examining whether the means of two samples tested are equivalent or whether they differ by a small pre-defined amount (Δ margin; for Δ, see Delta in this glossary. A variation of the classical hypothesis testing where the null hypothesis is defined as an effect large enough to be deemed interesting (specified by the margin), and the alternative hypothesis is any effect that is less extreme than specified margin.

Ethical Guidelines: Standards that must be met to conduct a research study according to specific ethical parameters.

Ethics Review Board (ERB) or Committee: A committee that oversees and reviews research proposals in terms of appropriate ethical conduct of researchers and treatment of participants over the course of a research study. Also see Institutional Review Board (IRB) and Research Ethics Committee (REC).

Ethics: Refers to both (1) a branch of moral philosophy that examines moral character and conduct that humans ought to engage in terms of rights, values, fairness, obligations, and benefits to society and (2) a standard of conduct that is founded upon moral philosophy. Professional organizations develop ethical guidelines to inform their own actions and evaluate actions of their members.

Ethnography: The scientific description of the customs of individual peoples and cultures determined through an in-depth systematic study.

Exclusion Criteria: Criteria for which participants from the target population are to be excluded from the current study sample.

Expectancy or demand characteristics: An experimental artifact where participants form an interpretation of the experiment's purpose and subconsciously change their behavior to fit that interpretation.

Experimental Analysis: Study design isolates the influence of an independent variable on a dependent variable.

Experimental Control: Techniques that researchers use to minimize the effects of extraneous experience and environmental variables as well as to strengthen the inference that changes in the dependent variable are due to the independent variable.

Explained Variance (r^2): The proportion of the variance in a dependent variable that can be determined or explained by one or more variables.

External Validity: The extent to which the results of a study can be generalized to the population from which the sample has been selected and across other situations, people, stimuli, and times represented in the study.

F: A value named after Sir Ronald Fisher and based on the ratio of two mean squares. Analysis of variance (ANOVA) tests use F-tests to assess the equality of means. An F value is displayed in an ANOVA output.

Familywise Error Rate: When multiple tests are carried out on a single sample it affects the assumed 95% confidence levels and increases the chance of Type I error.

Generalizability: A measure of how useful the results of a study are for a broader group of people or situations.

Goniometer: An instrument that measures the available range of motion at a joint.

Grounded Theory: A systematic methodology where the discovery of emerging patterns is "grounded" in actual data. Involves the construction of hypotheses and theories through the collecting and analysis of data.

H_0: See Null Hypothesis.

Hawthorne Effect: The tendency for participants to change their behavior as a result of being studied. The effect was first noticed in an industrial study conducted in the Hawthorne suburb of Chicago in the 1920 and '30s. The validity of this specific finding has been refuted, but researchers need to guard against the potential for such influence.

History: A threat to internal validity involving changes in variables other than the IV that may influence the DV.

Holistic: A process where meaning is constructed by examining the whole, taking into account mental and social factors, for example, rather than just the individual parts.

Homogeneity: The quality or state of being of a similar kind or of having a uniform structure or composition throughout.

Homogenous: Refers to consistency or uniformity of the group.

Hypothesis (plural hypotheses): The precise, testable statement of what researchers predict will be the outcome of the study.

IADMS: International Association for Dance Medicine & Science.

Inclusion Criteria: Criteria for which participants from the target population are to be included in the current study sample.

Independent Variable (IV): In scientific research, the variable the experimenter manipulates or changes, assumed to have a direct effect on the dependent variable.

Indirect Observations: Inferring the presence, absence, or state of a trait by observing a related behavior such as answering a survey question.

Inductive Reasoning: A method of reasoning in which a body of observations is synthesized to come up with a general principle.

Inferential Statistics: A way of making inferences about a population based on sample subset.

Informed Consent: The communication to research participants about the research purpose, procedures, risk, benefits, and rights of their participants. It ensures that the participants understand what is involved in their participation and that they are free of coercion or undue influence or demands by the researchers.

Instrumentation: A threat to internal validity involving a change in the accuracy with which measurements are made.

Inter-Observer Agreement (IOA): The degree to which two observers agree on independent observations of the same phenomena.

Inter-Rater Reliability: Two or more researchers working on the same analysis independently code some or all the data, then discuss and resolve differences in interpretations.

Internal Validity: The extent to which one can be confident that a cause-and-effect relationship established in a study cannot be explained by other factors.

Interpretivism: A theoretical stance proposing that the social realm cannot be studied with the scientific method of investigation utilized within the natural sciences, and that investigation of the social realm requires a different epistemology.

Interrupted Time-Series Design: A study design in which measurements are made repeatedly over time and an intervention is applied at a specific point in time to observe any systematic influence on a DV.

Intervention: Applying an IV systematically.

Intervention Mapping Protocol: The iterative path from problem identification to problem solving or mitigation.

Intra-class Correlation Coefficients: A descriptive quantitative statistic used to describe how strongly measurements that are grouped into units are related to each other.

Institutional Review Board (IRB): An institution-specific committee that oversees and reviews research proposals. See Ethics Review Board (ERB).

Justice: Fairness or equity and equal rights to all.

Kruskal-Wallis test: A non-parametric test that is the equivalent of an ANOVA.

Levene's Test for Equality of Variance: Measures the difference in variance between two independent groups.

Linear Regression: A statistical approach for modeling the relationship between a response and one or more explanatory variables (also known as dependent and independent variables). If there is one explanatory variable this process is called simple linear regression; for more than one, the process is called multiple linear regression.

Literature Review: A description of the peer-reviewed papers on a particular topic.

Literature Search: A process by which previously published information is gathered for review.

Lived Experiences: The representation of the experiences and choices of a given person, and the knowledge that they gain from these experiences and choices.

Longitudinal Study: A study in which observations are made at several points in time.

Mann-Whitney U Test: A non-parametric test that requires two independent groups and one dependent variable.

Marley Flooring: Dance performance surfaces made of sheet vinyl, also called dance floors and marley floors. They are called marley-type floors from a very popular reversible vinyl flooring manufactured for years until about 1978 by Marley Floors Ltd.

Mean: The mathematical average of multiple numbers.

Measurement Stability: The change in bias over time. It represents the total variation in measurements of the same part measured over time. This variation over time is called drift.

Median: The middle number in a sorted, ascending, or descending list of numbers.

Medical Subject Heading (MeSH) Terms: A comprehensive controlled vocabulary created and updated by the United States National Library of Medicine (NLM) to index journal articles and books in the life sciences; serves as a thesaurus to facilitate searching.

Member checking or member reflection: A procedure that can control for researcher bias by returning the data and/or interpretations to the participants for confirmation. An alternative is member reflection, in which researcher and participant engage in a cooperative dialogue about the findings to provide further insight. Braun, V., & Clarke, V. (2013). *Successful Qualitative Research: A Practical Guide for Beginners*. Sage Publications.

Mentor: An experienced and trusted adviser.

Mentorship: Guidance provided by a mentor, especially an experienced person.

Meta-Analysis: A quantitative, formal, epidemiological study design used to assess systematically the results of previous research to derive conclusions about that body of research.

Methodological Integrity: The demonstration of procedural rigor and quality in research. One of the most effective standards for measuring rigor that consists of two components: fidelity to the subject matter and utility of research contributions.

Minors: Individuals who do not meet the legal age of majority and therefore cannot consent to medical care or research participation, for example. Parents or guardians are responsible for minors and any decisions about such participation. This age is often 18 but differs by jurisdiction.

Mixed Methods: An emergent methodology of research that advances the systematic integration, or "mixing," of quantitative and qualitative data within a single investigation or sustained program of inquiry.

Mode: The value that appears most frequently in a set of data values.

Multiple-Baseline Design: A within-subject experimental design that uses repeated measurements of each participant's performance for a DV and the staggered application of an intervention (IV) at different points in time.

N=1 Trial: See Within-Subject Experiments.

Narrative Inquiry: The process of gathering information for the purpose of research through storytelling.

Naturalistic: A research method used in psychology and other social sciences in which research participants are observed in their natural environments.

Non-Experimental Research: The description of phenomena without random selection of participants or random assignment to groups; no independent variables are manipulated.

Non-Parametric Tests: Tests that do not make assumptions about the parameters of the population distribution from which the sample is drawn.

Non-Response Bias: When the participants who respond to a research questionnaire differ substantially from those who do not respond.

Normal Distribution: A probability distribution that is symmetric about the mean, with data near the mean being more frequent than data far from the mean. In graph form, normal distribution will appear as a bell curve. Also known as the Gaussian distribution.

Normative Research: A study that seeks to define a population and what are considered normal or ideal characteristics.

Null Hypothesis: A statement of no difference between a sample mean or proportion and a population mean or proportion. In other words, the difference equals 0. Denoted as H_0.

Objectivity: The quality or character of being objective and showing no favoritism toward one side or another.

Observation: The examination, measurement, or description of a phenomenon.

Ontology: The nature of personal growth and development.

Operational Definition: The statement of procedures the researcher is going to use in order to measure a specific dependent variable or apply a specific independent variable.

Outcome Variable: Usually the dependent variables that are observed and measured while changing an independent variable.

P-Value: The probability that a particular outcome may have occurred by chance. Short form of Probability-Value, usually reported at 95% level to accept or reject null hypothesis $p \leq 0.05$.

Paradigm: An overarching framework for conducting research that is aligned with a particular way of seeing the world.

Parametric Tests: Tests that make assumptions about the parameters of the population distribution from which the sample is drawn. Often the assumption is that the population data are normally distributed.

Parsimony: Using the simplest explanation available to describe a phenomenon.

Peer debriefing: The data and interpretations are reviewed by experts in either the research process, or the topic under inquiry, or both. The goal is not necessarily to reach a consensus but to reflect on interpretations and develop a more informed understanding.

Phenomenology: The study of phenomena as they manifest in our experience, of the way we perceive and understand phenomena, and of the meaning phenomena have in our subjective experience.

PICO Process (or model or framework): A mnemonic used in evidence-based practice to frame and answer a clinical or health care related question and develop literature search strategies. The acronym stands for Population, Intervention, Comparison, and Outcome.

Pilot Testing: Trying out measurements, treatment, training, and other components of a study before the study to be reported is initiated.

Pirouette: A dance movement that involves spinning while supported on one leg.

Population: A distinct group of individuals, whether that group comprises a nation or a group of people with a common characteristic.

Post hoc tests: Statistical analyses used only after researchers find a statistically significant result when examining multiple (usually more than two) groups; e.g., in ANOVA designs. These tests allow researchers to determine where the differences truly came from among the groups being tested by examining pairs of values while minimizing the Type I error rates. The phrase "post hoc" comes from the Latin for "after the event."

Post-Positivism: A metatheoretical perspective that argues against the underlying assumptions associated with logical positivism.

Postmodernism: An attitude toward the social world in its current state of historical development.

Power Analysis: A way to estimate the sample size required to obtain a statistically significant difference based on the anticipated differences.

Power Statistics: Calculations that help determine the minimum number of participants required to detect an effect of a given size.

Power, Physical: Force applied per unit of time.

Practical Significance: The extent to which a study result has meaningful applications in real-world settings.

Pre-experimental Designs: Preliminary research study designs where a participant or a group is observed after a treatment has been applied to test whether the treatment has the potential to cause change.

PRISMA: An evidence-based minimum set of items for reporting in systematic reviews and meta-analyses, primarily focusing on the reporting of reviews evaluating the effects of interventions, but can also be used to report systematic reviews with objectives other than evaluating interventions (e.g., evaluating etiology, prevalence, diagnosis, or prognosis). The acronym stands for Preferred Reporting Items for Systematic Reviews and Meta-Analyses.

Privacy: The state or quality of being free from intrusion from others. Individuals have a right to privacy that needs to be respected in the context of research participation.

Procedural Fidelity (IV Integrity): The degree to which the intended procedures are applied as intended.

Prospective: Relating to or effective in the future.

PubMed: A free resource supporting the search and retrieval of biomedical and life sciences literature with the aim of improving health, both globally and personally.

Purposive sampling: Participants are chosen in a purposeful fashion to suit the needs of the study.

Qualitative Analysis: The process of examining and interpreting data that are textual in nature.

Qualitative Data: Non-numerical descriptions of phenomena.

Qualitative Methods: Research methods that do not involve measurement or statistics but involve collecting and analyzing non-numerical data to understand concepts, opinions, or experiences.

Qualitative Research: Research that generates detailed descriptions of phenomena by reference to a particular worldview and paradigm of inquiry.

Quality Assessment Tool (QAT): Any tool designed to target one or more aspects of the quality of a research study, whether focused specifically on one aspect of study quality (usually risk of bias) or broadly covering aspects such as applicability, generalizability, and reporting quality of the study.

Quantitative Data: Observations that are described or summarized with numerical values.

Quasi-experimental Design: A study design that, generally due to pragmatic considerations, is unable to isolate the influence of an independent variable on a dependent variable.

Questionnaire: A set of questions with a choice of answers, devised for the purposes of a survey or statistical study.

Random Assignment: Using a chance procedure (e.g., coin-flip, random number table) to assign participants to experimental conditions; an essential element for assessing experimental control in Between-Group Experiments.

Random Selection: Using a chance procedure (e.g., coin-flip, random number table) to choose a sample of participants from a population; allows researchers to use statistical principles to generalize the results from a sample to the population from which the sample was taken.

Randomized Controlled Trial: A rigorous prospective study design where participants are randomly assigned to one of at least two groups: an experimental group that receives intervention being tested, and a control group that does not receive the intervention or receives an alternative (conventional) treatment. If the methods are correctly followed, this study design allows researchers to be confident that the only expected difference in the outcome variable between the experimental and control group is due to the intervention.

Range: Difference between the lowest and highest values.

Reactivity: Also known as the observer effect, takes place when the act of doing the research changes the behavior of participants, thereby making the findings of the research subject to error.

Reference Management Software: Software for scholars and authors to use for recording and utilizing bibliographic citations as well as managing project references either as a company or an individual.

Reference: To provide (a book or article) with citations of sources of information.

Reflexivity: The process of self-awareness and reflection about a researcher's role, perspectives, experiences, and biases related to the research endeavor, interpretation of findings, and the co-construction of meaning between the participants and researcher.

Regression Analyses: Statistical processes for estimating relationships between a dependent variable (often called the "outcome" or "response" variable) and one or more independent variables (often called "predictors," "covariates," "explanatory variables," or "features").

Reliability: The quality of being trustworthy or of performing consistently.

Reliable: Consistent, particularly measurements.

Replication of Effect: Applying an independent variable a second and subsequent times and observing a similar influence on the dependent variable.

Replication, Direct, Systematic: Repeating a study as similarly as possible to the original study (direct replication) or with some systematic change in an aspect of the study (e.g., different participants, setting, parameters of the intervention; systematic replication).

Representativeness: The level of how well or how accurately something reflects upon a sample.

Research: Systematic investigation of materials and sources in order to establish facts and reach new conclusions.

Research Degrees: Program in which the majority of credits are earned by engaging intensively in the execution of research.

Research Design: Arranging the elements of a study in a manner that facilitates answering the research question.

Research Ethics Committee (REC): Institution-specific committee that oversees and reviews research proposals to ensure ethical conduct by researcher and that ethical treatment of participants is ensured during the course of the research proposed. This group may be referred to as an Institutional Review Board (IRB), Ethics Review Board, Ethics Committee, or Human Subjects Review Committee.

Research Ethics: Pertaining to the ethical aspects of a research endeavor that must be addressed in a study.

Research Question: The question a study is designed to answer; it focuses the researcher's efforts.

Research, Applied: Seeks to answer a question that has immediate practical value.

Research, Basic: Seeks to answer a question of curiosity.

Response Rate: A calculation that describes how many participants responded to a survey compared with the total number of participants who were surveyed. Also used to describe the frequency with which participants are responding in a particular experimental condition.

Responsive: Able to distinguish differences in observations smaller than the magnitude of the change anticipated.

Retrospective: Relative to past event.

Reversal (Withdrawal) Design: A within-subject design that uses repeated measurements of each participant's performance while, after a stable baseline is established, an intervention is applied, removed, and re-applied.

RM-ANOVA: Repeated Measures Analysis of Variance for one or more groups tested two or more times.

Sample: A subgroup of a particular population from which data are collected.

Sampling: How participants are selected and recruited.

Sampling Frame: The source material or device from which a sample is drawn.

Saturation: The point at which the sample is large enough and no other participants are needed because no new meaning, concepts, or themes are identified in new transcripts.

Simple Comparison Design: A quasi-experimental, within-subject design that permits a preliminary assessment of the influence of an intervention on a repeatedly measured dependent variable.

Simple Random Sampling: Selecting participants based on the premise that everyone has the opportunity to be selected.

Single-Case Designs: See Within-Subject Experiments.

Single-Subject Design: See Within-Subject Experiments

Snow-ball sampling: Participants involved in the study suggest other participants.

Social Desirability: The tendency of people to present themselves in a socially acceptable fashion or provide responses that they deem to be desired by the researcher.

Social Validity (Significance): Allowing participants in a study to evaluate the importance, appropriateness, and cost-benefit value of a change in a performance following an intervention.

Source, Primary: Report of a study presented by the researchers who conducted it.

Source, Secondary: Report of a study, or its elements, by anyone other than the researcher who conducted it.

Specific Aims: Active verb-oriented statements that clearly state the objective in relationship to the central hypothesis of the study. Often these are the most important component of the entire study that describes the study strategy, the outcomes, and how it will help advance the knowledge in the field.

Stable Baseline: Repeated measurements made prior to intervention in which variability and trending are minimized.

Standard Deviation: A measure of the amount of variation or dispersion of a set of values.

Stratified Random Sampling: A process of identifying categories in the population and then randomly selecting participants from each category.

Student Dependent/Paired Samples t Test: An inferential statistical test that compares the same continuous, dependent variable on the same group twice to determine whether there is a statistically significant difference between these means.

Student Independent t Test: A type of inferential statistical test to determine if there is a significant difference between the means of two groups. Also referred to as t-tests.

Survey Research: Involves collecting large amounts of data from a sample of a specified population through tools such as questionnaires or interviews.

Systematic Random Sampling: A process of selecting participants according to a particular system, such as every tenth person in a telephone directory.

Systematic Replication: The practice of repeating an existing study with one small change, such as using a new research population, setting, methodology, or duration. It applies most of the protocols used in a prior study while changing a key element (e.g., population, targeted skill, intervention duration or intensity).

Systematic Review: A review of the evidence on a clearly formulated question that uses systematic methods to identify, select and appraise relevant primary research, and to extract and analyze data from the included studies.

t-test: A type of inferential statistical test to determine if there is a significant difference between the means of two groups.

t: A value determined by a t-test and provided as statistical output for independent and dependent sample student t-tests.

Taught Degrees: Programs where the majority of credits are earned by taking courses involving frequent interaction with the instructor.

Thematic analysis: A process of categorizing patterns observed in qualitative data.

Thick description: The detailed explanation of what has been observed.

Threats to Internal Validity: Features of an experiment that limit the ability to isolate the influence of an IV on a DV.

TOST: Two one-sided equivalence test.

Transferability: The extent to which findings can be generalized to other contexts.

Trending: Systematically increasing or decreasing values of a repeatedly measured DV.

Triangulation: Combining multiple data sources (including qualitative and quantitative) to establish commonalities and confirm findings from various perspectives.

Trustworthiness: How well the quality, credibility, and accuracy of research are assured.

Type I error: Concluding that there is a difference when there is no actual difference (i.e., not significant); also known as alpha (α).

Type II error: Concluding that there is no difference, when there is an actual difference (significant); also known as beta (β).

Validity: The quality of being logically or factually sound.

Variability: Degree to which measurements of a dependent variable vary across participants, or across repeated assessments of a particular participant.

Within Group Variability: Size of standard deviation.

Within-Subject Experiments: Research that examines changes in scores in each participant over time and across conditions.

Worldview: An individual's particular perception and understanding of the world.

X^2: This symbol identifies the chi-square statistic and is used in chi-square output. See Chi-square Test.

Z: This symbol identifies a test statistic call a z-score used in a Mann-Whitney U test output. See Mann-Whitney U test.

Contributors

Contributing Authors

Amit Abraham, PhD, MAPhty (Musculoskeletal), BPT, is Assistant Professor in the Faculty of Health Sciences, Department of Physical Therapy, Ariel University, Ariel, Israel.

Imogen Aujla, PhD, FHEA, is a freelance lecturer, researcher, and well-being coach with expertise in dance psychology, inclusive dance, and talent identification and development at Dance in Mind in Milton Keynes, England, United Kingdom.

Manuela Angioi, PhD, FHEA, FIADMS, is Senior Lecturer and Program Leader for the iBSc Sports and Exercise Medicine Course at William Harvey Research Institute, Queen Mary University of London, United Kingdom.

Liliana S. Araújo, PhD, C.Psych, SFHEA is Assistant Professor of Performance Science at the Schulich School of Music, McGill University, Montreal, Canada, and former Program leader in Dance Science and Senior Lecturer in Performance Psychology at Trinity Laban Conservatoire of Music and Dance, London, United Kingdom.

Wendy C. Coates, MD, FACEP, is Emeritus Professor of Emergency Medicine at the Geffen School of Medicine, University of California Los Angeles, United States of America.

Steven J. Chatfield, PhD, is Professor of Dance in the School of Music and Dance at the University of Oregon, Eugene, United States of America.

Boyi Dai, PhD, is Associate Professor in the Division of Kinesiology and Health at the University of Wyoming, United States of America.

Jennifer Deckert, MFA, is Associate Professor of Dance at the University of South Carolina, United States of America.

Claire E. Hiller, PhD, MAppSc, BAppSc, PT, is Associate Professor in the Sydney School of Health Sciences, Faculty of Medicine and Health at the University of Sydney, Australia.

Sarah J. Kenny, PhD, is an Assistant Professor on the Faculty of Kinesiology and the Division of Dance in the School of Creative and Performing Arts, Faculty of Arts, at the University of Calgary, Canada.

Nico Kolokythas, PhD, ASCC, is a strength and conditioning coach at Elmhurst Ballet School and Birmingham Royal Ballet and a visiting lecturer at the University of Wolverhampton, United Kingdom, and Bern University, Switzerland.

Yiannis Koutedakis, PhD, is Professor Emeritus in Applied Physiology, Thessaly University, Greece.

Donna H. Krasnow, PhD, is Professor Emerita and Senior Scholar at York University, Toronto, Canada, and creator of C-I Training™ (Conditioning with Imagery) for dancers.

Merry Lynn Morris, PhD, MFA, is Assistant Director of the Dance Program and coordinator of the Dance Medicine and Science certificate program at the University of South Florida, Tampa, United States of America.

Emma Redding, PhD, FSHEA, FIADMS, is Professor and Director of the Victorian College of the Arts at the University of Melbourne, Melbourne, Australia, and Visiting Professor in Performance Science at Trinity Laban Conservatoire of Music and Dance, London, United Kingdom.

K. Michael Rowley, PhD, is Assistant Professor in the Kinesiology Department at California State University East Bay in Hayward, and Adjunct Assistant Professor of Clinical Physical Therapy at the University of Southern California in Los Angeles, United States of America.

Jeffrey A. Russell, PhD, AT, FIADMS, is Associate Professor of Athletic Training and Director of Science and Health in Artistic Performance at Ohio University, Athens, United States of America.

Selina Shah, MD, FACP, FAMSSM, is the company physician for Diablo Ballet, AXIS Dance Company, and Contra Costa Ballet, and a sports medicine physician in private practice as part of the BASS Medical Group in Walnut Creek, California, United States of America.

Leah Reilly Sherman, MA, MLIS, is the Visual and Performing Arts Librarian at Florida State University in Tallahassee, United States of America.

Rogier M. van Rijn, PhD, is Associate Professor of Performing Arts Medicine at Codarts Rotterdam, and board member of the Performing Artist and Athlete Research Laboratory (PEARL), Rotterdam, The Netherlands.

Janine H. Stubbe, PhD, is Professor of Performing Arts Medicine at Codarts Rotterdam, Associate Professor at Erasmus MC Medical University Center, and Director of the Performing Artist and Athlete Research Laboratory (PEARL), Rotterdam, The Netherlands.

Virginia Wilmerding, PhD, is Research Professor in the Department of Theatre and Dance and the Department of Health, Education and Sports Sciences at the University of New Mexico in Albuquerque, United States of America.

Margaret Wilson, PhD, FIADMS, is a professor in the Department of Theatre and Dance at the University of Wyoming, Laramie, United States of America.

Matthew Wyon, PhD, CSCS, FIADMS, FHEA, is Professor in Dance Science at the School of Performing Arts, University of Wolverhampton, Walsall, United Kingdom.

Editors

Jatin P. Ambegaonkar, PhD, ATC, FIADMS, is Professor in the School of Kinesiology and the founding Operations Director of the Sports Medicine Assessment, Research, and Testing (SMART) Laboratory at George Mason University, Virginia, United States of America.

Lynda Mainwaring, PhD, C. Psych, is Associate Professor in the Faculty of Kinesiology and Physical Education at the University of Toronto, Canada.

Tom Welsh, PhD, is Dance Sciences Professor at Florida State University, Tallahassee, and Courtesy Research Professor in Dance at the University of Utah, Salt Lake City, United States of America.

Index

Page numbers followed by *f* and *t* indicate figures and tables.